The Pathobiology of Breast Cancer

Jose Russo

The Pathobiology of Breast Cancer

Springer

Jose Russo
Fox Chase Cancer Center
Philadelphia, PA, USA

ISBN 978-3-319-40813-2 ISBN 978-3-319-40815-6 (eBook)
DOI 10.1007/978-3-319-40815-6

Library of Congress Control Number: 2016945167

Printed on acid-free paper

This Springer imprint is published by Springer Nature
The registered company is Springer International Publishing AG Switzerland

Preface

This book is targeted to all those that are wishing to have a comprehensive view of breast cancer and will provide novel information to clinicians, researchers and academia.

This book provides the latest advances in the pathobiology of breast cancer including initiation and progression of the disease, the mechanisms of invasion and metastasis, the concept of stem cells in treatment and drug resistance. The role of personalized medicine and genomic testing provides a window to the future of cancer patient care in diagnosis, prognosis and treatment.

Altogether this book provides a new insight on the pathobiology of the breast using a meticulously researched process that has been lead from basis to translational research.

Philadelphia, PA, USA — Jose Russo, MD, FACP

Acknowledgments

My specific acknowledgement and thanks to *Ms. Patricia A. Russo* for her insightful editorial suggestions, critiques and the delightful moments spent with her discussing the manuscripts and its ideas.

A special thanks to Ms. *Linda Cathay* for helping me in the formatting and downloading of all the manuscripts.

My thanks also to *Pathology Consultation Service* in Rydal, PA who financed the writing and editing of this book.

Author's Bio

Jose Russo, MD is the Director of the Irma H, Russo, MD-Breast Cancer Research Laboratory and Director of the Breast Cancer and The Environmental Research Center at the Fox Chase Cancer Center, Temple Health in Philadelphia, PA. Dr. Russo is also an Adjunct Professor of Pathology and Cell Biology at Jefferson Medical School and Professor of Biochemistry at Temple Medical School. He has received numerous research awards from the National Cancer Institute (NCI) of the National Institute of Health (NIH), from the American Cancer Society and the Department of Defense for his original research in breast cancer. Throughout his career, Russo's work and research interests have had a broad base but with a focused goal: to understand the mechanisms that control cancer metastasis; and to develop strategies for breast cancer prevention.

The original version of this book was revised. An erratum to this book can be found at DOI 10.1007/978-3-319-40815-6_11

Contents

Chapter 1
The Windows of Susceptibility to Breast Cancer

1.1 Introduction

The occurrence of cancer of the breast has long been known [1–4] and the disease affects women of all races and nationalities and the incidence of has increased 30–40 % since the 1970s [2, 4–7]. This already dismal picture is worsened by the gradual increase in breast cancer incidence in most Western countries and in societies that recently became westernized or that are in the process of westernization [8, 9]. Epidemiological observations that daughters of women who migrate from low-incidence to high-incidence countries acquire the breast cancer risk prevailing in the new country [10], suggest that aspects of lifestyle or the environment are major determinants of breast cancer risk. A study of population-attributable risks has estimated that at least 45 % to 55 % of breast cancer cases in the United States may be explained by the following factors: advanced age at the time of the first full-term pregnancy, nulliparity, family history of breast cancer, higher socioeconomic status, earlier age at menarche, and prior benign breast disease [11]. Other statistical models appear to explain an even higher proportion of breast cancer on the basis of known risk factors [12]. Studies of atomic bomb survivors have shown that environmental exposures, such as ionizing radiation, are a risk factor for breast cancer [13]. Exposure to radiation at a young age (Fig. 1.1) has been identified as a causative agent of breast cancer in selected populations [14–17], but there is no definitive proof of what causes breast cancer in the population at large. The increased risk associated with exposure to environmental chemicals, such as alcohol [18] and cigarette smoke [19–25], makes these agents suspects for causing cancer in the human population (Fig. 1.1).

J. Russo, *The Pathobiology of Breast Cancer*,
DOI 10.1007/978-3-319-40815-6_1

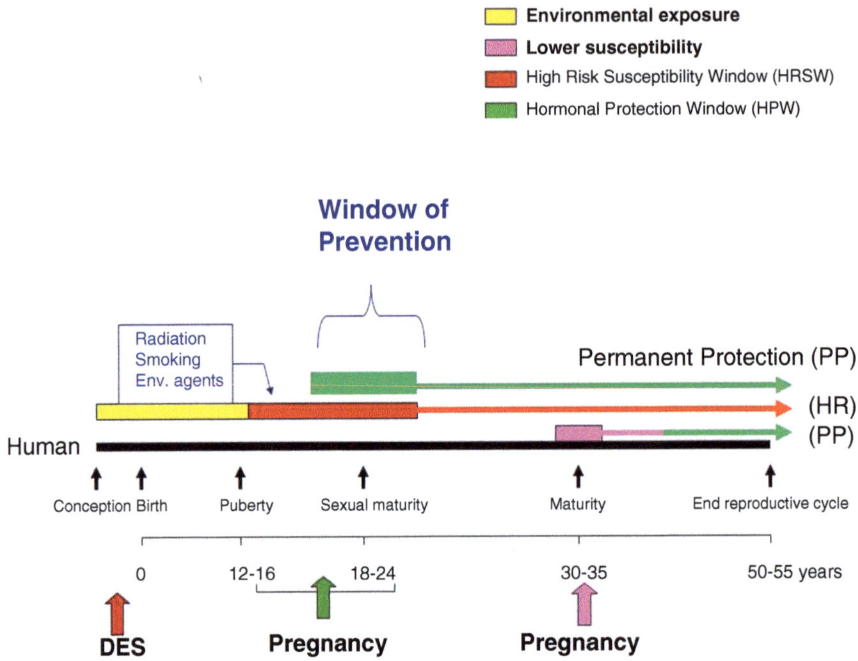

Fig. 1.1 Windows of risk and prevention

1.2 Risk Factor and Etiological Agents

Breast cancer rates among older women have been reported to be higher in the Northeast than in the South; they are also higher in urban than in rural areas [26, 27]. There is an increasing linear relationship between standardized incidence rates of breast cancer and population density [28]. There is a higher breast cancer incidence in the San Francisco Bay Area when compared to the other seven registries located throughout the country [29]. Regional differences in the prevalence of known breast cancer risk factors, such as low parity, higher education, and higher income, seem to play an important role in the elevated rates of breast cancer reported in affluent communities, such as in Marin County, California [30]. In addition some of the geographic variations in breast cancer have been attributed to differences in exposure to sunlight [31–33].

Radiation from natural sources is ubiquitous in the environment [34]. Only accidental or iatrogenic radiation has been demonstrated to exert a carcinogenic effect on the breast. High energy X- or γ rays from a bomb exposure in Hiroshima and Nagasaki in Japan [35], and chest radiation administered for the treatment of scoliosis, or repeated fluoroscopies for tuberculosis [36] have been well-documented causes of breast cancer (Fig. 1.1). Importantly radiation has a carcinogenic effect when exposure occurs at a young age. Only those women who were younger than 29 at the time of the bombing in Japan developed breast cancer [35], whereas older

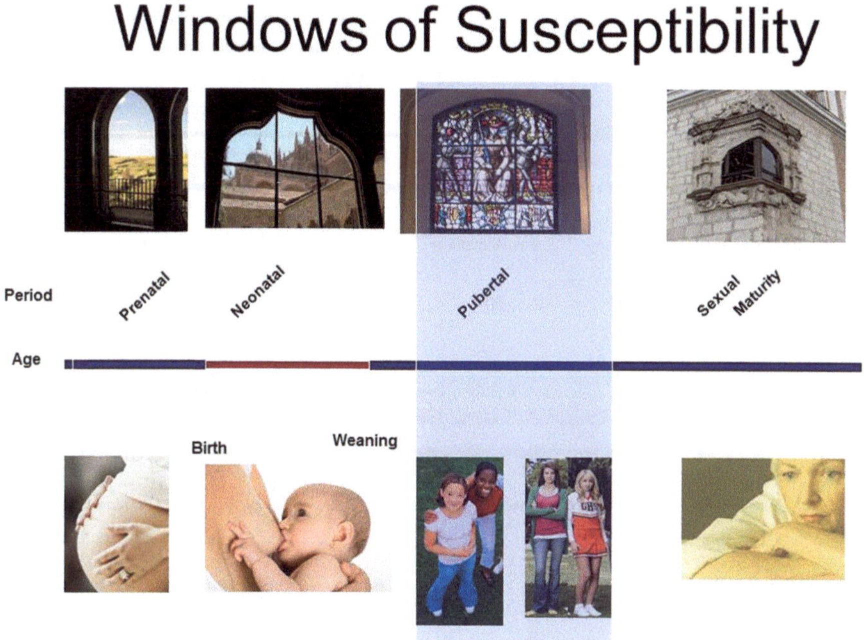

Fig. 1.2 The pubertal period is the windows that offer the highest risk for cancer initiation and is also the best windows for cancer prevention

women developed benign breast diseases. There is evidence that age at exposure influences the relative risk (Figs. 1.1 and 1.2). Excess risk decreases with increasing age at exposure; the highest relative risks are observed for women exposed between the ages of 10 and 20, and there is little risk for those greater than 40 [37–39] (Figs. 1.1 and 1.2). Parity, in addition to age at exposure, modifies the risk of developing radiation-induced breast cancer, since the risk is greater in nulliparous women, but no carcinogenic effect has been reported in women treated with radiation for postpartum mastitis. However, young women that are successfully treated with radiation for early-stage Hodgkin's disease [14–17] develop breast cancers after a median interval of 15 years. While the risk of recurrent Hodgkin's disease decreases as time from treatment elapses, the risk of radiation-induced breast cancer rises. Women irradiated between the time of puberty and the ages of 30 are at the highest risk of developing cancer. It would be of great interest to consider the possibility of preventing breast cancer in this patient population by administering a hormonal treatment that would differentiate their breast "before" or shortly after administering radiation, in order to reduce the susceptibility of the organ to be transformed by the treatment (Fig. 1.3).

Exposure to environmental lighting in the visible range of the spectrum [40] and low-level EMF [41] have been hypothesized to increase the risk of breast cancer due to a decrease in the secretion of the hormone melatonin and a subsequent increase in circulating estrogens [42–48]. The general population is exposed to EMF

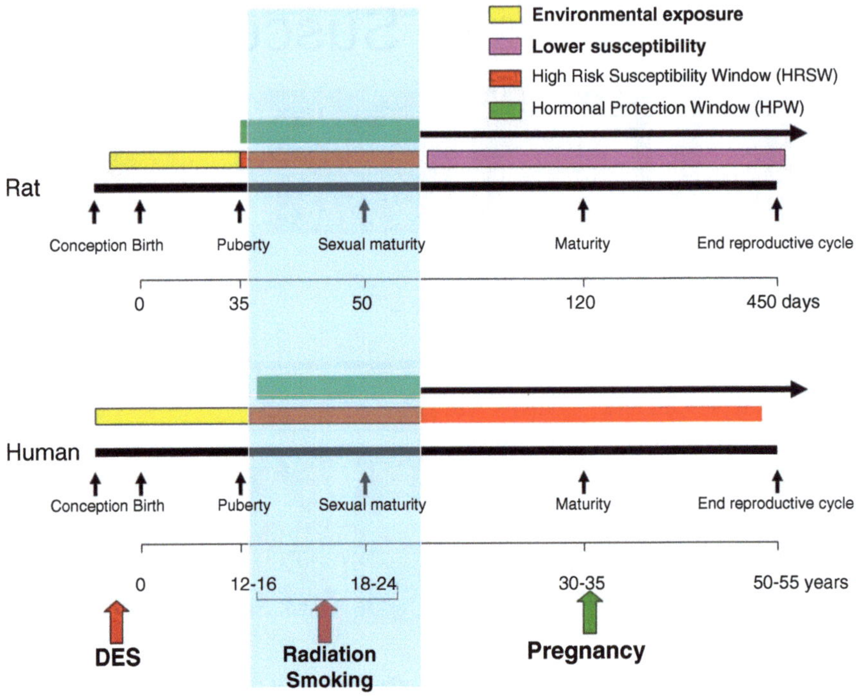

Fig. 1.3 Comparative windows of susceptibility for the rat mammary gland and the human breast

primarily from power lines, transformer substations, and electrical appliance use. Elevation in female breast cancer incidence has been associated with magnitude of exposure at the current residence [49–52].

Exposure to pesticides, e.g., 2,2-bis(p-chlorophenyl)-1,1,1-trichloromethane (DDT), chlordane, hexachloro-cyclohexane (HCH, lindane), hexachlorobenzene (HCB), kepone, and mirex; industrial chemicals, e.g., polychlorinated biphenyls (PCBs); and dioxins (polychlorinated dibenzo-furans (PCDFs), and polychlorinated dibenzodioxin (PCDDs), produced as combustion byproducts of PCBs or contaminants of pesticides have been postulated to increase the incidence of breast cancer . Most of the recent large studies, however, have not found evidence of increased breast cancer risk associated with blood levels of DDE or total PCBs. The possibility exists that a positive association might be limited to women with particular reproductive characteristics [53]. Exposure of the general population to environmental compounds occurs predominantly through ingestion of fish, dairy products, and meat [54]. The experimental and epidemiological evidence of potential links to cancer has been reviewed in detail elsewhere by Adami et al. [55], Ahlborg et al. [56], and Wolff and associates [57].

All women in the general population are exposed to similar environmental influences; yet, not all of them develop breast cancer. Among women with no family history of breast or breast/ovarian cancer [58, 59], or Li-Fraumeni

Syndrome [60], the higher risk of developing breast cancer is associated with a history of early menarche [61], nulliparity [62–65], late first full-term pregnancy [62, 66], and late menopause [62] (Figs. 1.1 and 1.3), all conditions that are under the direct control of the ovary. The central role played by the ovary in breast cancer development is further confirmed by the marked reduction in cancer incidence after surgical or chemical ovariectomy [67]. The indirect evidence that depression of gonadal function, attributed to elevated melatonin levels in profoundly blind women, decreases the risk of breast and other cancers [68–71] suggests that light acts as an important environmental factor modulating breast cancer risk through endocrine disruption [72]. The paradox that ovarian stimulation such as that induced by pregnancy [58–62] or by treatment of women with the pregnancy hormone human chorionic gonadotropin (hCG) [73] exerts a protective effect, highlights the importance of induction of complete breast differentiation for protecting the breast from developing cancer (Fig. 1.3). Differentiation, however, has to be induced during a specific period in the lifetime of a woman, as indicated by epidemiological observations that a full-term pregnancy that markedly reduces the lifetime breast cancer risk of a woman if it occurs before 24 years of age, increases the risk above that observed in nulliparous women when it is postponed beyond the 30th to 35th birthday [62–66] (Figs. 1.1 and 1.2).

1.3 The Concept of the Windows of Susceptibility to Carcinogenesis

Spontaneous mammary tumors are frequently observed in long term rodent studies [74, 75]. The induction of hormone dependent rat mammary tumors with chemical carcinogens, on the other hand, has become an essential model for testing the carcinogenic potential of specific chemicals, such as 3,4-benzopyrene, 3-methylcholanthrene (MCA) [76] and the polycyclic aromatic hydrocarbon (PAH) 7,12-dimethylbenz(a)anthracene (DMBA) [77], or the alkylating agent N-methyl-N-nitrosourea (MNU) [78, 79]. Chemically-induced tumors developed in mice strains of low spontaneous mammary cancer incidence or in transgenic mice are adenoacanthomas or type B adenocarcinomas that are in general estrogen receptor alpha (ERα) negative [80]. However, in p53 null mice hormonal stimulation by estrogen and/or progesterone or prolactin/progesterone, markedly enhances tumorigenesis, whereas blocking estrogen signaling through ovariectomy or tamoxifen treatment greatly reduces the tumorigenic capability of the mammary epithelium, an indication that normal mammary gland and preneoplastic lesions are responsive to estrogen [80]. The majority of rat mammary tumors induced by DMBA or MNU are ductal adenocarcinomas that are ER-α positive and reproduce the pathological features of the most frequent type of adenocarcinomas developed by women [81]. The characteristics of this model have opened a myriad of opportunities for dissecting the initiation, promotion, and progression steps of carcinogenesis and for translating these findings to the human situation [74, 82, 83].

The response of the mammary gland to specific carcinogenic stimuli depends upon the physiologic state of the mammary tree under the control of the endocrine system. The administration of optimal carcinogenic doses to young and sexually mature virgin rats induces maximal tumorigenic response [5–7, 84–93] (Fig. 1.3). This period of highest susceptibility of the mammary gland to be transformed by such stimulus represents the "high risk susceptibility window" (HRSW), which encompasses different stages of development, i.e., prenatal life, infancy, puberty and early adulthood (Figs. 1.1, 1.2 and 1.3). Thus, in addition to age, the tumorigenic response elicited by carcinogenic agents is modulated by the animal's endocrinological milieu prevailing at the time of exposure, as well as by endocrine and environmental influences occurring during the HRSW [94–97] (Fig. 1.3). The peak of cancer incidence occurring when virgin rats reach the age of 45 to 55 days and have had at least two ovulatory cycles after vaginal opening [98], represents the response of numerous mammary terminal end buds (TEBs) that are predominantly composed of progenitor mammary stem cells (PMSCs). These cells have been characterized by their size, nuclear-cytoplasmic ratio and euchromatin-heterochromatin ratio, number and distribution of organelles and proliferative activity [74, 99]. PMSCs that have been primed by ovarian hormones for the expansion the mammary parenchyma through branching and lobular formation, when they are exposed to a carcinogen, such as tritiated (^{3}H) DMBA they exhibit the highest rate of carcinogen uptake as well as a high rate of cell proliferation [74]. Within a few days transformed PMSCs expand and form intraductal proliferations (IDPs) that progress to ductal carcinomas *in situ* and invasive, confirming the transition of PMSCs to mammary cancer stem cells (MCSC) under the influence of a carcinogen [99, 100]. Morphologically similar cells have been isolated from DMBA-induced mammary tumors [101]. In the mouse the mammary gland continually undergoes postnatal developmental changes that are driven by signals from TEBs [102]. They direct ductal growth and elongation, producing a progeny of varied lineages that include luminal and myoepithelial cells under the influence of signals from the local tissue microenvironment [102].

1.4 The Windows of Susceptibility Apply to Human Breast Cancer

The comparison of events influencing the initiation of cancer in humans and animals and of the factors that influence both led us to postulate that a commonality exist. Central to this hypothesis is the initiation of puberty (Figs. 1.1 and 1.2). Menarche, or the first menstruation that marks the initiation of puberty, is an objective manifestation of ovarian function [69, 103]. The age at menarche has been observed to decrease in the Western world, with no clear explanation for this phenomenon. It is of great importance to take into consideration the facts that pubertal development is modulated by ovarian function, which in turn is under the control of the hypothalamic-pituitary axis under the control of two interacting timekeeping mechanisms in the central nervous system (CNS): endogenous circadian rhythmicity and sleep-wake homeostasis [72]. Circadian rhythmicity is an endogenous, near 24-h oscillation, generated in the suprachiasmatic nuclei (SCN) of the hypothalamus (H) that generate pulses transmitted to the pituitary-gonadal (PG) axis via neural and humoral mechanisms. The SCN, under the light-dark cycle, controls the pineal gland and the levels of circulating melatonin. The photoperiod via melatonin secretion determines the timing of puberty in some species and delays reproductive maturity in both males and females. The production of melatonin is inhibited by visible light, which alters the circadian rhythm, disrupting the body's physiology and metabolism [72]. The stimulus of the ovary by pituitary follicle stimulating hormone (FSH) results in follicular maturation and estrogen secretion, followed by a mid-cycle peak of luteinizing hormone (LH) that triggers ovulation and subsequent progesterone secretion. Ovarian stimulation per se is insufficient for driving the breast to the completely differentiated condition that should be reached for achieving protection from cancer development. Additional hormonal supplementation, such as that provided by full-term pregnancy, or specific hormonal regimens, are required for that purpose [73] (Fig. 1.3).

In the last two decades, approximately 3 million women have died prematurely from smoking-related diseases, including cancer. In 1998, 22 % of all women in the USA smoked cigarettes, with a higher percentage of high school senior girls smoking (Figs. 1.1 and 1.3). Lung cancer surpassed breast cancer as the leading cause of cancer death in women and it killed nearly 68,000 women in 2000 [20, 104]. Tobacco smoke is a complex mixture of several thousand chemicals that include carcinogens, namely polycyclic hydrocarbons (PAHs) such as benzo(a)pyrene (BP), which are metabolically activated, forming carcinogen-DNA adducts in human breast tissues. BP selectively binds to deoxyguanine at CpG dinucleotides within codons of the gene, making them mutational hotspots. The resultant G:C to T:A transversions in the p53 gene show a dose-response relationship in lung cancers of smokers [20, 104]. The need to determine whether tobacco smoking is a causative agent in breast cancer has stimulated numerous studies at both epidemiological and basic research levels. The fact that several studies support the hypothesis that "women are more susceptible than men to smoking-induced lung cancer," and that estradiol regulates activities that enhance lung carcinogenesis and tumor progression, as supported by the detection of ER and progesterone receptors (PR) in lung cancer [104], indicate that there similarities at least in hormone dependence between breast and lung cancer. The use of smokeless tobacco has been reported to elevate significantly the risk of developing younger-onset (<55 years) breast cancer (OR = 7.79, 95 % CI = 1.05--66.0) for ever-users of smokeless tobacco [19]. Additional support to the etiologic role of chemical carcinogens in the initiation of breast cancer has been obtained from experimental studies of in vitro transformation of human breast epithelial cells with BP [105].

A clear association has been found between breast cancer occurrence and drinking alcohol. Any history of drinking alcohol increases the risk of breast cancer 1.2-fold above that of women who never drank alcohol (95 % confidence interval 0.7--1.8). A greater than average consumption of alcohol for 6 months or more increases the relative risk of breast cancer to 2.6 (95 % confidence interval 1.1--5.8) [18]. Positive dose-response trends have been also reported in pooled analysis of large cohort studies and meta-analyses of a broader spectrum of studies. Although the ultimate mechanism through which alcohol increases breast cancer risk has not been clarified, it has been suggested that alcohol consumption alters hormonal levels by affecting the opioid hypothalamic activity, altering LH secretion [103]. These changes have been associated with an advance in the onset of sexual behavior when alcohol is consumed before puberty [103] (Fig. 1.2). An effect of alcohol consumption on the hypothalamic-pituitary-gonadal axis is also supported by the reported observations that the nocturnal urinary concentration of 6-sulfatoxymelatonin, the primary metabolite of melatonin, decreases in a dose-dependent manner with increasing consumption of alcoholic beverages in the preceding 24-h period [106]. A categorical analysis revealed no effect of one drink, but a 9 % reduction with two drinks, a 15 % reduction with three drinks, and a 17 % reduction with four or more drinks [106].

A higher breast cancer incidence has been reported in women that work at night, whereas the risk is lower in profound bilateral blind women [69–71]. The rate of breast cancer risk reduction in blind women is proportional to the degree of blindness [72]. Experimental studies have demonstrated that exposure to constant light from birth enhances mammary carcinogenesis in rodents [70]. Pinealectomy, that suppress nocturnal melatonin production and increases prolactin levels, also stimulates tumor progression in rodents [71]. Exposure to constant light initiated at the age of 26 days in rat, on the other hand, induces lactational changes and inhibits rat mammary carcinogenesis. A suggested mechanism for the effect of light on mammary carcinogenesis is its suppressive effect on nocturnal melatonin, in association with increased levels of DNA synthesis and elevated circulating levels of prolactin [70–72]. Discrepancies in results obtained when constant light exposure is initiated at birth and those initiated at 26 days of age might reflect age-related differences in the maturation of the suprachiasmatic nucleus (SCN) and hypothalamic-pituitary-gonadal axis in response to environmental stimuli that deserve further investigation. During the past years, significant progress has been made in identifying the molecular components of the mammalian circadian clock system [107]. An autoregulatory transcriptional feedback loop similar to that described in Drosophila appears to form the core circadian rhythm generating mechanism in mammals. Two basic helix-loop-helix (bHLH) PAS (PER-ARNT-SIM) transcription factors, CLOCK and BMAL1, form the positive elements of the system and drive transcription of three Period and two Cryptochrome genes. The protein products of these genes are components of a negative feedback complex that inhibits CLOCK and BMAL1 to close the circadian loop. The novel findings that environmental carcinogens bind to epithelial cells through the aryl hydrocarbon receptor (AhR), which upon translocation to the nucleus it dimerizes with the co-factor AhR nuclear translocator (ARNT), a member of the Per-ARNT-Sim (PAS) protein-containing the same transcription factors CLOCK and BMAL1 found in the suprachiasmatic nucleus (SCN), suggest that in addition to being involved in the initiation of puberty, the SCN might be affected by environmental carcinogens that are traditionally considered to exert their carcinogenic effects on peripheral organs by acting directly on mammary epithelial receptors.

1.5 Effect of Hormones on Breast Cancer

The hormone dependence of breast cancer that had been established by Beatson in 1896 [108] was not recognized in laboratory animals until Huggins et al. [76] demonstrated that all 3-MC treated rats exhibited a deep reduction of tumor size after hypophysectomy. Ovariectomy also reduced mammary cancer incidence by 40 %; administration of daily injections of 0.1 or 0.2 μg 17-β estradiol increased mammary cancer incidence to 100 %, whereas rats receiving 20 μg 17-β estradiol daily had a 70 % reduction in incidence. Dihydrotestosterone treatment also decreased tumor size; whereas progesterone or diethylstilbestrol administered to ovariectomized rats increased tumor incidence and enhanced the speed of tumor growth. Blocking the action of estrogens by antiestrogens that bind to the estrogen receptor alpha (ERα), such as tamoxifen [109] has demonstrated a long-lasting chemopreventive effect on mammary tumors both benign and malignant. Nevertheless, hormone-independent tumors continue growing after ovariectomy as well after prolonged treatment with tamoxifen [76, 109].

Numerous treatments have been developed for the extinction of chemically induced tumor in rodents. DMBA-treated Sprague-Dawley rats that begin receiving a daily injection of 100 IU hCG 20 days after carcinogen administration exhibit a significant reduction in mammary adenocarcinoma incidence and number of tumors per animals, an effect that becomes evident as early as 10 days after initiation of the hormonal treatment and persisted for 40 days after its termination [83, 110, 111] (Fig. 1.3). Treatment of various strains of rats with hormonal combinations, i.e., ethinyl estradiol-megestrol acetate; ethinyl estradiol-norethindrone [74, 93], or 17β estradiol-progesterone [112] two weeks after NMU administration significantly inhibits tumor progression. Protection conferred by 17β-estradiol and progesterone to BALB/c mice after treatment with DMBA administration is associated with activation of p53 in response to the hormonal treatment, which is sustained to induce p21 upon carcinogen challenge [113, 114].

1.6 Hormones as Carcinogens

The dependence of breast cancer from estrogens has been demonstrated through the induction of mammary cancer in female August/Copenhagen/Irish (ACI) rats in which administration of 17β-estradiol induces tumors that are similar to the *in situ* and invasive ductal carcinomas developed by women [115]. Estrogen-induced lesions are completely prevented by concomitant treatment with tamoxifen citrate (TAMc), confirming their estrogen receptor dependence [116]. The carcinogenicity of 17β–estradiol has been confirmed by *in vitro* experiments through the induction of neoplastic transformation of the human breast epithelial cells MCF-10 F, supporting the concept that this hormone could act as an initiator of breast cancer in women [117]. The biphasic effect of estradiol on breast cancer risk is highlighted by the findings that elevated levels of estradiol in maternal serum during the first-trimester of pregnancy are positively associated with risk of breast cancer before age 40, but inversely associated with risk in women whose breast cancer was diagnosed after the age 40 [118, 119]. Among the hormones used for contraception, medroxyprogesterone acetate (MPA) also has a biphasic effect when administered to rats of different ages before DMBA inoculation. MPA given for 21 days moderately increase the incidence of adenocarcinomas in 45 day-old rats, and more significantly in 75 day-old rats, whereas the same doses significantly reduced mammary cancer incidence in the 55 day-old rats, and indication that the same hormone and same dose exerts a biphasic effect that is modulated by the age of the rats at the time of initiation of treatment [120]. MPA administered to BALB/c female mice induces mammary ductal carcinomas in 80 % of treated animals. The tumors are hormone-dependent ERα and progesterone receptor positive that metastasize to lymph nodes and lungs; the tumors evolve through different stages of hormone dependence that are progesterone receptor dependent [121].

1.7 Role of Breast Development and Cancer

The development of the breast from birth to puberty follows a general pattern common for all normally cycling women, with the formation of Lob 1, Lob 2 and Lob 3 [81, 122]. The progression of lobular development under the cyclic influence of ovarian hormones is rapidly accelerated during the first pregnancy, which for being successful requires the timely fertilization of an ovocyte followed by its uterine implantation. The embryo drives a process that establishes a collaboration of the newly formed placenta with the maternal environment [123]. The placenta alone elaborates a myriad of proteins, glycoproteins, steroid hormones, growth factors, tumor suppressor factors and cytokines that control the local environment of the fetus and regulate the metabolic activities of both the mother and the fetus [124]. In addition to estrogen and progesterone, newly secreted hormones, such as human growth hormone (hGH); hCG, human placental lactogen (hPL), inhibin [125, 126] stimulate breast development and differentiation. Elevated serum levels of Metastin (KISS1) have been detected during pregnancy [127], but the role of this hormone in breast development has not been identified as yet. LH, progesterone and hCG are the main hormones driving the initial phase of growth, which is followed by the secretion of the pituitary hormone prolactin (PRL) that stimulates milk secretion and contributes to the development of the fully differentiated Lob 4 during the last trimester of pregnancy and lactation. After weaning Lob 4 regress to Lob 3, which persists in the breast as long as women continue cycling. At peri-menopause the number of Lob 3 progressively decreases due to their involution to Lob 2 and Lob 1 [83].

There exists consensus on the fact that an early pregnancy and increasing number of full-term pregnancies are associated with greater risk reduction for invasive breast cancer compared with never-pregnant women and women whose first full-term pregnancy occurred after 30 years of age [2, 3] (Fig. 1.1). Data on the mechanisms mediating the protection conferred by early pregnancy or the increased risk resulting from delayed first birth in women are discussed in Chap. 10. Epidemiological data have provided useful information on cancer causation through the identification of deleterious environmental exposures, such as radiation exposure, either accidental [35] or diagnostic and therapeutic [15]; and passive or active smoking at an early age [128] (Fig. 1.1). The association of these early exposures with the development of breast cancer are an indication that the initiation of breast epithelial cell transformation took place several years before a first FTP during a window of susceptibility similar to that described in the rodent experimental model (Fig. 1.3). The hormonal milieu of pregnancy might stimulate the progression of pre-existing pre-neoplastic lesions or small tumors that would be diagnosed during pregnancy or within one or two years following delivery as pregnancy associated breast cancer (PABC) [129], whose incidence is likely to increase because of the continuous trend towards postponement of childbearing [130, 131]. PABC diagnosed shortly after delivery has a worse prognosis and more pronounce mortality than breast cancer in women with no PABC. Unfortunately, the knowledge of the response of the breast to the hormones of pregnancy is scant and insufficient for understanding the complex interactions between pregnancy hormones required for stimulating fetal growth and breast development and those that activate differentiation pathways exerting a cancer risk reduction effect [132]. This is a field that requires extensive investigation for clarifying these hormonal-carcinogen relationships.

Current knowledge on the development of the breast before and during pregnancy has been acquired through the study of *in vivo* and *in vitro* experimental models, autopsy material, or cancer-free breast tissues obtained from biopsies or reduction mammoplasties [83, 84, 110, 111]. Conclusions drawn through these studies had to be extrapolated to the conditions prevailing in the breast during pregnancy due to the difficulties in obtaining breast tissue specimens during the gestational process. In order to understand how the dramatic modifications that occur during pregnancy in the pattern of lobular development and differentiation, cell proliferation and hormone receptors influence a woman's lifetime cancer risk, it has been necessary to analyse the pattern of gene expression in the epithelium of Lob 1 in breast biopsies of parous and nulliparous postmenopausal women with a history of invasive breast cancer (cases) or benign breast disease without hyperplasia or atypia (controls). Genomic and Gene Ontology (GO) analyses revealed that the mRNA obtained from the breast of parous women free of cancer had a higher level of gene expression in processes that included proteolysis and ubiquitination, cell adhesion, response to exogenous agents, metabolism, DNA repair and replication, RNA processing, apoptosis, anti-apoptosis, and chromatin modifications [133, 134]. These data indicate that parous women after menopause who had not developed breast cancer exhibit a genomic "signature" that differs from that present in the breast of parous postmenopausal women with cancer or in nulliparous women, who traditionally represent a high breast cancer risk group [133, 134] (Fig. 1.1). The importance of these findings is that they support the concept that for being protective, parity should occur during a specific time of breast development, which is represented by the interplay of the HRSW/HPW and that the stem progenitor cells must be programmed for differentiation before any carcinogenic insult has activated them for becoming cancer stem cells (Figs. 1.1, 1.2 and 1.3). If a carcinogenic insult reaches the progenitor stem cells of the breast before the first FTP, then pregnancy, instead of being protective, might stimulate the progression of transformed cells.

1.8 Fertility and Breast Cancer Risk

In Chap. 10 we described the genomic changes taking place in the human breast that explain why women that have completed their first pregnancy before age 24 have significantly reduced risk of developing breast cancer after menopause, whereas women that become first pregnant after age 30 or that are younger than 40 years of age at the time of breast cancer diagnosis are not protected by parity. Studies of the association of cohort fertility and breast cancer incidence have revealed that childbearing trends account at least in part for cohort variations in breast cancer incidence in the Western world. In the United States the average age of mothers at the first birth was 21.4 in 1970, increasing 3.6 years by 2006 [135]. During this period the proportion of first births to women aged 35 years and over increased nearly eight times; whereas the percentage of first time mothers under age 20 was reduced from 36 % to 21 %, and the number of invasive breast cancer cancers increased 2.71 fold (171 %), in contrast with an increase in the female population of 1.29 fold (29 %), an indication that delayed first birth played a greater role in the increased breast cancer incidence than population growth [135, 136]. These epidemiological data support our hypothesis that in order for a first FTP to be protective it should occur during specific periods of time, the HRSW/HPW interface.

The understanding of the mechanisms that determine the variability in breast cancer risk in relation to age at the first full term pregnancy requires to take into consideration not only the development and differentiation of the breast at various ages, but of the organs that control both breast development and the capabilities of women to be fertile, namely the ovary and the HPG axis [137, 138]. The fertility of a sexually mature woman is established as early as the 4th month of her fetal life, when the number of ovarian primordial follicles is set. At birth they number approximately 5 million; progressively decreasing to 500,000 when the girl reaches menarche [139]. Between the initiation of ovulation and up to 30 years of age, the monthly fecundity rate (MFR), i.e., the ability of a woman to become pregnant, ranges from 20 to 25 % [139]. Continuous ovulation and associated follicular atresia and apoptosis further reduce the natural MFR to below 10 % by age 35 and even more thereafter, resulting in either sub fertility or clinical infertility, which affects approximately 15 % of women in developed countries [140]. With aging of the ovary the size of recruited follicles declines, with a concomitant reduction in the number of follicles entering maturation and of pre-ovulatory oocytes. Under these conditions the quality of the oocytes declines, leading to lower pregnancy rates even after infertility treatments. In addition, embryo aneuploidy and miscarriage rates increase, ultimately resulting in poorer delivery rates [141].

Although follicular depletion is a critical factor mediating both fertility and menopausal transition, the age-related decline in reproductive function is affected by changes in all levels of the HPG axis [137]. The major hormonal elements that hierarchically define this system are the hypothalamic GnRH, and the pituitary gonadotropins LH and follicle stimulating hormone (FSH) that controls the synthesis and release of ovarian sex steroids and peptides. Accumulating evidence indicates that during the earliest stages of reproductive decline, changes at the level of the brain play an important role in the initiation of reproductive senescence, which is manifested as a dampened and delayed GnRH-driven pro-estrous LH surge [137, 141]. These data emphasize the need to understand how endocrinologic changes occurring in response to aging affect on the ability of the breast to respond to the hormonal effects of pregnancy or of exogenous hormones that might be used for the purposes of preventing breast cancer.

1.9 Conclusions

The identification of a "high risk window" for every individual will provide a physiological framework for minimizing exposures to environmental carcinogens during the high susceptibility periods (Figs. 1.1, 1.2 and 1.3). The ultimate goal would be to narrow the "high risk window" by inducing full differentiation of the breast before it is "hit" by one or many of the known or suspected environmental carcinogens (see Sect. 1.4). The mechanisms described above emphasize the importance of a timely differentiation of the breast for preventing breast cancer. The fact that only full term pregnancy occurring during a narrow window in the development of the breast would succeed in permanently inducing the molecular changes that will make the breast resistant to develop cancer imposes a heavy burden on reproductive biologists, breast specialists, endocrinologists and molecular biologists for developing the adequate strategies for preventing breast cancer in continuously changing populations (Fig. 1.3). A hope is offered by the observations that it is possible to protect the breast from cancer initiation and progression with hCG, one of the hormones produced by the human placenta (see Chap. 10). The implications of these observations are two-fold: on one hand, they indicate that hCG may induce early genomic changes that control the progression of the differentiation pathway, and that these changes are permanently imprinted in the genome, regulating the long-lasting refractoriness of the breast to develop cancer. The permanence of these changes would, in turn, make them ideal surrogate markers for the evaluation of novel chemopreventive agents designed for preventing breast cancer in the immediate future.

References

1. Breasted JH (ed) (1991) The Edwin Smith Surgical Papyrus: published in facsimile and hieroglyphic transliteration with translation and commentary in two volumes, vol 1. University of Chicago Press, Chicago
2. Clarke CA, Purdie DM, Glaser SL (2006) Population attributable risk of breast cancer in white women associated with immediately modifiable risk factors. BMC Cancer 6:170
3. Botha JL, Bray F, Sanlika R, Parkin DM (2003) Breast cancer incidence and mortality trends in 16 European countries. Eur J Cancer 39:1718–1729
4. Jemal A, Bray F, Center MM, Ferlay J, Ward E, Forman D (2011) Global cancer statistics. CA Cancer J Clin 61:69–90
5. Nagata C, Mizoue T, Tanaka K, Tsuji I, Wakai K, Inoue M et al (2006) Tobacco smoking and breast cancer risk: an evaluation based on a systematic review of epidemiological evidence among the Japanese population. Jpn J Clin Oncol 36:387–394
6. Maskarinec G, Pagano I, Chen Z, Nagata C, Gram IT (2007) Ethnic and geographic differences in mammographic density and their association with breast cancer incidence. Breast Cancer Res Treat 104:47–56
7. Althuis MD, Dozier JM, Anderson WF, Devesa SS, Brinton LA (2005) Global trends in breast cancer incidence and mortality 1973-1997. Int J Epidemiol 34:405–412
8. Jemal A, Murray T, Samuels A, Ghafoor A, Ward E, Thun MJ (2003) Cancer Statistics, 2003. CA Cancer J Clin 53:5–26
9. Howe HL, Wingo PA, Thun MJ, Ries LA, Rosenberg HM, Feigal EG, Edwards BK (2001) Annual report to the nation on the status of cancer (1973 through 1998), featuring cancers with recent increasing trends. J Natl Cancer Inst 93:824–842
10. Buell P (1973) Changing incidence of breast cancer in Japanese-American women. J Natl Cancer Inst 51:1479–1483
11. Madigan MP, Ziegler RG, Benichou J, Byrne C, Hooper RN (1995) Proportion of breast cancer cases in the United States explained by well-established risk factors. J Natl Cancer Inst 87:1681–1685
12. Colditz GA, Frazier AL (1995) Models of breast cancer show that risk is set of events of early life: prevention efforts must shift focus. Cancer Epidemiol Biomarkers Prev 4:567–571
13. Segaloff A, Maxfield WS (1971) The synergism between radiation and estrogen in the production of mammary cancer in the rat. Cancer Res 31:166
14. Hancock SL, Tucker MA, Hoppe RT (1993) Breast cancer after treatment of Hodgkin's disease. J Natl Cancer Inst 85:25–31
15. Cutuli B, Borel C, Dhermain F, Magrini SM, Wasserman TH, Bogart JA, Provencio M, de Lafontan B, de la Rochefordiere A, Cellai E, Graic Y, Kerbrat P, Alzieu C, Teissier E, Dilhuydy J, Mignotte H, Velten M (2001) Breast cancer occurred after treatment for Hodgkin's disease: analysis of 133 cases. Radiother Oncol 59:247–255
16. Gaffney DK, Hemmersmeier J, Holden J, Marshall J, Smith LM, Avizonis V, Tran T, Neuhausen SL (2001) Breast cancer after mantle irradiation for Hodgkin's disease: correlation of clinical, pathologic, and molecular features including loss of heterozygosity at BRCA1 and BRCA2. Int J Radiat Oncol Biol Phys 49:539–546
17. Janov AJ, Tulecke M, O'Neill A, Lester S, Mauch PM, Harris J, Schnitt SJ, Shapiro CL (2001) Clinical and pathologic features of breast cancers in women treated for Hodgkin's disease: a case-control study. Breast J 7:46–52
18. Lash TL, Aschengrau A (2000) Alcohol drinking and risk of breast cancer. Breast J 6:396–399
19. Spangler JG, Michielutte R, Bell RA, Dignan MB (2001) Association between smokeless tobacco use and breast cancer among Native-American women in North Carolina. Ethn Dis 11:36–43
20. Satcher D (2001) Women and health: a report of the Surgeon General - 2001. U.S. Dept. Health and Human Serv., PHS Office of the Surgeon General, 1–4

21. Russo IH, Tahin Q, Huang Y, Russo J (2001) Cellular and molecular changes induced by the chemical carcinogen benzo(a)pyrene in human breast epithelial cells and its association with smoking and breast cancer. J Women Cancer 3:29–36
22. Baron JA (2001) Breast cancer, hormone-related disorders and cigarette smoking. J Women Cancer 3:23–28
23. Phillips DH, Martin FL, Grover PL, Williams JA (2001) Toxicological basis for a possible association of breast cancer with smoking and other sources of environmental carcinogens. J Women Cancer 3:9–16
24. Ambrosone CB (2001) Impact of genetics on the relationship between smoking and breast cancer risk. J Women Cancer 3:17–22
25. Morabia A, Ambrosone CB, Baron JA, Phillips DH, Russo IH (2001) What do we currently know about the epidemiological and biologic plausibility of the association of smoking and breast cancer? J Women Cancer 3:5–8
26. Blot WJ, Fraumeni JF, Stone BJ (1977) Geographic patterns of breast cancer in the United States. J Natl Cancer Inst 59:1407–1411
27. Sturgeon SR, Schairer C, Gail M, McAdams M, Brinton LA, Hoover RN (1995) Geographic variation in mortality from breast cancer among white women in the United States. J Natl Cancer Inst 87:1846–1853
28. Nasca PC, Mahoney MC, Wolfgang PE (1992) Population density and cancer incidence differentials in New York State. Cancer Causes Control 3:7–15
29. Robbins AS, Brescianini S, Kelsey JL (1997) Regional differences in known risk factors and the higher incidence of breast cancer in San Francisco. J Natl Cancer Inst 89:960–965
30. Clarke CA, Glaser SL, West DW, Ereman RR, Erdmann CA, Barlow JM, Wrensch MR (2002) Breast cancer incidence and mortality trends in an affluent population: Marin County, California, USA, 1990–1996. Breast Cancer Res 4:R13
31. Garland FC, Garland CF, Gorham ED, Young JF (1990) Geographic variation in breast cancer mortality in the United States: a hypothesis involving exposure in solar radiation. Prev Med 19:614–622
32. Eisman JA, McIntyre I, Marti TJ, Frampton RJ, King RJ, King RJ (1980) Normal and malignant breast tissue is a target organ for 1,25-$(OH)_2$ vitamin D_3. Clin Endocrinol 13:267–272
33. Gorham ED, Garland FC, Garland CF (1990) Sunlight and breast cancer incidence in the USSR. Int J Epidemiol 19:820–824
34. National Council on Radiation Protection and Measurements (1987) Ionizing radiation exposures of the population of the United States. Rep. 93, Natl. Counc. Radiation. Prot. Meas., Washington, D.C.
35. McGregor DH, Land CE, Choi K, Tokuoka S, Liu PI, Wakabayashi I, Beebe GW (1977) Breast cancer incidence among atomic bomb survivors, Hiroshima and Nagasaki 1950–1989. J Natl Cancer Inst 59:799–811
36. Boice JD Jr, Preston D, Davis FG, Monson RR (1991) Frequent chest X-ray fluoroscopy and breast cancer incidence among tuberculosis patients in Massachusetts. Radiat Res 125:214–222
37. Boice JD Jr, Land CE, Preston DL (1996) Ionizing radiation. In: Scottenfeld DL, Fraumedni JF Jr (eds) Cancer epidemiology and prevention, 2nd edn. Oxford Univ. Press, New York, pp 319–372
38. National Academy Press (1990) Health effects of exposure to low levels of ionizing radiation (BIER V). National Academy Press, Washington, D.C., 421 pp
39. Brenner DJ, Sawant SG, Hande MP, Miller RC, Elliston CD, Fu Z, Randers-Pehrson G, Marino SA (2002) Routine screening mammography: how important is the radiation-risk side of the benefit-risk equation? Int J Radiat Biol 78:1065–1067
40. U.S. Department of Energy (1996) Energy Information Administration. The changing structure of the Electric Power Industry; An Updata Rep. DOE/EIA-0562(96). U.S. Dept. Energy, Washington, D.C.
41. Stevens RG (1987) Electric power use and breast cancer: a hypothesis. Am J Epidemiol 125:556–561

42. Tamarkin L, Baird CJ, Almeida OF (1985) Melatonin: a coordinating signal for mammalian reproduction? Science 227:714–720
43. Wurtman RJ, Axelrod J (1965) The pineal gland. Sci Am 213:50–60
44. Reiter RJ (1994) Melatonin suppression by static and extremely low frequency electromagnetic fields: relationship to the reported increased incidence of cancer. Rev Environ Health 10:171–186
45. Baum A, Mevissen M, Kamino K, Mohr U, Loscher W (1995) A histopathological study on alterations in DMBA-induced mammary carcinogenesis in rats with 50 Hz, 100 muT magnetic field exposure. Carcinogenesis 16:119–125
46. Liburdy RP, Loscher W (1997) Laboratory studies on extremely low frequency (50/60-Hz) magnetic fields and carcinogenesis. In: Stevens RG, Wilson BW, Anderson LE (eds) The melatonin hypothesis, breast cancer and use of electric power. Batelle Press, Columbus, pp 585–668
47. Liburdy RP, Sloma TR, Sokolic R, Yaswen P (1993) ELF magnetic fields, breast cancer, and melatonin: 60 Hz fields block melatonin's oncostatic action on ER+breast cancer cell proliferation. J Pineal Res 14:84–97
48. Loscher W, Mevissen M (1994) Mini-review: animal studies on the role of 50/60-Hertz magnetic fields in carcinogenesis. Life Sci 54:1531–1543
49. Wertheimer N, Leeper E (1987) Magnetic field exposure related to cancer subtypes. Ann NY Acad Sci 502:43–54
50. Florig HK, Hoburg JF (1990) Power-frequency magnetic fields from electric blankets. Health Phys 58:493–502
51. Preston-Martin S, Peters JM, Yu MC, Garabrant DH, Bowman JD (1988) Myelogenous leukemia and electric blanket use. Bioelectromagnetics 9:207–213
52. Vena JE, Freudenheim JL, Marshall R Jr, Laughlin R, Swanson M, Graham S (1994) Risk of premenopausal breast cancer and use of electric blankets. Am J Epidemiol 140:974–979
53. Moysich KB, Ambrosone CB, Vena JE, Kostyniak P, Shields PG et al (1997) Serum polychlorinated biphenyls (PCBs) and postmenopausal breast cancer risk. Proc Am Assoc Cancer Res 38:627a
54. Kutz FW, Wood PH, Bottimore DP (1991) Organochlorine pesticides and polychlorinated biphenyls in human adipose tissue. Rev Environ Contam Toxicol 120:1–82
55. Adami H-O, Lipworth L, Titus-Ernstoff L, Hsieh C-C, Hanberg A et al (1995) Organochlorine compounds and estrogen-related cancers in women. Cancer Causes Control 6:551–566
56. Ahlborg UG, Lipworth L, Titus-Ernstoff L, Hsieh C-C, Hanberg A et al (1995) Organochlorine compounds in relation to breast cancer, endometrial cancer, and endometriosis: an assessment of the biological and epidemiological evidence. Crit Rev Toxicol 25:463–531
57. Wolff MS, Collman GW, Barrett JC, Huff J (1996) Breast cancer and environmental risk factors: epidemiological and experimental findings. Annu Rev Pharmacol Toxicol 36:573–596
58. Bergthorsson JT, Ejlertsen B, Olsen JH, Borg A, Nielsen KV, Barkardottir RB, Klausen S, Mouridsen HT, Winther K, Fenger K, Niebuhr A, Harboe TL, Niebuhr E (2001) BRCA1 and BRCA2 mutation status and cancer family history of Danish women affected with multifocal or bilateral breast cancer at a young age. J Med Genet 38:361–368
59. Greenblatt MS, Chappuis PO, Bond JP, Hamel N, Foulkes WD (2001) TP53 mutations in breast cancer associated with BRCA1 or BRCA2 germ-line mutations: distinctive spectrum and structural distribution. Cancer Res 61:4092–4097
60. Lehman TA, Haffty BG, Carbone CJ, Bishop LR, Gumbs AA, Krishnan S, Shields PG, Modali R, Turner BC (2000) Elevated frequency and functional activity of a specific germline p53 intron mutation in familial breast cancer. Cancer Res 60:1062–1069
61. Kelsey JL, Horn-Ross PL (1993) Breast cancer: magnitude of the problem and descriptive epidemiology. Epidemiol Rev 15:7–16
62. MacMahon B, Cole P, Lin TM et al (1970) Age at first birth and breast cancer risk. Bull World Health Organ 43:209–217

63. Vessey MD, McPherson K, Roberts MM, Neil A, Jones L (1985) Fertility and the risk of breast cancer. Br J Cancer 52:625–628
64. Trapido EJ (1983) Age at first birth, parity and breast cancer risk. Cancer 51:946–948
65. De Waard F, Trichopoulos D (1988) A unifying concept of the etiology of breast cancer. Int J Cancer 41:666–669
66. Lambe M, Hsieh CC, Chan HW, Ekbom A, Trichopoulos D, Adami HO (1996) Parity, age at first and last birth, and risk of breast cancer: a population-based study in Sweden. Breast Cancer Res Treat 38:305–311
67. Meiser B, Butow P, Barratt A, Friedlander M, Gattas M, Kirk J, Suthers G, Walpole I, Tucker K (1999) Attitudes toward prophylactic oophorectomy and screening utilization in women at increased risk of developing hereditary breast/ovarian cancer. Gynecol Oncol 75:122–129
68. Feychting M, Osterlund B, Ahlbom A (1998) Reduced cancer incidence among the blind. Epidemiology 9:490–494
69. Wurtman RJ (1984) Fall of nocturnal melatonin during puberty and prepubescence. Lancet 362:85
70. Dauchy RT, Blask DE, Sauer LA, Brainard GC, Krause JA (1999) Dim light during darkness stimulates tumor progression by enhancing tumor fatty acid uptake and metabolism. Cancer Lett 144:131–136
71. Tamarkin L, Cohen M, Roselle D, Reichert C, Lippman M, Chabner B (1981) Melatonin inhibition and pinealectomy enhancement of 7,12-dimethylbenz(a)anthracene-induced mammary tumors in the rat. Cancer Res 44:4432–4436
72. Pillittere D, Miller M (2000) Researchers search for link between circadian rhythms, breast cancer. J Natl Cancer Inst 92:686–689
73. Bernstein L, Hanisch R, Sullivan-Halley J, Ross RK (1995) Treatment with human chorionic gonadotropin and risk of breast cancer. Cancer Epidemiol Biomarkers Prev 4:437–440
74. Russo IH, Russo J (1996) Mammary gland neoplasia in long-term rodent studies. Environ Health Perspect 104:938–967
75. Rao GN, Piegorsch WW, Haseman JK (1987) Influence of body weight on the incidence of spontaneous tumors in rats and mice of long term studies. Am J Clin Nutr 45:252–260
76. Huggins C, Briziarelli G, Sutton H (1959) Rapid induction of mammary carcinoma in the rat and the influence of hormones on the tumors. J Exp Med 709:25–42
77. Huggins C, Grand L, Fukunishi R (1964) Aromatic influences in the yields of mammary cancers following administration of 7,12-dimethylbenzanthracene. Proc Natl Acad Sci U S A 57:737–742
78. Gullino PM, Pettigrew HM, Grantham FH (1975) N-nitrosomethylurea as mammary gland carcinogen in rats. J Natl Cancer Inst 54:401–414
79. Thordarson G, Lee AV, McCarty M, Van Horn K, Chu O, Chou YC et al (2001) Growth and characterization of N-methyl-N-nitrosourea-induced mammary tumors in intact and ovariectomized rats. Carcinogenesis 22:2039–2047
80. Medina D, Kittrell FS, Hill J, Shepard A, Thordarson G, Brown P (2005) Tamoxifen inhibition of estrogen receptor-α–negative mouse mammary tumorigenesis. Cancer Res 65:3493–3496
81. Russo J, Russo IH (1987) Development of human mammary gland. In: Neville MC, Daniel C (eds) The mammary gland development, regulation and function. Plenum Publishing Corp, New York, pp 67–93
82. Russo J, Gusterson BA, Rogers AE, Russo IH, Wellings SR, Van Zwieten MJ (1990) Comparative study of human and rat mammary tumorigenesis. Lab Invest 62:1–32
83. Russo J, Russo IH (eds) (2004) Molecular basis of breast cancer: prevention and treatment. Springer Verlag, Berlin
84. Phipps AI, Chlebowski RT, Prentice R, McTiernan A, Wactawski-Wende J, Kuller LH et al (2011) Reproductive history and oral contraceptive use in relation to risk of triple-negative breast cancer. J Natl Cancer Inst 103:1–8
85. Jemal A, Bray F, Center MM, Ferlay J, Ward E, Forman D (2011) Global cancer statistics. CA Cancer J Clin 61:69–90

86. MacMahon B, Cole P, Lin TM, Lowe CR, Mirra AP, Ravnihar B et al (1970) Age at first birth and breast cancer risk. Bull World Health Organ 43:209–221
87. Hinkula M, Pukkala E, Kyyrönen P, Kauppila A (2001) Grand multiparity and the risk of breast cancer: population-based study in Finland. Cancer Causes Control 12:491–500
88. Ma H, Henderson KD, Sullivan-Halley J, Duan L, Marshall SF, Ursin G et al (2010) Pregnancy-related factors and the risk of breast carcinoma in situ and invasive breast cancer among postmenopausal women in the California Teachers Study cohort. Breast Cancer Res 12:R35
89. Toniolo P, Grankvist K, Wulff M, Chen T, Johansson R, Schock H et al (2010) Human chorionic gonadotropin in pregnancy and maternal risk of breast cancer. Cancer Res 70:6779–6786
90. Mustacchi P (1961) Ramazzini and Rigoni-Stern on parity and breast cancer. Clinical impression and statistical corroboration. Arch Intern Med 108:639–642
91. Kroman N, Mouridsen HT (2003) Prognostic influence of pregnancy before, around, and after diagnosis of breast cancer. Breast 12:516–521
92. Kroman N, Melbye M, Mouridsen HT (2002) Prognostic influence of age at diagnosis in premenopausal breast cancer patients. Scand J Surg 91:305–308
93. Welsch CW (1985) Host factors affecting the growth of carcinogen-induced rat mammary carcinomas: a review and tribute to Charles Brenton Huggins. Cancer Res 45:3415–3443
94. Moral R, Wang R, Russo IH, Lamartiniere CA, Pereira J, Russo J (2008) Effect of prenatal exposure to the endocrine disruptor bisphenol A on mammary gland morphology and gene expression signature. J Endocrinol 196:101–112
95. Kawaguchi H, Miyoshi N, Miyamoto Y, Souda M, Umekita Y, Yasuda N et al (2009) Effects of fetal exposure to diethylstilbestrol on mammary tumorigenesis in rats. J Vet Med Sci 71:1599–1608
96. Umekita Y, Souda M, Hatanaka K, Hamada T, Yoshioka T, Kawaguchi H et al (2011) Gene expression profile of terminal end buds in rat mammary glands exposed to diethylstilbestrol in neonatal period. Toxicol Lett 205:15–25
97. Goodman A, Schorge J, Greene MF (2011) The long-term effects of in utero exposures–the DES story. N Engl J Med 364:2083–2084
98. Russo IH, Russo J (2007) Primary prevention of breast cancer by hormone-induced differentiation. Recent Results Cancer Res 174:111–130
99. Russo J, Tait L, Russo IH (1983) Susceptibility of the mammary gland to carcinogenesis: III the cell of origin of rat mammary carcinoma. Am J Pathol 113:50–66
100. Russo J, Balogh GA, Chen J, Fernandez SV, Fernbaugh R, Heulings R et al (2006) The concept of stem cell in the mammary gland and its implication in morphogenesis, cancer and prevention. Front Biosci 11:151–172
101. Bennett DC, Peachey LA, Durbin H, Rudland PS (1978) A possible mammary stem cell line. Cell 15:283–298
102. Bussard KM, Smith GH (2011) The mammary gland microenvironment directs progenitor cell fate in vivo. Int J Cell Biol. doi:10.1155/2011/451676
103. Hernandez-Gonzalez M, Juarez J (2000) Alcohol before puberty produces an advance in the onset of sexual behavior in male rats. Alcohol 21(2):133–140
104. Harris C (2001) Tobacco smoking and cancer risk in women. J Women Cancer 3:1–4
105. Russo J, Barnabas N, Higgy N, Salicioni AM, Wu YL, Russo IH (1996) Molecular basis of human breast cell transformation. In: Calvo F, Crepin M, Magdelenat H (eds) Advances in breast cancer research. Eurotext, Paris, pp 33–43
106. Stevens RG, Davis S, Mirick DK, Kheifets L, Kaune W (2000) Alcohol consumption and urinary concentration of 6-sulfatoxymelatonin in healthy women. Epidemiology 11(6):660–665
107. Falcon J (1999) Cellular circadian clocks in the pineal [Review]. Prog Neurobiol 58:121–162
108. Beatson GT (1896) On the treatment of inoperable cases of carcinoma of the mamma: suggestions for a new method of treatment with illustrative cases. Lancet ii:104

109. Jordan VC (1976) Effect of tamoxifen (ICI 46,474) on the initiation and growth of DMBA-induced rat mammary carcinoma. Eur J Cancer 12:419–424
110. Russo IH, Koszalka M, Gimotty PA, Russo J (1990) Protective effect of chorionic gonadotropin on DMBA-induced mammary carcinogenesis. Br J Cancer 62:243–247
111. Russo IH, Koszalka M, Russo J (1991) Comparative study of the influence of pregnancy and hormonal treatment on mammary carcinogenesis. Br J Cancer 64:481–484
112. Blakely CM, Stoddard AJ, Belka GK, Dugan KD, Notarfrancesco KL, Moody SE et al (2006) Hormone-induced protection against mammary tumorigenesis is conserved in multiple rat strains and identifies a core gene expression signature induced by pregnancy. Cancer Res 66:6421–6431
113. Medina D, Kittrell FS (2003) p53 function is required for hormone-mediated protection of mouse mammary tumorigenesis. Cancer Res 63:6140–6143
114. Rajkumar L, Kittrell FS, Guzman RC, Brown PH, Nandi S, Medina D (2007) Hormone-induced protection of mammary tumorigenesis in genetically engineered mouse models. Breast Cancer Res 9:R12
115. Weroha SJ, Li SA, Tawfik O, Li JJ (2006) Overexpression of cyclins D1 and D3 during estrogen-induced breast oncogenesis in female ACI rats. Carcinogenesis 27:491–498
116. Li SA, Weroha SJ, Tawfik O, Li JJ (2002) Prevention of solely estrogen-induced mammary tumors in female aci rats by tamoxifen: evidence for estrogen r eceptor mediation. J Endocrinol 175:297–305
117. Russo J, Fernandez SV, Russo PA, Fernbaugh R, Sheriff FS, Lareef HM, Garber J, Russo IH (2006) 17-Beta-estradiol induces transformation and tumorigenesis in human breast epithelial cells. FASEB J 20:1622–1634
118. Chen T, Lundin E, Grankvist K, Zeleniuch-Jacquotte A, Wulff M, Afanasyeva Y et al (2010) Maternal hormones during early pregnancy: a cross-sectional study. Cancer Causes Control 21:719–727
119. Lukanova A, Surcel HM, Lundin E, Kaasila M, Lakso HA, Schock H et al (2011) Circulating estrogens and progesterone during primiparous. pregnancies and risk of maternal breast cancer. Int J Cancer 130:910–920
120. Russo IH, Gimotty P, Dupuis M, Russo J (1989) Effect of medroxyprogesterone acetate on the response of the rat mammary gland to carcinogenesis. Br J Cancer 59:210–216
121. Lanari C, Lamb CA, Fabris VT, Helguero LA, Soldati R, Bottino MC et al (2009) The MPA mouse breast cancer model: evidence for a role of progesterone receptors in breast cancer. Endocr Relat Cancer 16:333–350
122. Russo J, Rivera R, Russo IH (1992) Influence of age and parity on the development of the human breast. Breast Cancer Res Treat 23:211–218
123. Lathi RB, Fisher SJ, Giudice LC (2006) Implantation and placental physiology in early human pregnancy: the role of the maternal decidua and the trophoblast. In: De Groot L, Jameson LJ (eds) Endocrinology. Elsevier Saunders, Philadelphia, pp 3341–3351
124. Parry S, Strauss F III (2006) Placental hormones. In: De Groot L, Jameson LJ (eds) Endocrinology. Elsevier Saunders, Philadelphia, pp 3353–3367
125. Alvarado MV, Ho T-Y, Russo J, Russo IH (1994) Human chorionic gonadotropin regulates the synthesis of inhibin in the ovary and the mammary gland of rats. Endocrine 2:1–10
126. Alvarado ME, Alvarado NE, Russo J, Russo IH (1994) Human chorionic gonadotropin inhibits proliferation and induces expression of inhibin in human breast epithelial cells *in vitro*. In Vitro 30A:4–8
127. Horikoshi Y, Matsumoto H, Takatsu Y, Ohtaki T, Kitada C, Usuki S et al (2003) Dramatic elevation of plasma metastin concentrations in human pregnancy: metastin as a novel placenta-derived hormone in humans. J Clin Endocrinol Metab 88:914–919
128. Johnson KC, Miller AB, Collishaw NE, Palmer JR, Hammond SK, Salmon AG et al (2011) Active smoking and secondhand smoke increase breast cancer risk: the report of the Canadian Expert Panel on Tobacco Smoke and Breast Cancer Risk (2009). Tob Control 20, e2
129. Johansson AL, Andersson TM, Hsieh CC, Cnattingius S, Lambe M (2011) Increased mortality in women with breast cancer detected during pregnancy and different periods postpartum. Cancer Epidemiol Biomarkers Prev 20:1865–1872

130. Hahn RA, Moolgavkar SH (1989) Nulliparity, decade of first birth, and breast cancer in Connecticut cohorts, 1855 to 1945: an ecological study. Am J Public Health 79:1503–1507
131. Mathews TJ, Hamilton BE (2009) Delayed childbearing: more women are having their first child later in life. National Center for Health Statistics data brief No. 21, Hyattsville, MD
132. Lakshmanaswamy R, Guzman RC, Nandi S (2008) Hormonal prevention of breast cancer: significance of promotional environment. Adv Exp Med Biol 617:469–475
133. Balogh GA, Heulings R, Mailo DA, Russo PA, Sheriff F, Russo IH et al (2006) Genomic signature induced by pregnancy in the human breast. Int J Oncol 28:399–410
134. Russo J, Balogh GA, Russo IH (2008) Full-term pregnancy induces a specific genomic signature in the human breast. Cancer Epidemiol Biomarkers Prev 17:51–66
135. George K, Kamath MS (2010) Fertility and age. J Hum Reprod Sci 3:121–123
136. Homan GF, Davies M, Norman R (2007) The impact of lifestyle factors on reproductive performance in the general population and those undergoing infertility treatment. Hum Reprod Update 13:209–223
137. Downs JL, Wise PM (2009) The role of the brain in female reproductive aging. Mol Cell Endocrinol 299:32–38
138. Russo IH, Medado J, Russo J (1989) Endocrine influences on mammary gland structure and development. In: Jones TC, Mohr U, Hunt RD (eds) Integument and mammary gland of laboratory animals. Springer Verlag, Berlin, pp 252–266
139. Gleicher N, Weghofer A, Barad DH (2011) Defining ovarian reserve to better understand ovarian aging. Reprod Biol Endocrinol 9:23
140. Danforth DR, Arbogast LK, Mroueh J, Kim MH, Kennard EA, Seifer DB et al (1998) Dimeric inhibin: a direct marker of ovarian ageing. Fertil Steril 70:119–123
141. Rance NE (2009) Menopause and the human hypothalamus: evidence for the role of kisspeptin/neurokinin B neurons in the regulation of estrogen negative feedback. Peptides 30:111–122

Chapter 2
The So Called Pre-Neoplastic Lesions and Carcinoma In Situ

2.1 Introduction

In situ breast cancer represents 15 % to 30 % of all diagnosed cancer, from all the *in situ* breast cancer 80 % of them are ductal carcinoma or DCIS [1, 2]. Similar to invasive breast cancer, DCIS is not a single disease but rather many different diseases, each with its own clinical, morphologic, and molecular characteristics [3]. Ductal carcinoma *in situ* of the breast is characterized by malignant epithelial cells confined to the ductal system of the breast without evidence of invasion through the basement membrane into the surrounding stroma [4, 5].

DCIS constitutes 30 % to 40 % of the breast cancer cases diagnosed mammographically, however, only 1 out 1300 screening mammograms are carcinoma in situ. The most prominent feature in the mammogram is the presence of micro calcifications or as non-palpable masses or combination of both [6–8].

2.2 The So Called Pre-neoplastic Lesions

Although the sequence from normal to ductal hyperplasia, atypical ductal hyperplasia, carcinoma invasive and invasive has been considered the natural progression of the disease [6], there are some evidence that the ductal hyperplasia has few similarities to ADH, DCIS, or invasive cancer. Whereas ADH was shown to have many similarities to low-grade DCIS, such as losses at 16q and 17p and gains at 1q.7 [9, 10] In contrast, low-grade DCIS appears to be genetically distinct from high-grade DCIS [9, 10].

J. Russo, *The Pathobiology of Breast Cancer*,
DOI 10.1007/978-3-319-40815-6_2

2.2.1 Ductal Hyperplasia

Epithelial hyperplasia of ductal type, has been classified as *mild* (when made up of three or four epithelial cells in thickness), *moderate to florid* (when more pronounced), and *atypical*. Nuclei are oval, normochromatic and with slight overlap; small, single, indistinct nucleoli; scanty or no mitotic activity (Figs. 2.1, 2.2 and 2.3). The cytoplasm is acidophilic and finely granular (Figs. 2.4 and 2.5). An interesting feature is that cytoplasmic borders are not well demarcated giving a syncytial appearance. The intratubular lumina of ductal hyperplasia tend to be irregular in

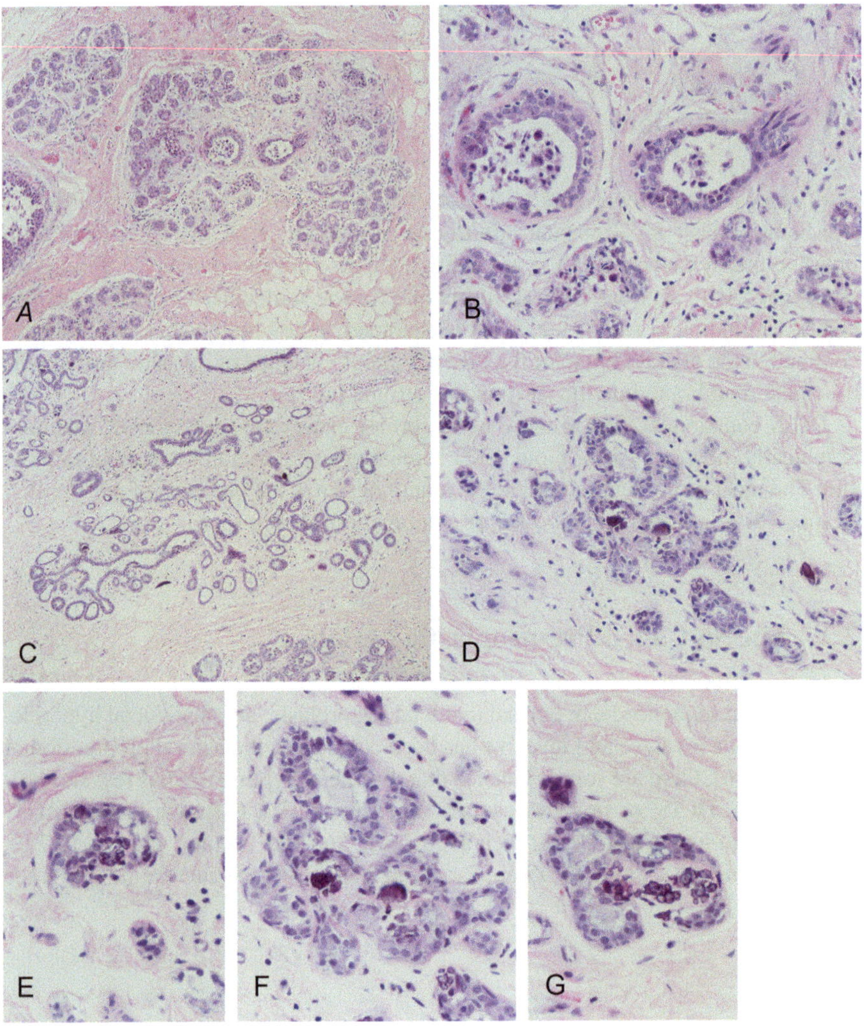

Fig. 2.1 Mild ductal hyperplasia. (**a**): 4×; (**b**): 10×; (**c**): 4×; Microcalcificatons in the lumen are found in (**d**): 10×; (**e, f** and **g**): 10×

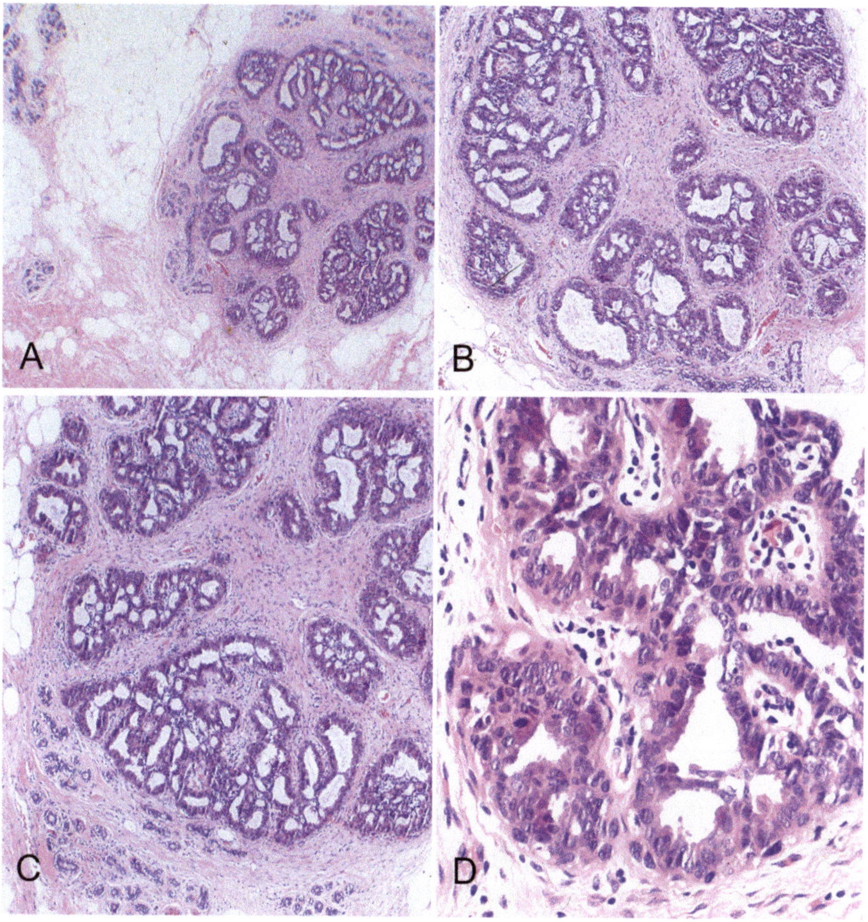

Fig. 2.2 Mid to moderate ductal hyperplasia. (**a**), 4×; (**b** and **c**): 10× ; (**d**) 40×

size, and more elongated rather than rounded, and most often are located in the periphery. The cells have a "Tufts" and "mounds" projecting into the lumen [11] (Figs. 2.4, 2.5 and 2.6). This must not be confused with the cytoplasmic blebbing of the apocrine metaplasia. Presence of irregularly shaped bridges connecting opposite portions of the wall formed by cells with oval nuclei arranged parallel to the long axis of the bridge (Fig. 2.6). Their appearance is very different from that seen in the rigid trabecular bars and Roman bridges of intraductal carcinoma. The luminal cells of the ductal hyperplasia are surrounded by myoepithelial cells either forming a continuous layer or scattered in the basal surface (Figs. 2.3, 2.4, 2.5 and 2.6). There is absence of necrosis but is not uncommon the presence of calcifications either in the lumina or in the stroma (Figs. 2.1, 2.6, 2.7, and 2.8).

Ductal hyperplasia at difference of atypical ductal hyperplasia express high-molecular-weight (HMW) keratin associated with S-100 protein expression (Fig. 2.9) [12], whereas atypical ductal hyperplasia lack reactivity for HMW keratin.

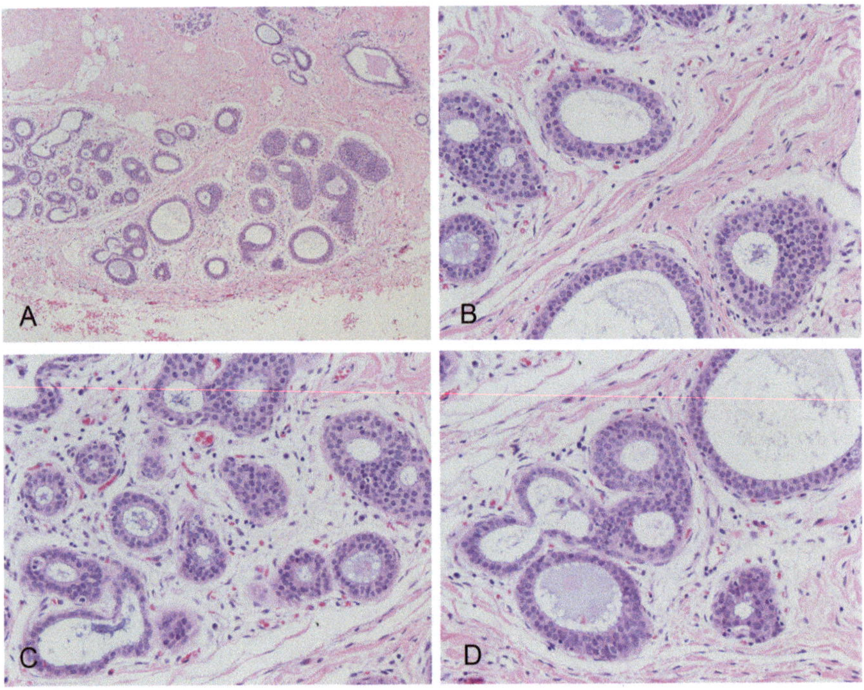

Fig. 2.3 Mild ductal hyperplasia with cystic changes. (**a**): 4×; (**b**, **c** and **d**) 10×

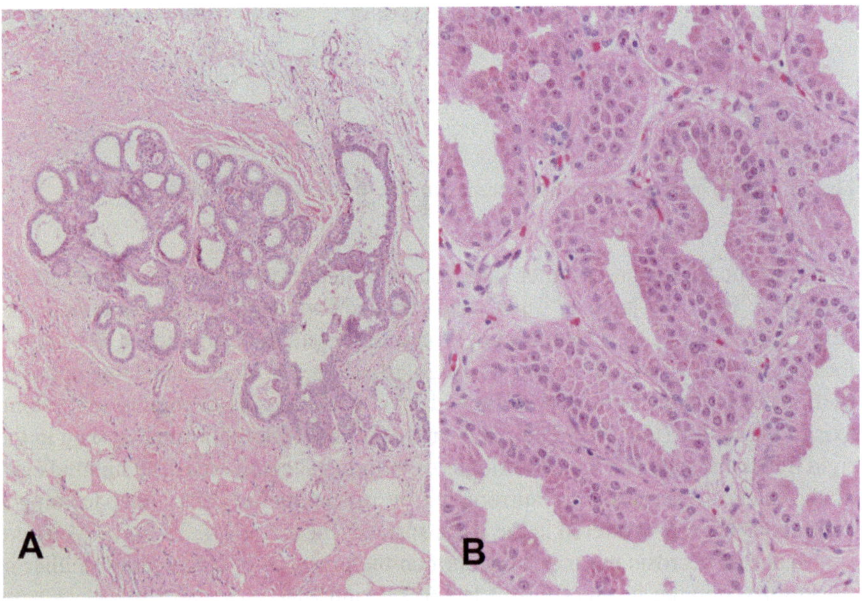

Fig. 2.4 Mild ductal hyperplasia. The cytoplasm is acidophilic and finely granular. (**a**): 4× and (**b**) 40×

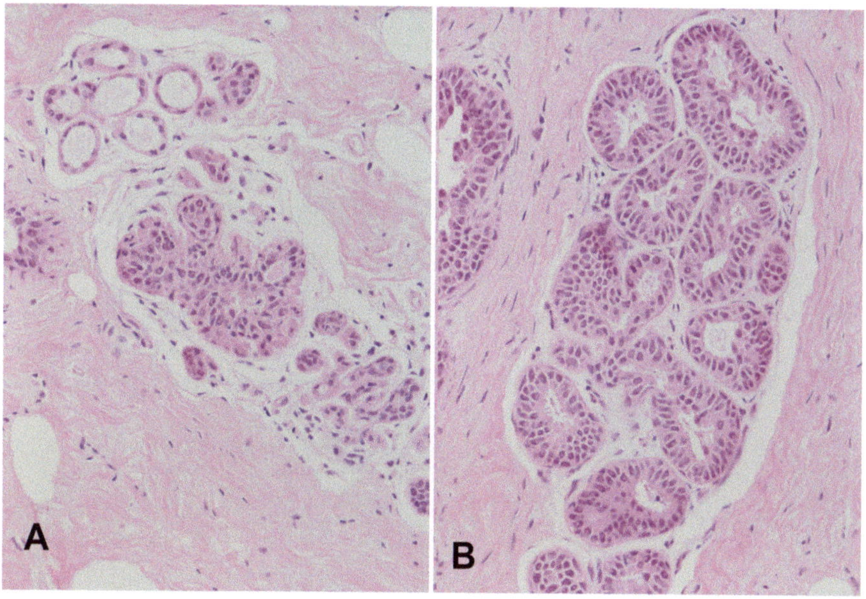

Fig. 2.5 Moderate ductal hyperplasia. (**a**): 4× and (**b**) 40×

2.2.2 *Lobular Hyperplasia*

The histological pattern of this lesion is characterized by abundant lobular formation and more cellular than usual (Fig. 2.10). These lesions do not fulfill the criteria for lobular CIS or even for atypical lobular hyperplasia (ALH). According to Rosai [11] the definition of ALH is rather vague itself.

2.2.3 *Atypical Ductal and Lobular Hyperplasia*

Page-Dupont [13–18] proposed the terms atypical ductal hyperplasia (ADH) and atypical lobular hyperplasia (ALH) for proliferative lesions in which some but not all of the features of intraductal carcinoma or lobular CIS, respectively, are present. Using these criteria in a retrospective study, they diagnosed ADH and / or ALH in 3.6 % of the cases and concluded that these patients had a risk of invasive breast carcinoma that was four to five times that of the general population. The currently accepted definition of ADH is that of a lesion with cytologic (monomorphic cells with ovoid to rounded nuclei) and architectural (micro papillae, tufts, fronds, bridges, solid and cribriform patterns) features indistinguishable from those of low-grade DCIS, but (1) intimately admixed with usual ductal hyperplasia, and/ or (2) showing only partial involvement of the terminal ductal lobular

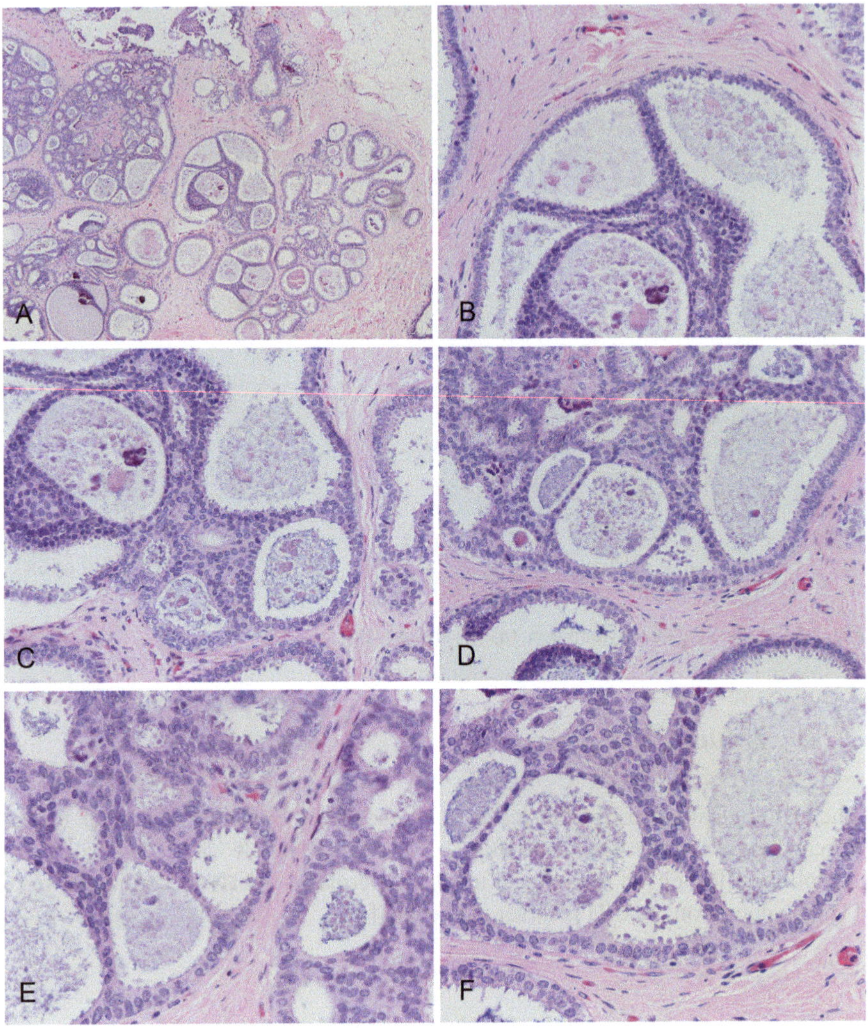

Fig. 2.6 Moderate ductal hyperplasia. (**a**): 4×, (**b**, **c**, **d**, **e** and **f**), 40×. The intratubular lumina of ductal hyperplasia tend to be irregular in size. The cells have a "Tufts" and "mounds" projecting into the lumen. Presence of irregularly shaped bridges connecting opposite portions of the wall formed by cells with oval nuclei arranged parallel to the long axis of the bridge

unit (TDLU) (Figs. 2.11 and 2.12). Quantitative requirements have been proposed (to measure <2 mm in aggregate or to be present in two spaces), but these have not been agreed upon [19–21]. The diagnosis of this type of lesions carries a significant subjectivity in the microscopic interpretation [22–24]. The intraductal proliferative lesions of the breast have been reformulated in the WHO book on Tumors of the Breast and Female Genital Organs and they are part of fibrocystic

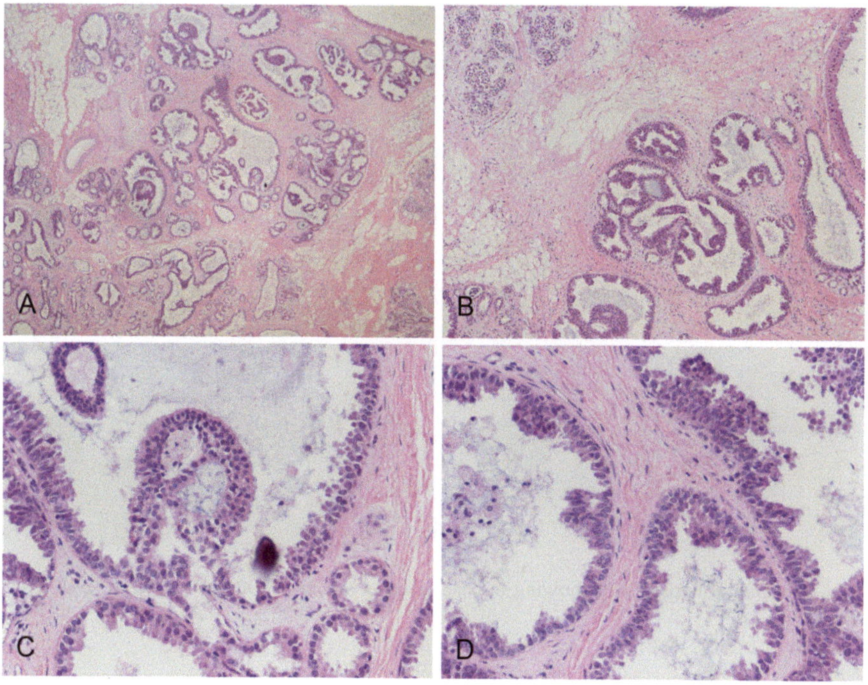

Fig. 2.7 Moderate ductal hyperplasia. (**a**): 4×; (**b**): 10×; (**c**): and (**d**): 40×. There is absence of necrosis but is not uncommon the presence of calcifications in the lumina

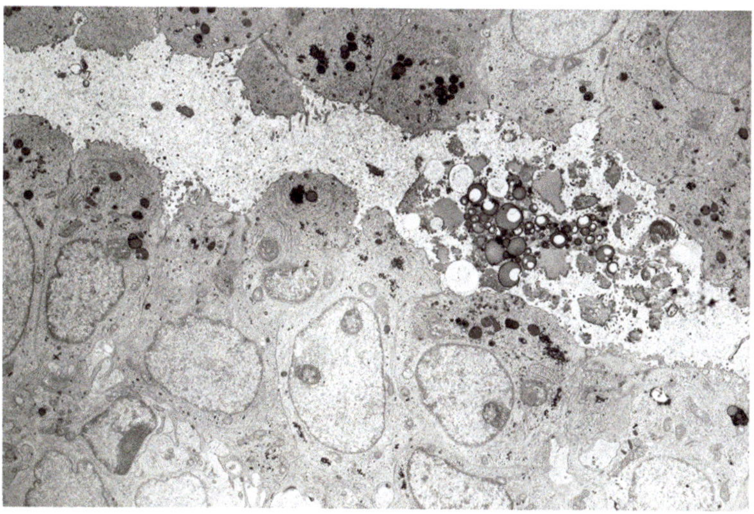

Fig. 2.8 Electron macrograph showing the multilayer epithelium and the presence of micro calcifications in the lumen. Stained with lead citrate and uranyl acetate; 5000×

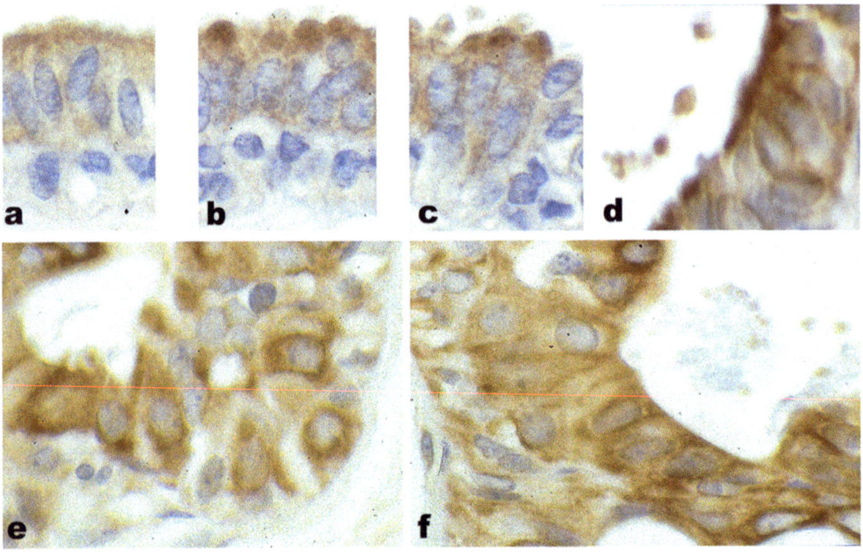

Fig. 2.9 Ductal hyperplasia showing intense reaction against S100 protein. (**a**, **b**, **c**, **d**, **e** and **f**): 40×

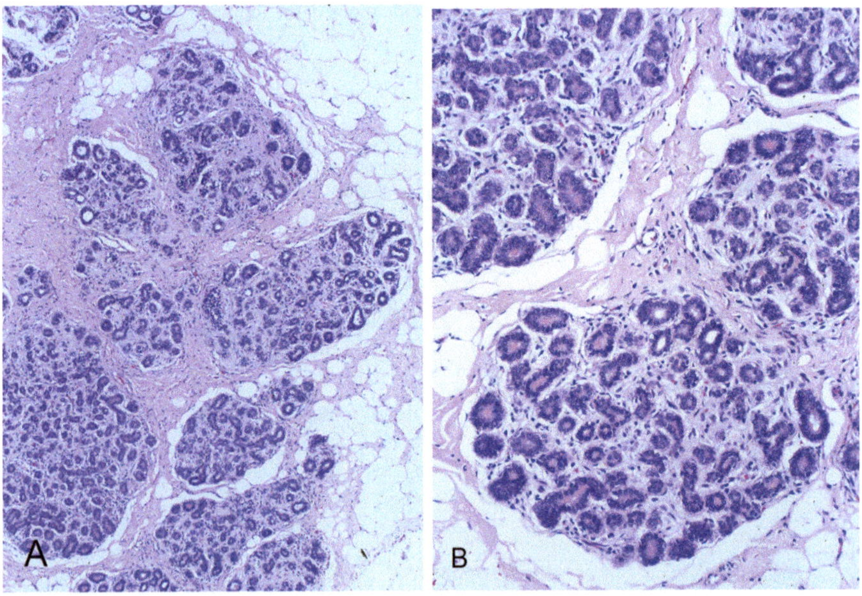

Fig. 2.10 Lobular hyperplasia is characterized by abundant lobular formation and more cellular than usual. (**a**): 4×, (**b**): 10×

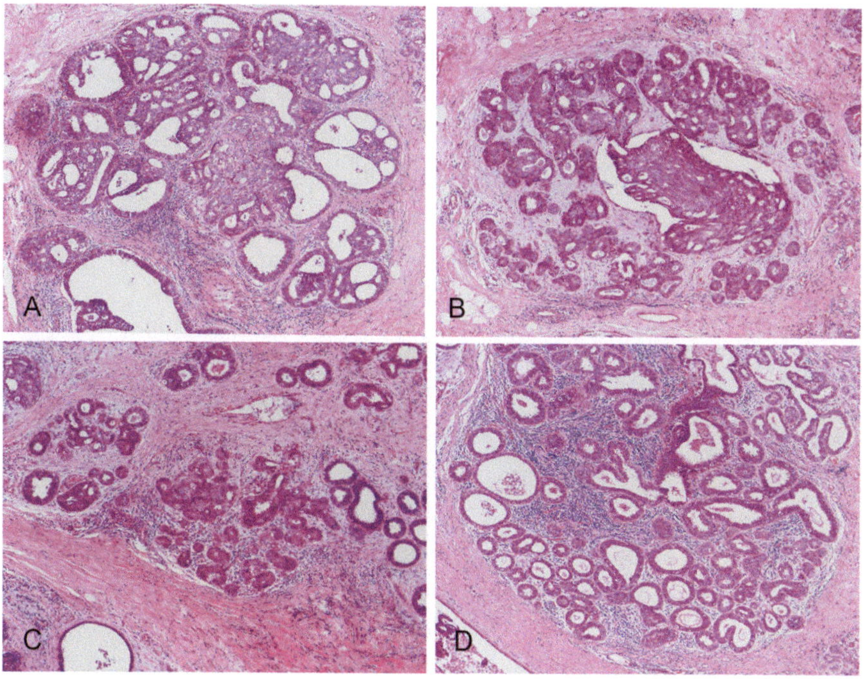

Fig. 2.11 Atypical ductal hyperplasia. It is characterized by architectural (micropapillae, tufts, fronds, bridges, solid and cribriform patterns) features indistinguishable from those of low-grade DCIS, but intimately admixed with usual ductal hyperplasia, and showing only partial involvement of the TDLU. (**a**, **b**, **c** and **d**): 4×

disease [25–28]. An important agreement is that the presence and type of proliferative epithelial disease determines the risk for subsequent carcinoma and that this risk seems to range from one to five times that of the control population [16, 22, 29–33].

2.3 The Histopathology of DCIS

The architectural subtypes of DCIS were classically divided into non-comedo (Fig. 2.13) and comedo subtypes (2.14); non-comedo subtypes were further subdivided into cribriform, micro papillary, solid and papillary, while the comedo subtype was defined by high-grade cells, prominent central necrosis, and associated pleomorphic micro calcifications [15, 34].

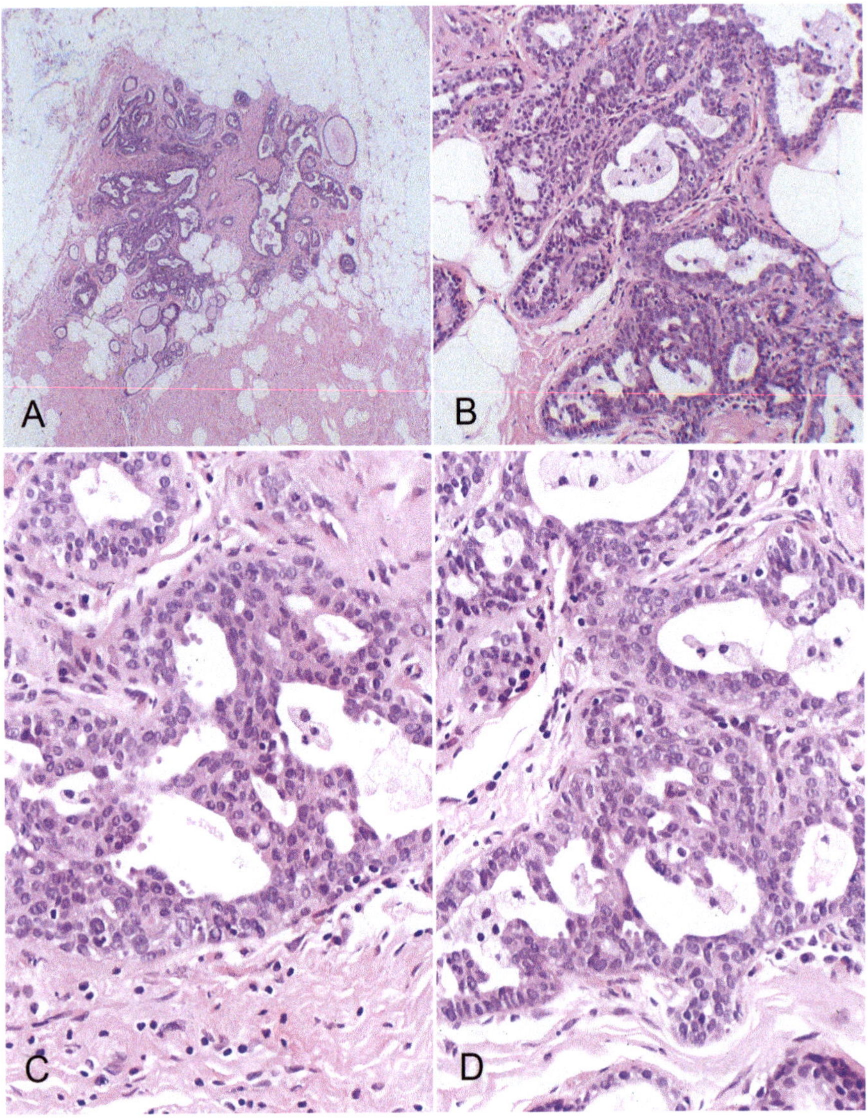

Fig. 2.12 Atypical ductal hyperplasia is characterized by atypical cytologic (monomorphic cells with ovoid to rounded nuclei) and architectural (micropapillae, tufts, fronds, bridges, solid and cribriform patterns) features indistinguishable from those of low-grade DCIS and showing only partial involvement of the TDLU. (**a**): 4×; (**b**) 10×; (**c** and **d**): 40×

2.3.1 Comedocarcinoma

Although comedocarcinoma are carcinoma in situ they may reach a relatively large size and become palpable [35]. They also can be multicentric and in 10 % of the cases could be bilateral [36, 37]. The term comedo is derived from the extruction of necrotic

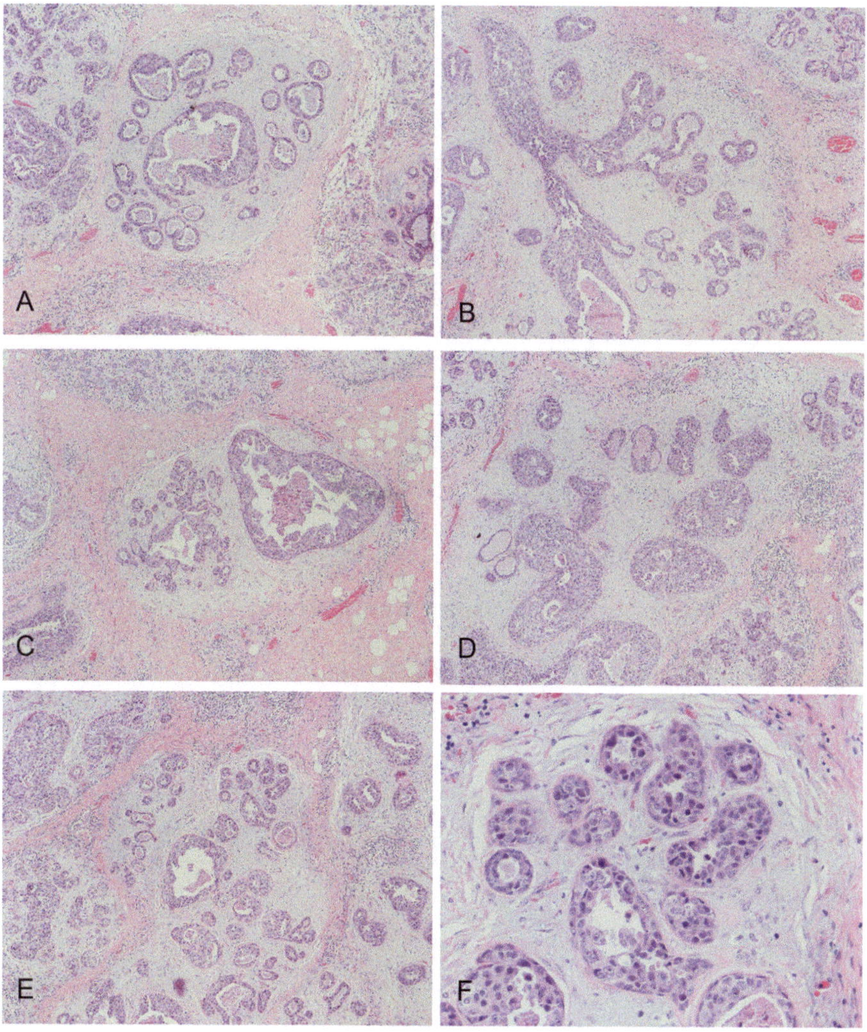

Fig. 2.13 DCIS noncomedo type. (**a**, **b**, **c**, **d** and **e**): 4×, (**f**): 10×

material or comedones upon compression of the lesion. Under the microscope the ducts show a solid growth of large pleomorphic tumor cells accompanied by generally abundant mitotic activity and lacking connective tissue support (Figs. 2.14 and 2.15). Most of these lesions are negative for hormone receptors and are expressing c-erbB-2 growth factors, P cadherin and mutation in P53 is a frequent finding [38–54]. In contrast, noncomedo subtypes are composed of cells with low-grade cytology, are very frequently positive for ER, negative for HER2/neu amplification, negative for p53 mutations, are not aneuploid, and have low proliferation rates [46, 49–54]. In the comedo carcinoma necrosis is always present and constitutes an important diagnostic sign, whether in the form of a large central focus or of individual tumor cells (Figs. 2.15 and 2.16).

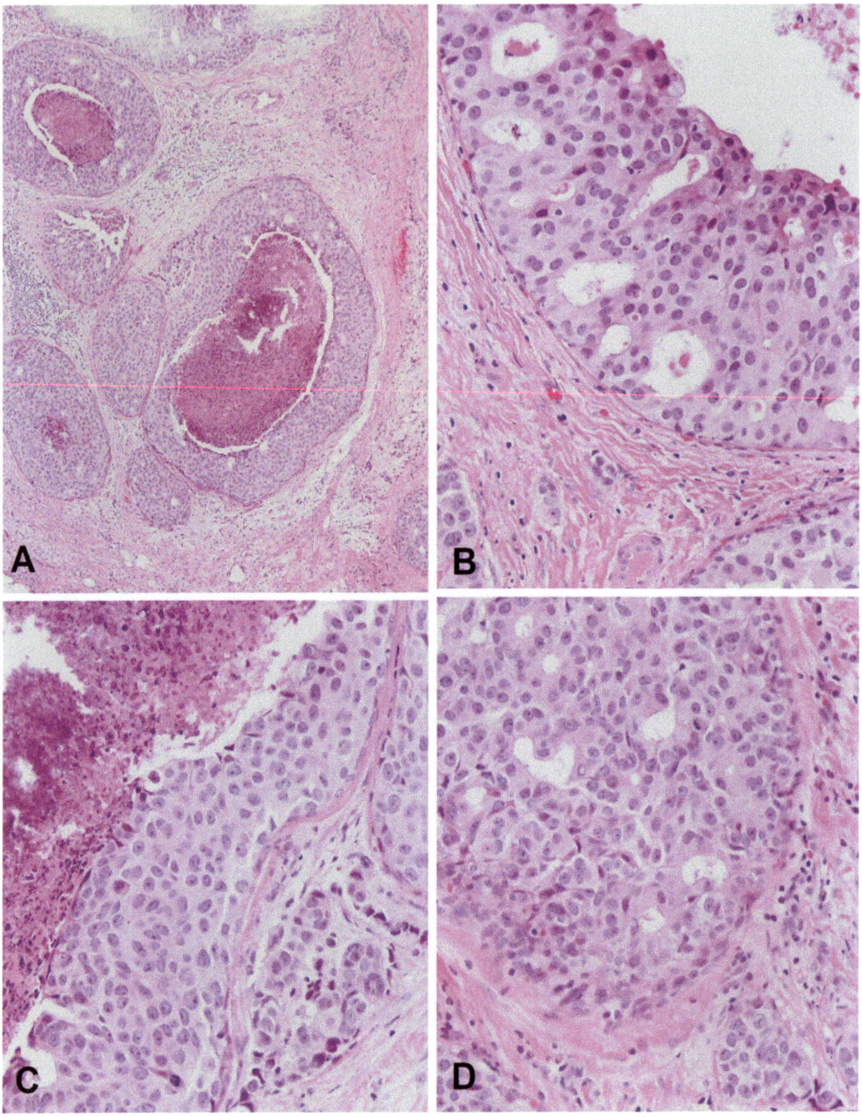

Fig. 2.14 DCIS comedo subtype. (**a**): 10×; (**b**, **c** and **d**): 40×

Calcification is often found in the center of the necrotic areas. The stroma around the involved ducts shows a characteristic concentric fibrosis accompanied by a mild to-moderate mononuclear inflammatory reaction.

In the European classification the pathologic report of comedocarcinoma is taking into account the degree of atypia of the nuclei that has a good correlation with clinical outcomes [55, 56]. In this system, the nuclear grade of the DCIS lesions is defined as low grade (grade 1), intermediate grade (grade 2), and high grade (grade 3) (Fig. 2.16),

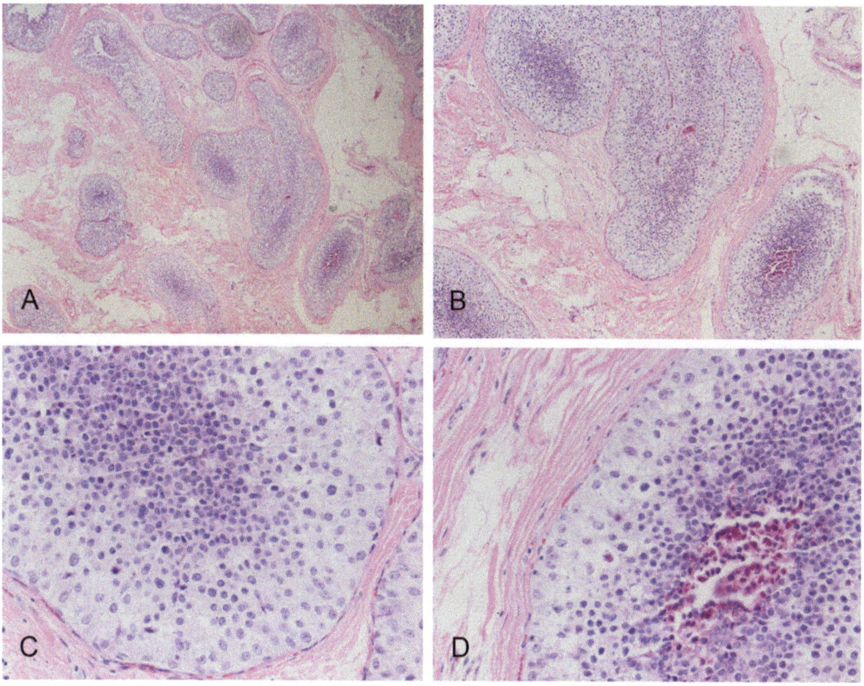

Fig. 2.15 DCIS comedo subtype. The ducts show a solid growth of large pleomorphic tumor cells accompanied by generally abundant mitotic activity and lacking connective tissue support. In the comedo carcinoma necrosis is always present and constitutes an important diagnostic sign, whether in the form of a large central focus or of individual tumor cells. (**a**): 4×; (**b**): 10×; (**c** and **d**): 40×

and this information is now one of the necessary components of a breast pathology report for DCIS, as emphasized in the 2009 College of American Pathologists-American Society for Clinical Oncology protocol for reporting of DCIS lesions [3].

2.3.2 Papillary Carcinoma in Situ

Papillary carcinomas occur in an older age group and are larger than papillomas. Microscopically, features favoring carcinoma are uniformity in size and shape of the epithelial cells, presence of one cell type only, nuclear hyperchromasia and high nucleo-cytoplasmic ratio, high mitotic activity, lack of apocrine metaplasia, cribriform and trabecular patterns, scanty or absent stroma, and lack of benign proliferative disease in the adjacent breast are the main features of this lesion (Figs. 2.17, 2.18 and 2.19).

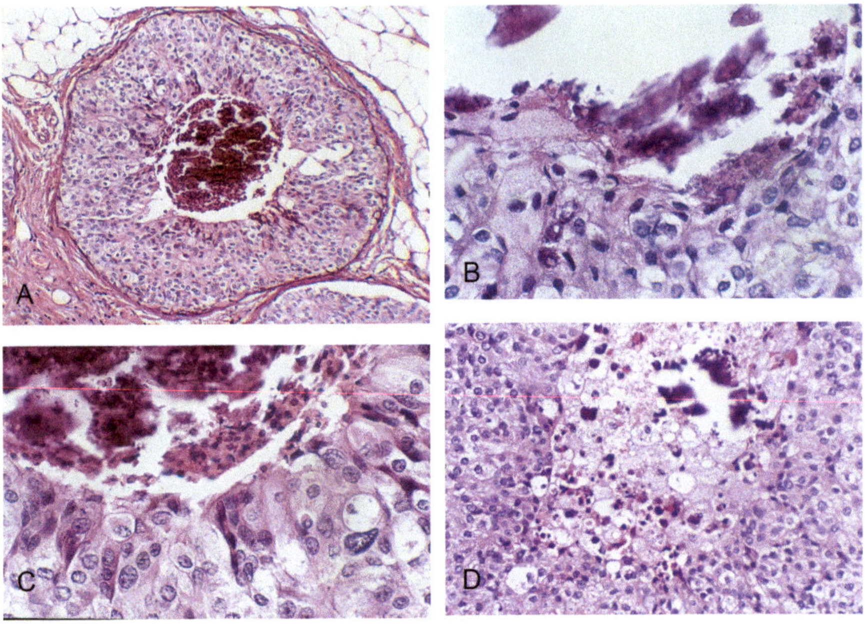

Fig. 2.16 DCIS comedo subtype high grade (grade 3). (**a**): 10×; (**b**, **c** and **d**): 40×

2.3.3 Solid Form of DCIS

In this type of carcinoma in situ, the glandular lumen is filled by the proliferation of medium-sized cells, which are larger than those found in lobular carcinoma in situ but smaller and more uniform than those of comedocarcinoma [57] (Fig. 2.20).

2.3.4 Cribriform Carcinoma In Situ

In this variety, round regular spaces are formed within the glands; the more regular these spaces are in terms of distribution, size, and shape, the more likely the lesion is to be malignant (Figs. 2.21 and 2.22). These spaces are often associated with formations of Roman bridges that are curvilinear trabecular bars connecting two portions of the epithelial lining (Figs. 2.23 and 2.24).

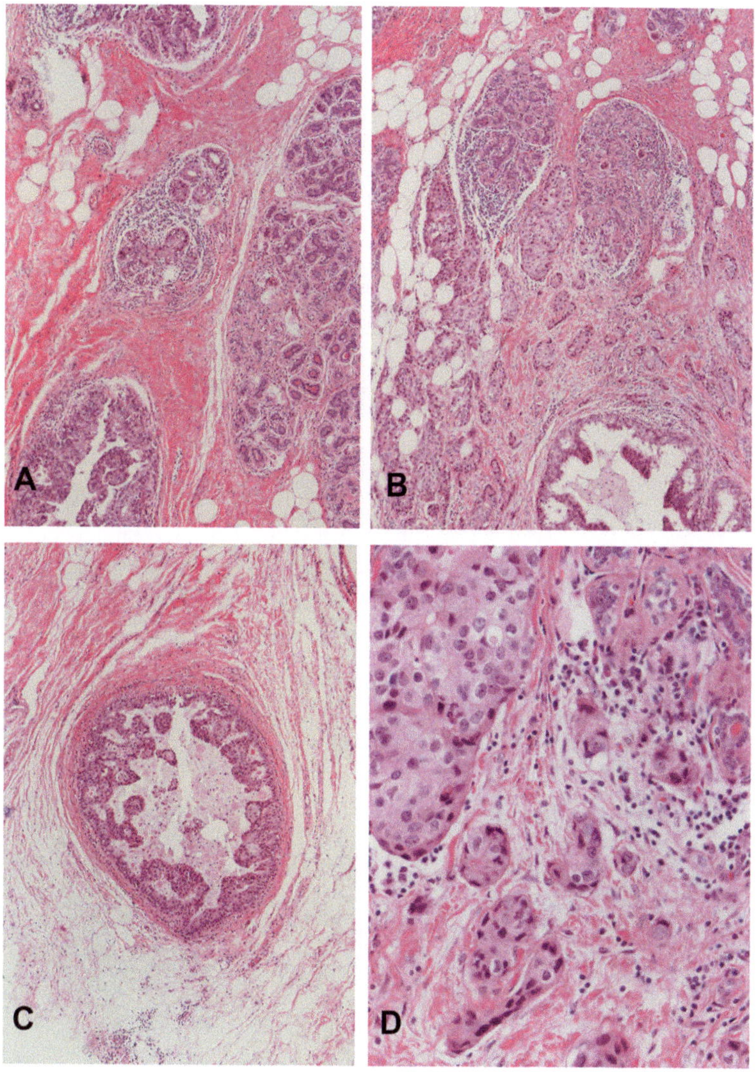

Fig. 2.17 Papillary carcinoma in situ with uniformity in size and shape of the epithelial cells, presence of one cell type only, nuclear hyperchromasia and high nucleocytoplasmic ratio, high mitotic activity, lack of apocrine metaplasia, cribriform and trabecular patterns, scanty or absent stroma. (**a** and **b**): 4×; (**c**): 10×. (**d**) shows an area of invasive cells in and adjacent area of a solid DCIS in the same woman that shows the areas (**a**), (**b**), and (**c**)

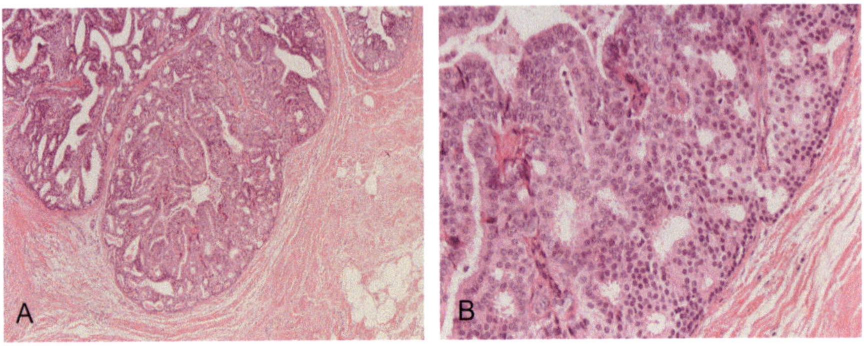

Fig. 2.18 Papillary carcinoma in situ. (**a**): 4× and (**b**): 10×

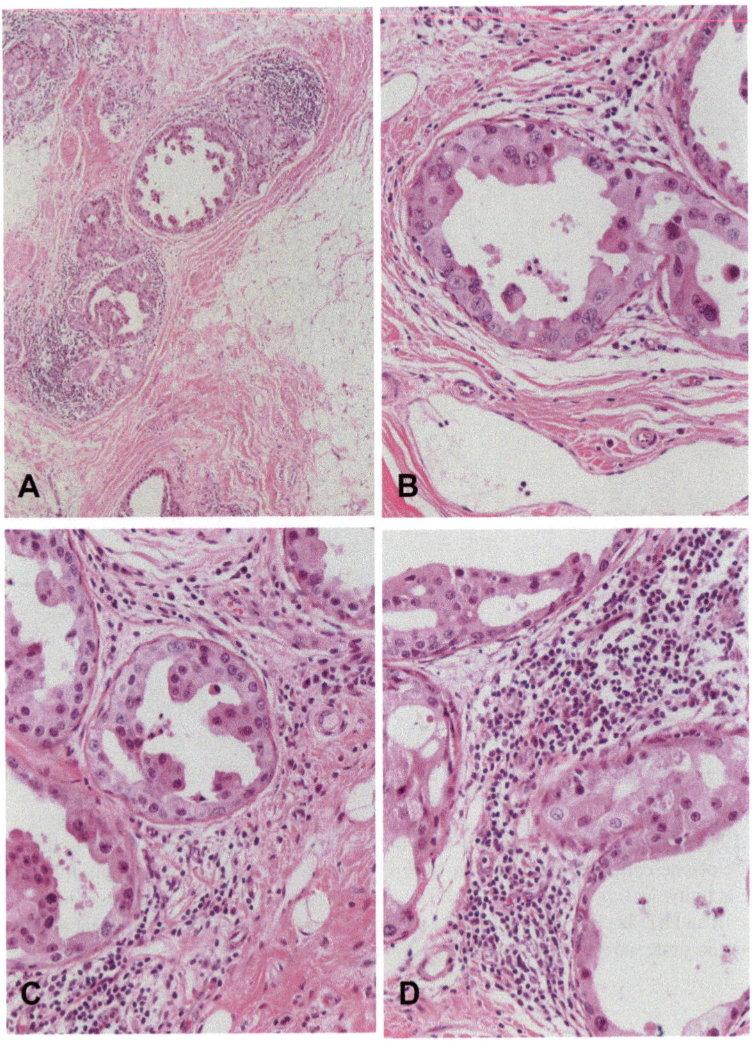

Fig. 2.19 Papillary carcinoma in situ. (**a**): 4×; (**b**, **c** and **d**): 40×. Observe the lymphocytic infiltration in (**c**) and in (**d**). In the figure (**d**) a tongue of invasive cells are seen

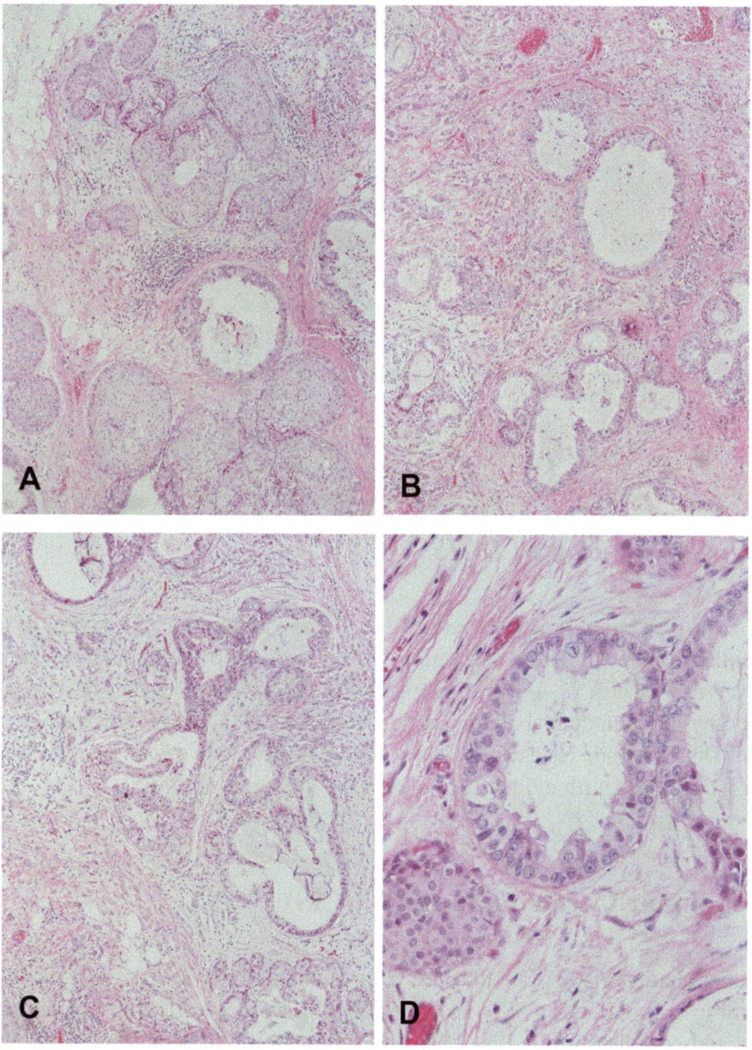

Fig. 2.20 Solid form of DCIS in which the glandular lumen is filled by the proliferation of medium-sized cells, which are larger than those found in lobular carcinoma in situ but smaller and more uniform than those of comedocarcinoma. (**a**, **b** and **c**): 10×; (**d**): 40×

2.3.5 Micropapillary Carcinoma In Situ

This variety could be associated with the cribriform type (Fig. 2.25). Histologically shows elongated epithelial projections projecting into the glandular lumen; these lack connective tissue support, may have a space at the base, and often show a bulbous expansion at the tip. The micro papillary carcinoma may involve multiple quadrants of the breast.

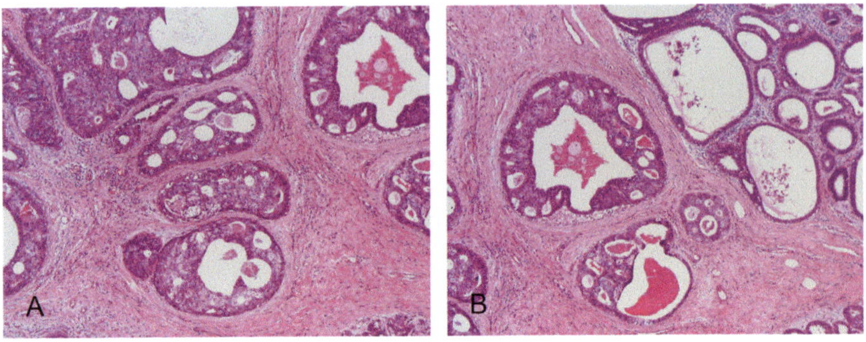

Fig. 2.21 Cribriform carcinoma in situ. (**a** and **b**): 4×

2.3.6 *Other Forms of DCIS*

The *clinging carcinoma* is a variety of DCIS showing one or two layers of malignant cells lining a glandular formation with a large empty lumen [57]. The *cystic hypersecretory* form is a variation of DCIS characterized by cystic formations induced by the abundant secretory material present [58]. Other morphologic variations of DCIS include cases with *signet ring cells* [59], *with apocrine differentiation* [60–62] *and* those with evidence of *endocrine differentiation* [63].

2.4 Lobular Carcinoma In Situ (lCIS)

The major characteristic of lobular CIS is its multicentricity in 70 % of cases [64] and bilateral in approximately 30 % to 40 % [65]. Microscopically, the lobules are distended and completely filled by relatively uniform, round, small- to medium-sized cells with round and normochromatic nuclei (Figs. 2.26, 2.27, 2.28). The *pleomorphic LCIS has* tumor cells of medium to large size, with moderate to marked pleomorphism, occasional prominent nucleoli, and moderate to abundant cytoplasm. According to Rosai [11] the diagnosis of LCIS should be made only in those cases in which the cellular proliferation has resulted in the formation of solid nests that have expanded the lobules, whereas the designation of atypical lobular hyperplasia is to be given to those lesions accompanied by normal-sized lobules in which central lumina are still identifiable. Staining for mucin show positivity in scattered tumor cells in about three fourths of cases [66, 67]. One immuno-cytochemical features of LCIS are the lack of reactivity for E-cadherin and the positivity for HMW keratin by contrast, DCIS is consistently positive for E-cadherin and shows significantly reduced or absent HMW keratin [68].

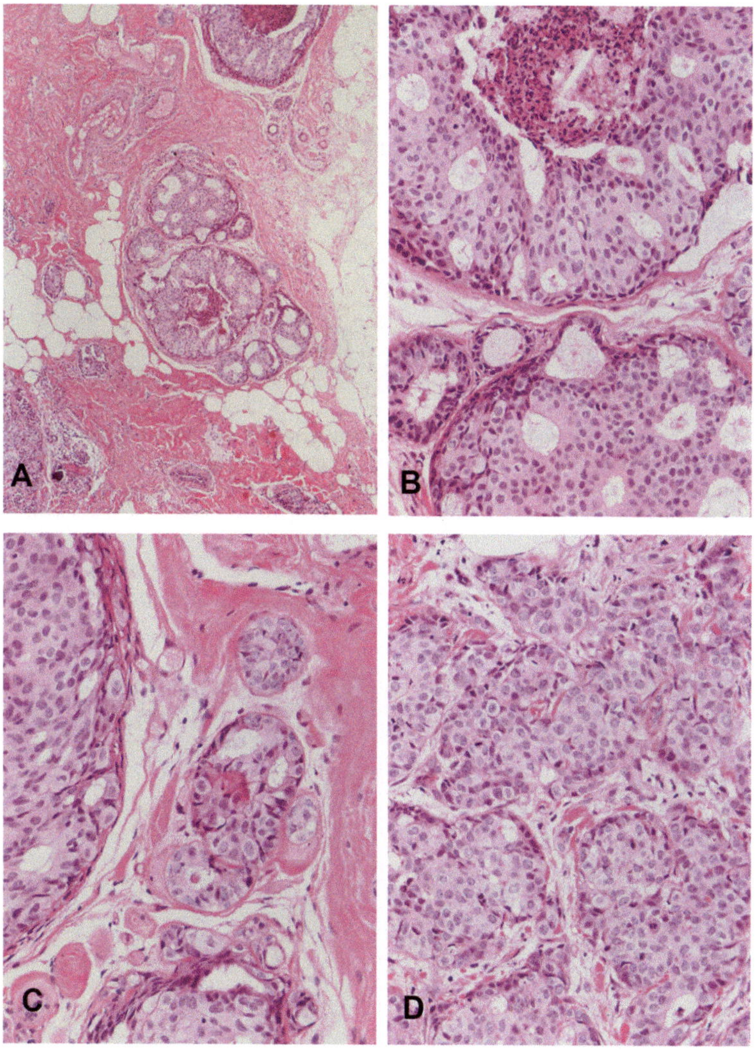

Fig. 2.22 Cribriform carcinoma in situ. Round regular spaces are formed within the glands; the more regular these spaces are in terms of distribution, size, and shape. (**a**): 10×; (**b** and **c**): 40×. Area of invasion is observed in a cribriform carcinoma subtype 40×

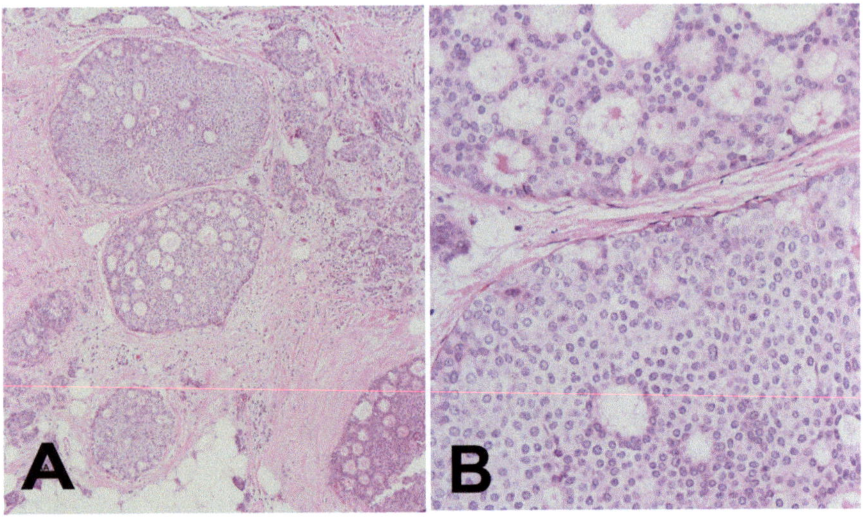

Fig. 2.23 Cribriform carcinoma in situ. Round regular spaces are formed within the glands and these spaces are often associated with formations of Roman bridges that are curvilinear trabecular bars connecting two portions of the epithelial lining. (**a**): 10× and (**b**): 40×

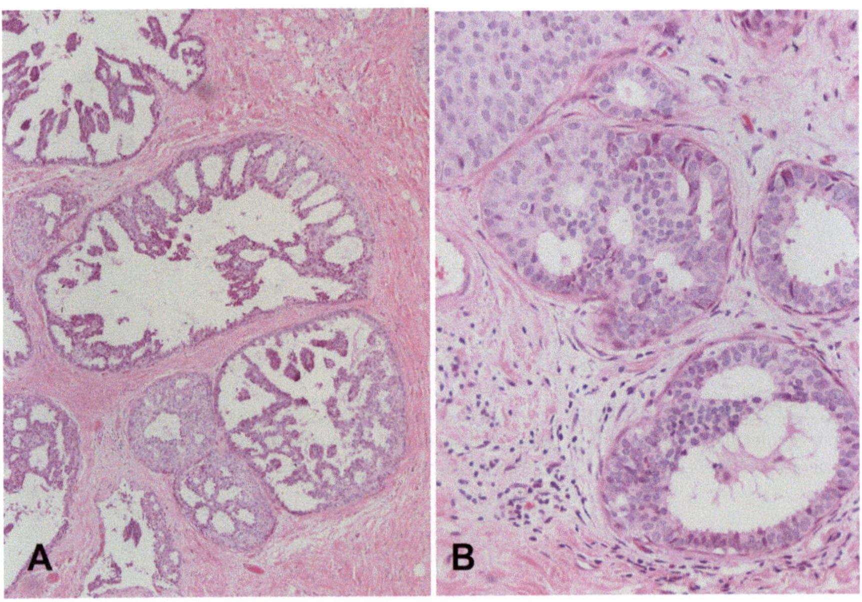

Fig. 2.24 Cribriform carcinoma in situ. (**a**): 10×; (**b**): 40×

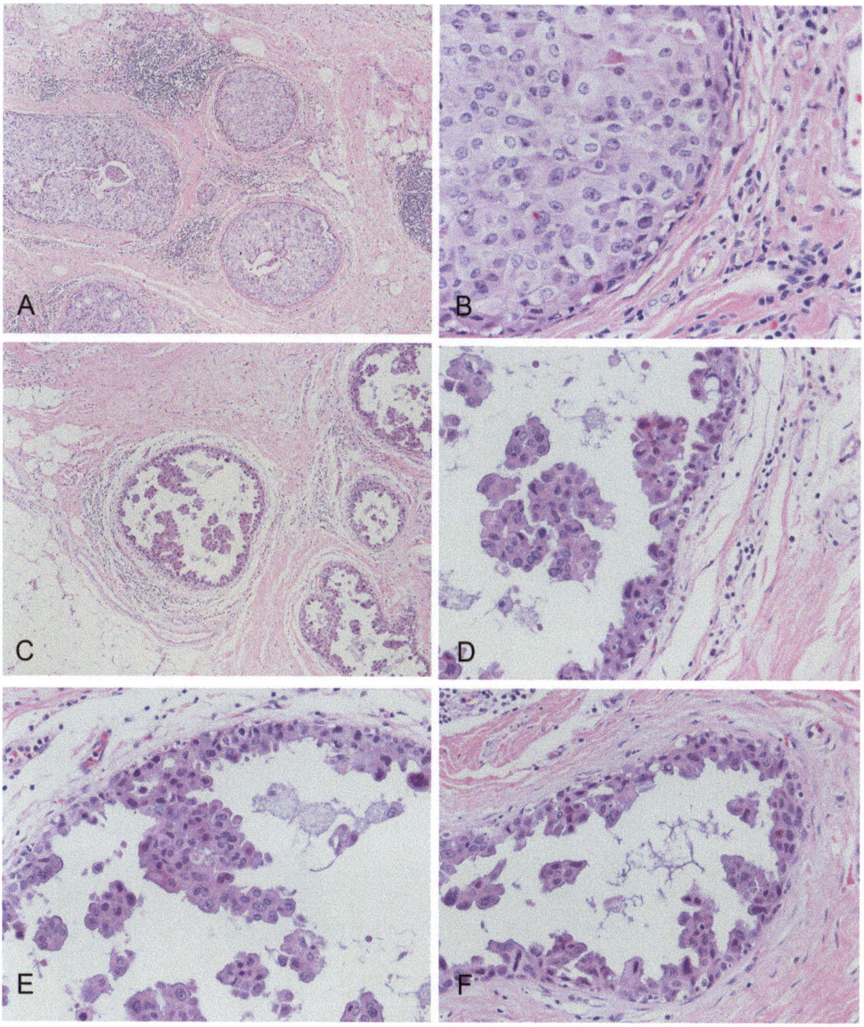

Fig. 2.25 Solid carcinoma in situ. (**a**): 4× and (**b**): 40×. Micropapillary carcinoma in situ showing elongated epithelial projections projecting into the glandular lumen; there is a lack of connective tissue support. (**c**): 4×; (**d**, **e** and **f**): 40×

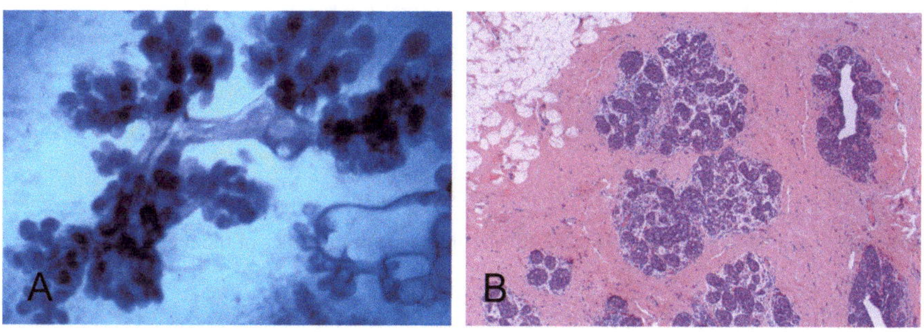

Fig. 2.26 (**a**): Whole mount of lobular carcinoma in situ (LClS) originated in the lobules type 2 of the breast, 4×. (**b**): Lobular carcinoma in situ win which the lobules are distended and completely filled by neoplasic cells, 4×

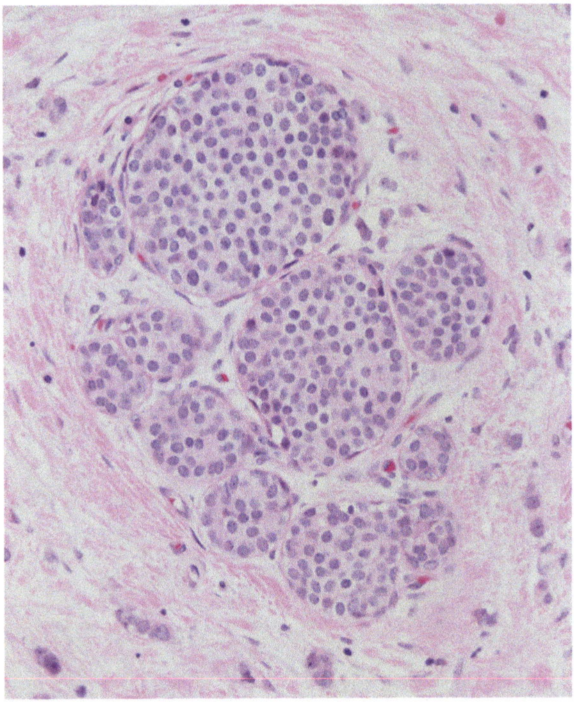

Fig. 2.27 Lobular carcinoma in situ (LCIS) clearly showing the lobules completely filled by relatively uniform, round, small- to medium-sized cells with round and normochromatic nuclei, 40×

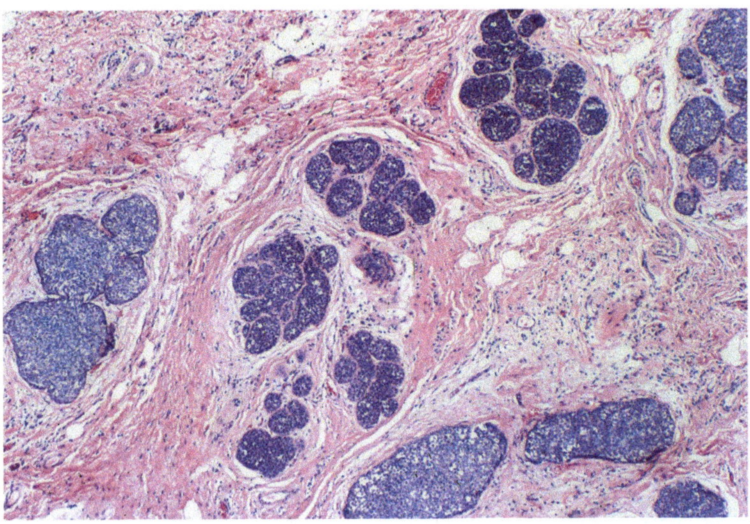

Fig. 2.28 Lobular carcinoma in situ (LCIS) involving several lobules of the breast, 4×

2.5 Differential Diagnosis

DCIS needs to be distinguished from atypical ductal hyperplasia (ADH) [7, 16, 29]. The difference lies in the extent of involvement of the ducts; specifically, ADH lesions occupy only part of the involved space, while low-grade DCIS occupies the entire duct space and often adjacent duct spaces as well [8, 17, 69]. Page et al. [17, 69] proposed that at least 2 spaces of uniformly present atypical cells should be seen in order to call a low-grade atypical epithelial lesion DCIS instead of ADH, while Tavassoli and Norris [21] proposed the 2-mm rule, namely, any low-grade atypical epithelial lesion smaller than 2 mm should be placed in the ADH category and larger than 2 mm, in the low-grade DCIS

DCIS lesions also need to be distinguished from invasive carcinomas; a frequent problem is invasive cribriform carcinoma that needs to be distinguished mostly from cribriform DCIS. Myoepithelial markers may help identify a basement membrane around cribriform DCIS and the absence of such barrier in invasive cribriform carcinomas is extremely helpful [8, 70–72]. Extension of cancer cells beyond the basement membrane with no focus larger than 0.1 cm in diameter is considered microinvasion. The presence of microinvasion is a frequent finding according to some authors [73]. Another problem is the differential diagnosis in which the DCIS extend in a benign lesion such as sclerosing adenosis, giving the morphologic impression of microinvasion [8, 70–72], The use of immuno-cytochemical markers like myosin heavy chain or p63 are useful [8, 70–72] The same confusion may occur when foci of cancer cells are in lymphatic and vascular spaces mimicking a carcinoma in situ, The use of markers such as CD31, CD34, or classic factor VIII immunostain are helpful to differentiate a DCIS from an intra- lymphatic or vascular invasion [8, 70–72].

Comedo-type DCIS lesions often need to be distinguished from the pleomorphic subtype of lobular carcinoma in situ lesions [42, 73–75]. The main difference between these two types of lesions is that LCIS subtype is the lobulocentric appearance of the lesion and the discohesive nature of the large atypical cells. Pleomorphic LCIS, as is the case with all other lesions of lobular histology are negative for E-cadherin expression [8, 70, 73–75].

References

1. SEER Cancer Statistics Review (2012) National Cancer Institute Web site. http://seer.cancer.gov/. Accessed April 16, 2012
2. Cancer Facts and Figures (2008) American Cancer Society Web site. www.cancer.org. Accessed June 1, 2012
3. Siziopikou Arch Pathol Lab Med—Vol 137, April 2013 Ductal Carcinoma In Situ of the Breast
4. Schnitt SJ, Silen W, Sadowsky NL et al (1988) Ductal carcinoma in situ (intraductal carcinoma) of the breast. N Engl J Med 318(14):898–903
5. Lester SC, Bose S, Chen Y et al (2009) Cancer Committee of the College of American Pathologists. Protocol for the examination of specimens from patients with ductal carcinoma in situ (DCIS) of the breast. Arch Pathol Lab Med 133(1):15–25

6. Harris J, Lippman ME, Morrow M, Osborne CK (2004) Diseases of the breast, 3rd edn. Lippincott Williams & Wilkins, Philadelphia
7. Stomper PC, Margolin FR (1994) Ductal carcinoma in situ: the mammographer's perspective. AJR Am J Roentgenol 162(3):585–591
8. Schnitt SJ, Collins LC (2009) Biopsy interpretation of the breast. Biopsy Interpretation Series. Lippincott Williams & Wilkins, Philadelphia, pp 51–95
9. Simpson PT, Reis-Filho JS, Gale T et al (2005) Molecular evolution of breast cancer. J Pathol 205(2):248–254
10. Burstein HJ, Polyak K, Wong JS et al (2004) Ductal carcinoma in situ of the breast. N Engl J Med 350(14):1430–1441
11. Rosai J (ed) (2004) Breast. Chapter 20. In: Rosai and Ackerman's Surgical Pathology. Mosby, New York, pp 1802–1818
12. Guerreiro Da Silva ID, Hu YF, Russo IH, Ao X, Salicioni AM, Yang X, Russo J (2000) S100P Ca^{+2}-binding protein overexpression is associated with immortalization and neoplastic transformation of human breast epithelial cells *in vitro* and tumor progression *in vivo*. Int J Oncol 16:231–240
13. Dupont WD, Page DL (1985) Risk factors for breast cancer in women with proliferative breast disease. N Engl JMed 312:146–151
14. Fitzgibbons PL, Henson DE, Hutter RV (1998) Benign breast changes and the risk for subsequent breast cancer: an update of the 1985 consensus statement. Cancer Committee of the College of American Pathologists. Arch Pathol Lab Med 122:1053–1055
15. Page DL (1986) Cancer risk assessment in benign breast biopsies. Hum Pathol 17:871–874
16. Page DL, Dupont WD, Rogers LW, Rados MS (1985) Atypical hyperplastic lesions of the female breast. A long-term follow-up study. Cancer 55:2698–2708
17. Page DL, Kidd IE, Dupont WD, Rogers LW (1988) Lobular neoplasia of the breast (LN) has varying magnitudes of risk for subsequent invasive carcinoma (IBC) (abstract). Lab Invest 58:69A
18. Schnitt SJ, Connolly JL, Tavassoli FA, Fechner RE, Kempson RL, Gelman R, Page DL (1992) Interobserver reproducibility in the diagnosis of ductal proliferative breast lesions using standardized criteria. Am J Surg Pathol 16:1133–1143
19. Page DL, Rogers LW (1992) Combined histologic and cytologic criteria for the diagnosis of mammary atypical ductal hyperplasia. Hum Pathol 23:1095–1097
20. Purcell CA, Norris HJ (1998) Intraductal proliferations of the breast: a review of histologic criteria for atypical intraductal hyperplasia and ductal carcinoma in situ, including apocrine and papillary lesions. Ann Diagn Pathol 2:135–145
21. Tavassoli FA, Norris HJ (1990) A comparison of the results of long-term follow-up for atypical intraductal hyperplasia and intraductal hyperplasia of the breast. Cancer 65:518–529
22. Rosai J (1991) Borderline epithelial lesions of the breast. Am J Surg Pathol 15:209–221
23. Bratthauer GL, Tavassoli FA (2002) Lobular intraepithelial neoplasia: previously unexplored aspects assessed in 775 cases and their clinical implications. Virchows Arch 440:134–138
24. Tavassoli FA, Hoefler H, Rosai J, Holland R, Ellis I, Schnitt S (2003) Intraductal proliferative lesions. Pathology and genetics of tumours of the breast and female genital organs. IARC Press, Lyon, pp 14–20
25. McDivitt RW (1978) Breast carcinoma. Hum Pathol 9:3–21
26. Schnitt SJ, Jimi A, Kojiro M (1993) The increasing prevalence of benign proliferative breast lesions in Japanese women. Cancer 71:2528–2531
27. Skolnick MH, Cannon-Albright LA, Goldgar DE, Ward JH, Marshall CJ, Schumann GB, Hogle H, McWhorter WP, Wright EC, Tran TD et al (1990) Inheritance of proliferative breast disease in breast cancer kindreds. Science 250:1715–1720
28. Steinhoff NG, Black WC (1970) Florid cystic disease preceding mammary cancer. Ann Surg 171:501–508
29. Dupont WD, Parl FF, Hartmann WH, Brinton LA, Winfield AC, Worrell JA, Schuyler PA, Plummer WD (1993) Breast cancer risk associated with proliferative breast disease and atypical hyperplasia. Cancer 71:1258–1265

30. Bianchi S, Palli O, Galli M, Zampi G (1993) Benign breast disease and cancer risk. Crit Rev Oncol Hematol 15:221–242
31. Bodian CA, Perzin KH, Lattes R, Hoffmann P, Abernathy TG (1993) Prognostic significance of benign proliferative breast disease. Cancer 71:3896–3907
32. Page DL, Dupont WD (1990) Anatomic markers of human premalignancy and risk of breast cancer. Cancer 66:1326–1335
33. Rosen PP (1993) Proliferative breast "disease." An unresolved diagnostic dilemma. Cancer 71:3798–3807
34. The Consensus Conference Committee (1997) Consensus conference on the classification of ductal carcinoma in situ. Cancer 80(9):1798–1802
35. Millis RR, Thynne GSI (1975) In situ intraduct carcinoma of the breast. A long-term follow-up study. Br J Surg 62:957–962
36. Brown PW, Silverman J, Owens E, Tabor DC, Terz JJ, Lawrence W Jr (1976) Intraductal "non-infiltrating" carcinoma of the breast. Arch Surg 111:1063–1067
37. Lagios MD, Westdahl PR, Margolin FR, Rose MR (1982) Duct carcinoma in situ. Relationship of extent of noninvasive disease to the frequency of occult invasion, multicentricity, lymph node metastases, and short-term treatment failures. Cancer 50:1309–1314
38. Bacus SS, Ruby SG, Weinberg DS, Chin D, Ortiz R, Bacus JW (1990) HER-2/ neu oncogene expression and proliferation in breast cancers. Am J Pathol 137:103–111
39. Bhoola S, DeRose PB, Cohen C (1999) Ductal carcinoma in situ of the breast: frequency of biomarkers according to histologic subtype. Applmmunohistochem 7:108–115
40. Bobrow LG, Happerfield LC, Gregory WM, Springall RD, Millis RR (1994) The classification of ductal carcinoma in situ and its association with biological markers. Semin Diagn Pathol 11:199–207
41. Bose S, Lesser ML, Norton L, Rosen PP (1996) lmmunophenotype of intraductal carcinoma. Arch Pathol Lab Med 120:81–85
42. Douglas-Jones AG, Schmid KW, Bier B, Horgan K, Lyons K, Dallimore ND, Moneypenny IJ, Jasani B (1995) Metallothionein expression in duct carcinoma in situ of the breast. Hum Pathol 26:217–222
43. Killeen JL, Namiki H (1991) DNA analysis of ductal carcinoma in situ of the breast. A comparison with histologic features. Cancer 68:2602–2607
44. Leal CB, Schmitt FC, Bento MJ, Maia NC, Lopes CS (1995) Ductal carcinoma in situ of the breast. Histologic categorization and its relationship to ploidy and immunohistochemical expression of hormone receptors, p53, and *c-erbB-2* protein. Cancer 75:2123–2131
45. Lodato RF, Maguire HC Jr, Greene MI, Weiner DB, Li Volsi VA (1990) Immunohistochemical evaluation of *c-erbB-2* oncogene expression in ductal carcinoma in situ and atypical ductal hyperplasia of the breast. Mod Pathol 3:449–454
46. O'Malley FP, Vnencak-Jones CL, Dupont WD, Parl F, Manning S, Page DL (1994) p53 mutations are confined to the comedo type ductal carcinoma in situ of the breast. Immunohistochemical and sequencing data. Lab Invest 71:67–72
47. Paredes J, Milanezi F, Viegas L, Amendoeira I, Schmitt F (2002) P-cadherin expression is associated with high-grade ductal carcinoma in situ of the breast. Virchows Arch 440:16–21
48. Poller ON, Silverstein MJ, Galea M, Locker AP, Elston CW, Blamey RW (1994) Ellis 10. Ideas in pathology. Ductal carcinoma in situ of the breast. A proposal for a new simplified histological classification association between cellular proliferation and *c-erbB-2* protein expression. Mod Pathol 7:257–262
49. Bur ME, Zimarowski MJ, Schnitt SJ et al (1992) Estrogen receptor immunohistochemistry in carcinoma in situ of the breast. Cancer 69(5):1174–1181
50. Rudas M, Neumayer R, Gnant MF et al (1997) p53 protein expression, cell proliferation and steroid hormone receptors in ductal and lobular in situ carcinomas of the breast. Eur J Cancer 33:39–44
51. Mack L, Doig G, O'Malley FP (1997) Relationship of a new histological categorization of ductal carcinoma in situ of the breast with size and the immunohistochemical expression of p53, c-erbB-2, bcl-2 and ki-67. Hum Pathol 28(8):974–979

52. van de Vijver MJ, Peterse JL, Mooi WJ et al (1988) Neu-protein overexpression in breast cancer: association with comedo-type ductal carcinoma in situ and limited prognostic value in stage II breast cancer. N Engl J Med 319(19):1239–1245
53. Bartkova J, Barnes DM, Millis RR et al (1990) Immunohistochemical demonstration of c-erbB-2 protein in mammary ductal carcinoma in situ. Hum Pathol 21(11):1164–1167
54. Poller DN, Roberts EC, Bell JA et al (1993) p53 protein expression in mammary ductal carcinoma in situ: relationship to immunohistochemical expression of estrogen receptor and c-erbB-2 protein. Hum Pathol 24(5):463–468
55. Holland R, Peterse JL, Millis RR et al (1994) Ductal carcinoma in situ: a proposal for a new classification. Semin Diagn Pathol 11(3):167–180
56. European Commission Working Group on Breast Screening Pathology (1998) Consistency achieved by 23 European pathologists in categorizing ductal carcinoma in situ of the breast using five classifications. Hum Pathol 29(10):1056–1062
57. Azzopardi JG (1979) Problems in breast pathology. In: Bennington JL (consulting ed.) Major problems in pathology, vol. 11. W.B. Saunders, Philadelphia
58. Guerry P, Erlandson RA, Rosen PP (1988) Cystic hypersecretory hyperplasia and cystic hypersecretory duct carcinoma of the breast. Pathology, therapy, and follow-up of 39 patients. Cancer 61:1611–1620
59. Andersen JA (1974) Invasive breast carcinoma with lobular involvement. Frequency and location of lobular carcinoma in situ. Acta Pathol Microbiol Scand (A) 82:719–729
60. Leal C, Henrique R, Monteiro P, Lopes C, Bento MI, De Sousa SP, Lopes P, Olson S, Silva MD, Page DL (2001) Apocrine ductal carcinoma in situ of the breast: histologic classification and expression of biologic markers. Hum Pathol 32:487–493
61. O'Malley FP, Page DL, Nelson EH, Dupont WD (1994) Ductal carcinoma in situ of the breast with apocrine cytology. Definition of a borderline category. Hum Pathol 25:164–168
62. Tavassoli FA, Norris HJ (1994) Intraductal apocrine carcinoma. A clinicopathologic study of 37 cases. Mod Pathol 7:813–818
63. Cross AS, Azzopardi JG, Krausz T, Van Noorden S, Polak JM (1985) A morphological and immunocytochemical study of a distinctive variant of ductal carcinoma in situ of the breast. Histopathology 9:21–37
64. Warner NE (1969) Lobular carcinoma of the breast. Cancer 23:840–846
65. Carter D, Smith RRL (1977) Carcinoma in situ of the breast. Cancer 40:1189–1193
66. Andersen JA, Vendelboe ML (1981) Cytoplasmic mucous globules in lobular carcinoma in situ. Diagnosis and prognosis. Am J Surg Pathol 5:251–255
67. Breslow A, Brancaccio ME (1976) Intracellular mucin production by lobular breast carcinoma cells. Arch Pathol Lab Med 100:620–621
68. Acs G, Lawton D, Rebbeck TR, LiVolsi VA, Zhang PJ (2001) Differential expression of E-cadherin in lobular and ductal neoplasms of the breast and its biologic and diagnostic implications. Am J Clin Pathol 115:85–98
69. Page DL, Anderson TJ (1987) Diagnostic histopathology of the breast. Churchill Livingstone, New York
70. Maluf HM (2004) Differential diagnosis of solid carcinoma in situ. Semin Diagn Pathol 21(1):25–31
71. Yaziji H, Gown AM, Sneige N (2000) Detection of stromal invasion in breast cancer: the myoepithelial markers. Adv Anat Pathol 7(2):100–109
72. Lerwill MF (2004) Current practical applications of diagnostic immunohistochemistry in breast pathology. Am J Surg Pathol 28(8):1079–1091
73. Sneige N, Wang J, Baker BA et al (2002) Clinical, histopathologic and biologic features of pleomorphic lobular (ductal-lobular) carcinomain situ of the breast: a report of 24 cases. Mod Pathol 15(10):1044–1050
74. Jacobs JW (2003) Recently recognized variants of lobular carcinoma in situ (LCIS) with an emphasis on management of LCIS on core needle biopsy. Pathol Case Rev 8(5):211–219
75. Fulford LG, Reis-Filho JS, Lakhani SR (2004) Lobular in situ neoplasia. Curr Diagn Pathol 10(3):183–192

Chapter 3
The Pathobiology of the Breast Cancer Invasive Process

3.1 Introduction

The progression of carcinoma cells into highly malignant states is thought to depend upon activation of the cell-biological program termed the epithelial-mesenchymal transition (EMT). This program, which plays key roles in normal embryogenesis, wound healing, and fibrosis, is also exploited by carcinoma cells, since it confers on them malignancy-associated traits, such as motility, invasiveness, heightened resistance to apoptosis, and an ability to disseminate from sites of primary tumor.

3.2 Epithelial Mesenchymal Transition (EMT)

The epithelial to mesenchymal transition (EMT) leads to exacerbation of motility and invasiveness in many cell types and is often considered a prerequisite for tumor infiltration and metastasis [1] (Fig. 3.1). EMT involves a shedding by epithelial cells of their characteristic morphology and gene expression pattern and the assumption of a shape and transcriptional program characteristic of mesenchymal cells. For understanding EMT is important to briefly describe the genomic classification of human breast cancer that will be discussed in Chap. 5 of this book. Breast cancer has been subdivided into five major subtypes: basal-like, Her2 (ERBB2)–overexpressing, normal breast tissue-like, and two subtypes of luminal-like, luminal A and luminal B [3]. The luminal-like subtypes display moderate to high expression of ERα and luminal cytokeratins, whereas the basal-like subtype is negative for both ERα and ERBB2, with high expression of basal cytokeratins 5 and 17. The ERBB2-overexpressing subtype is also ERα negative and, like the basal-like tumors, is associated with poorer prognosis as measured as time to development of distal metastasis [1, 3, 4]. Altogether, these data support the concept that ER-positive and ER-negative tumors may originate from two different cell populations, as postulated earlier [5]. In addition to differences inherent

J. Russo, *The Pathobiology of Breast Cancer*,
DOI 10.1007/978-3-319-40815-6_3

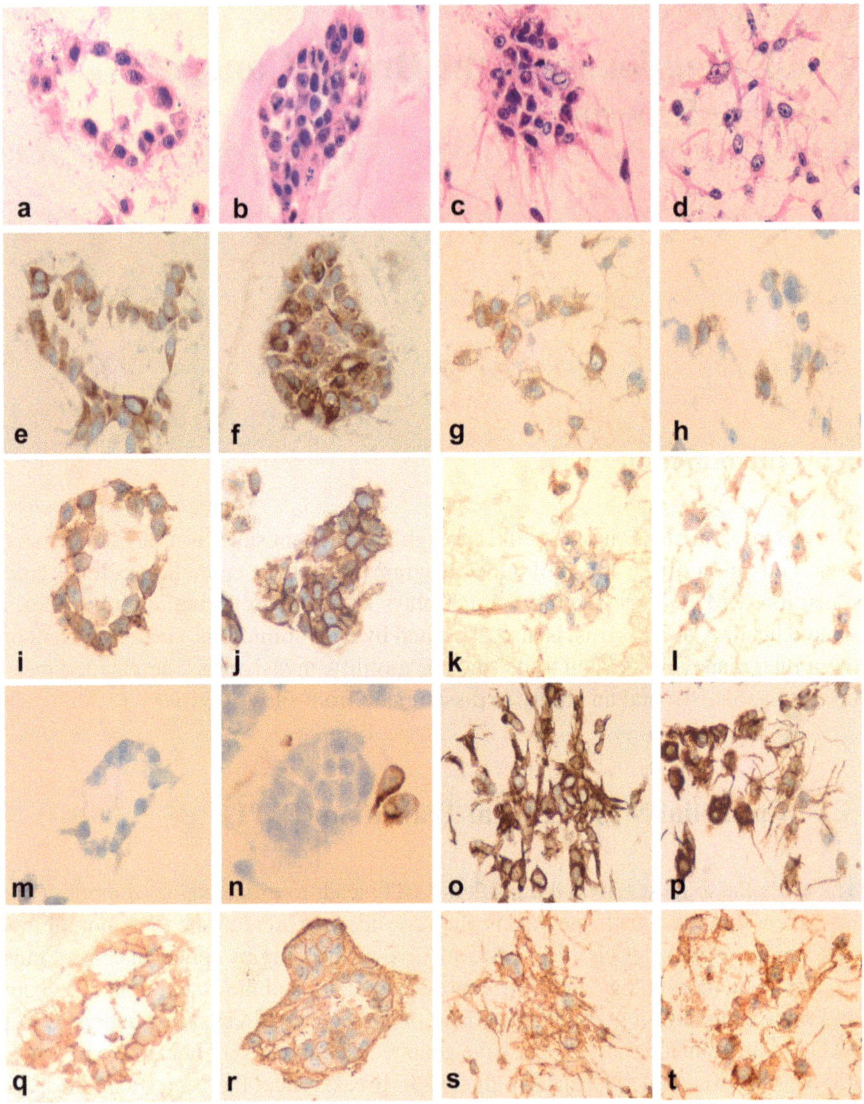

Fig. 3.1 (**a**) Histological section of MCF10F cells growing in collagen matrix, H&E ×40; (**b**) E2 70 cells growing in collagen matrix, H&E ×40; (**c** and **d**) C5-T8 cells growing in collagen matrix, H&E ×40; (**e**, **i**, **m** and **q**) MCF10F cells reacted with EMA, E-Cadherin, Vimentin, and Fibronectin respectively (40×); (**f**, **j**, **n** and **r**) E2 70 transformed cells reacted EMA, E-cadherin, Vimentin and fibronectin respectively (40×); (**g**, **h**, **k**, **l**, **o**, **p**, **s**, and **t**) C5-T8 cells reacted with EMA, E-cadherin, Vimentin and fibronectin respectively (40×) (From Tiezzi et al. [2])

to the type of cell in which cancer originates, neoplasticallly initiated cells lose specific characteristics of epithelial differentiation as the result of their progression toward malignancy. As the epithelial cells lose their polarity and cell-to-cell junctions, regulated in part by the expression of E-cadherin, they acquire characteristics of mesenchymal cells, which lack stable intercellular junctions [5] (Fig. 3.1).

3.3 A Human Breast Cancer Cell Model of EMT

A better understanding of the EMT has been obtained by the use an in vitro-in vivo system in which the spontaneously immortalized ERα-negative human breast epithelial cell (HBEC) line MCF-10F was transformed by treatment with 17-beta estradiol (E2) [5, 6]. E2-transformed cells progressively express phenotypes of in vitro cell transformation, including colony formation in agar methocel, decreased ductulogenesis, increased invasiveness in a Matrigel invasion system, and tumorigenesis in a heterologous host (Fig. 3.2). Tumors formed in severe combined immunodeficient (SCID) mice by invasive cells and by cell lines derived from those tumors were poorly differentiated ERα negative (–), progesterone receptor negative (–), and ERBB2-negative adenocarcinomas [6]. These characteristics are similar to those of basal cell type primary carcinomas previously described [7]. Using this in vitro in vivo model [8] it has been possible to identify associations between copy number changes, LOH, and tumorigenic phenotype, as well as the related changes in transcript expression. Functional analyses of these data identified several dysregulated pathways associated with progressive tumorigenic and invasive capacity (Fig. 3.3). This study integrates structural and functional genomic data analyses to elucidate the progressive molecular events in the EMT of ER (–) HBECs. Genomic aberrations progressively accumulated as the cells expressed more aggressive phenotypes (i.e.,

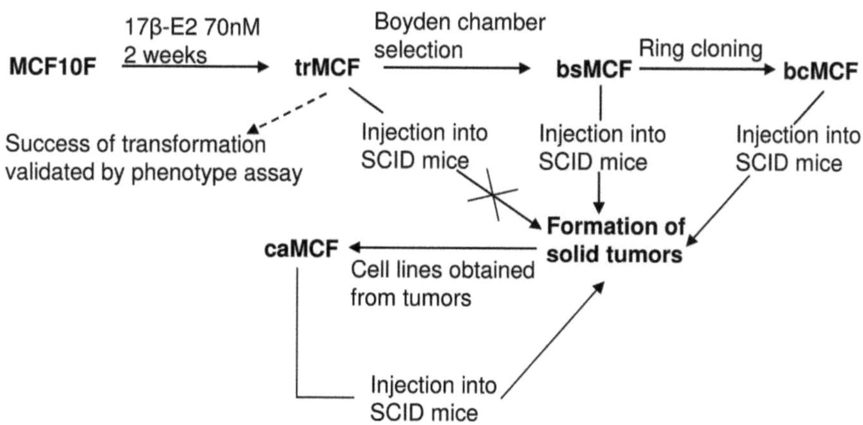

Fig. 3.2 Transformation of MCF-10F cells by E_2 treatment. Experimental protocol: MCF-10F cells treated with 70 nmol/L E_2 that expressed high colony efficiency and loss of ductulogenic capacity in collagen-matrix were classified as transformed (*trMCF*). Transformed cells that were invasive in a Matrigel Boyden-type invasion chambers were selected (*bsMCF*) and plated at low density for cloning (*bcMCF*). MCF-10F, trMCF, bsMCF, and bcMCF were tested for carcinogenicity by injecting them into the mammary fat pad of 45-d-old female SCID mice. MCF-10F and trMCF cells did not induce tumors (*canceled arrow*); bsMCF and bcMCF formed solid tumors from which four cell lines, identified as caMCF, were derived and proven to be tumorigenic in SCID mice (From Tiezzi et al. [2])

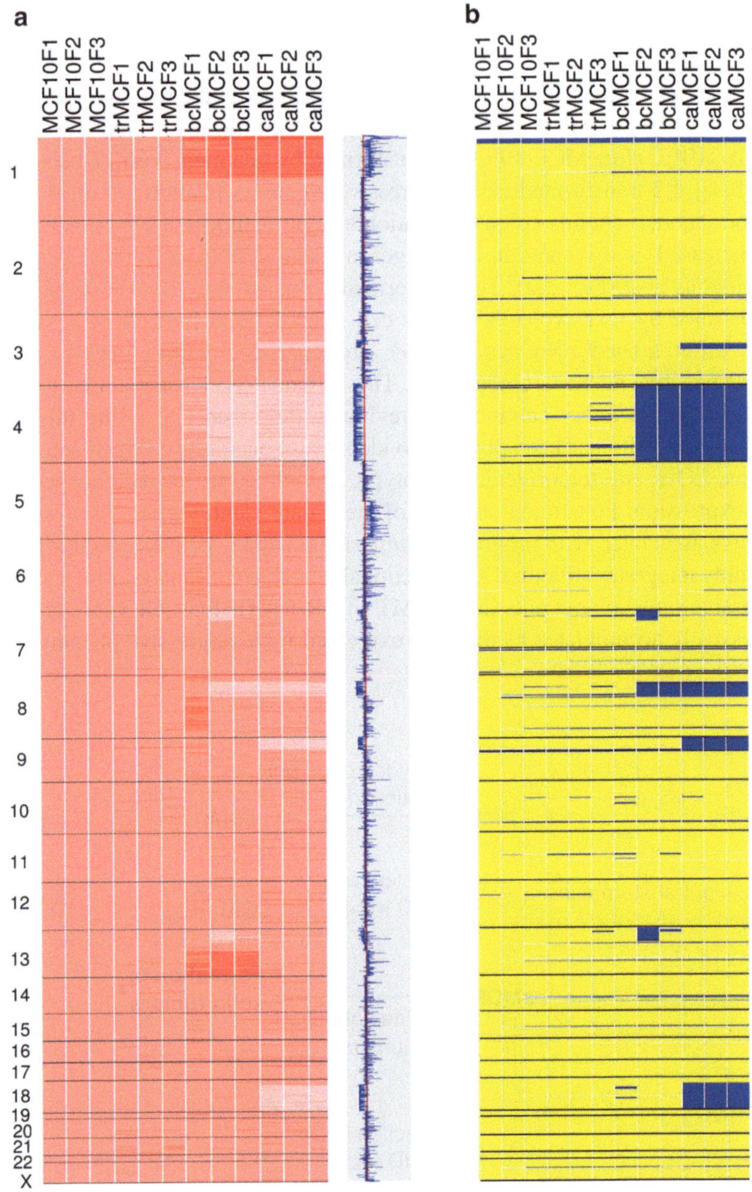

Fig. 3.3 *A* and *B,* chromosome copy number analysis using Affymetrix 100k SNP chips and dChip software. *B,* display of inferred copy number. MCF-10F at three different passages serves as diploid reference. *Pink shade,* diploidy; *darker red* and *lighter pink,* regions of copy number amplification and deletion, respectively. *Gray box,* range from 0 to 4 copies (*blue curve*); *red line,* baseline for diploidy. *C,* complete genome view of LOH: *yellow,* retention of heterozygosity; *blue,* LOH; *white,* no information due to lack of SNPs. Each column in *B* and *C* represent the different chromosomes; the chromosome number, from 1 to 22 and chromosome X are at the top of *B* and, at the left of these panels, the different cells are indicated (From Huang et al. [8])

in the tumorigenic bcMCF and caMCF) in comparison with the non-tumorigenic trMCF cells (Fig. 3.2). Accordingly, the number of genes with altered levels of expression was greater in the tumorigenic cells, as where the chromosomes enriched with up- or down-regulated genes. Importantly, the neoplasticallly transformed human breast epithelial cells were correctly classified into tumorigenic or non-tumorigenic groups based on the profile of copy number changes, indicating that in this model changes in copy number provide a genomic signature for the tumorigenic phenotype. Together, these findings revealed an intrinsic link between E2-induced copy number changes, gene expression alterations, and tumorigenesis in ERa (–) HBECs. For example, changes in CN (Fig. 3.3A) and LOH (Fig. 3.3B) were progressive in these cells. Only small and scattered chromosome gains were observed in trMCF cells and most alterations were observed in the bcMCF and caMCF (Fig. 3.3A). Large fragments of chromosome gain were observed only in the telomeric ends at 1p36.12 – pter and 5q21.1 – qter in both bcMCF and caMCF, and 13q21.31 – qter in bcMCF (Fig. 3.3A, dark pink areas). caMCF cell lines exhibited loss of the whole chromosome 4 and loss of 8p11.21 – 23.1 (Fig. 3.3A, pale pink areas). Loss of 3p12.1 – 14.1, 9p22.1 – pter, and 18q11.2 – qter was observed in the three lines of caMCF (Figs. 3.3A and 3.4a–d). Deletion was also observed in 7pter and 13pter of bcMCF2, but not in the rest of tumorigenic cells, indicating that such deletion is not required for the expression of tumorigenic capacity. Sample clustering based on CN profile grouped the bcMCF1 in the tumorigenic class (Fig. 3.5), and this was confirmed by sample clustering based on gene expression profile (Fig. 3.6). Furthermore, all bcMCF clones were shown to form tumors in SCID mice (Table 3.1). Therefore, the isolation of subclones may afford the possibility to detect minimal regions of CN change associated with neoplastic cell transformation. As major CN changes and LOH were observed in chromosome 4 (Fig. 3.3A, b). Interestingly, the CN loss in chromosome 4 detected in bcMCF cells can be observed in trMCF cells (Fig. 3.7 arrows), albeit at a level that does not reach statistical significance.

The chromosomes enriched with up- (Fig. 3.8a) or down- (Fig. 3.8b) -regulated genes also demonstrated large regions of amplification or deletion (Fig. 3.3A), respectively. For example, chromosomes 18, 4 and 8 were significantly overrepresented by down-regulated genes in caMCF cells (Fig. 3.3A, caMCF), and correspondently, demonstrated large regions of deletion in caMCF (Fig. 3.3A). Similarly, chromosome 4 was overrepresented by down-regulated genes in bcMCF cells (Fig. 3.3A, bcMCF), and showed the largest CN loss in bcMCF cells (Fig. 3.3A). Chromosomes 1 and 5 were significantly overrepresented by up-regulated genes in both bcMCF and caMCF cells (Fig. 3.3A, bcMCF and caMCF), and correspondently, demonstrated largest CN amplification (Fig. 3.3B). However, some exceptions were noteworthy. Chromosome 13 showed a high frequency of SNP amplification (50 %) in bcMCF (Fig. 3.5), but was not enriched with up-regulated genes. In fact, there were only 6 genes up-regulated in this region. Moreover, this amplification was not maintained in caMCF cells (Fig. 3.3B), indicating this aberration was not required for tumorigenesis. Also, chromosome 11 was over represented by down-regulated genes in trMCF cells (Fig. 3.8b, trMCF), although trMCF

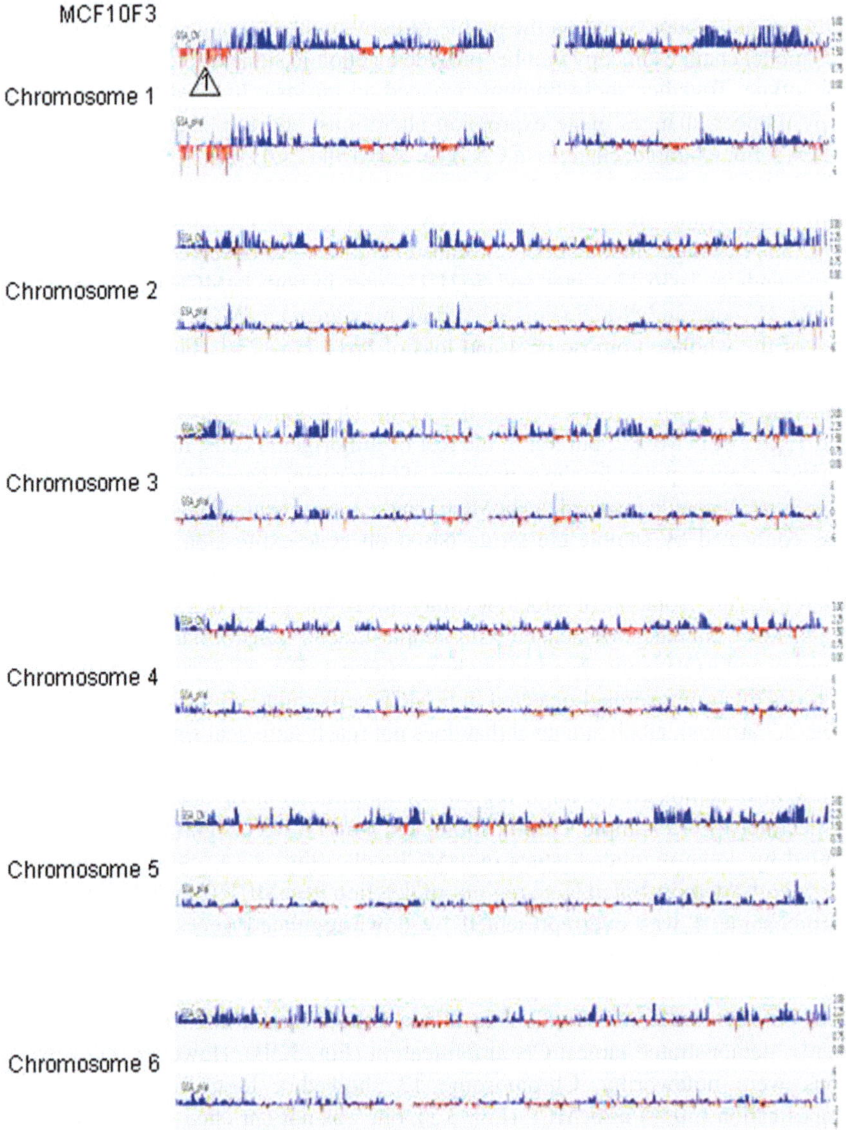

Fig. 3.4 The frequency of copy number (CN)

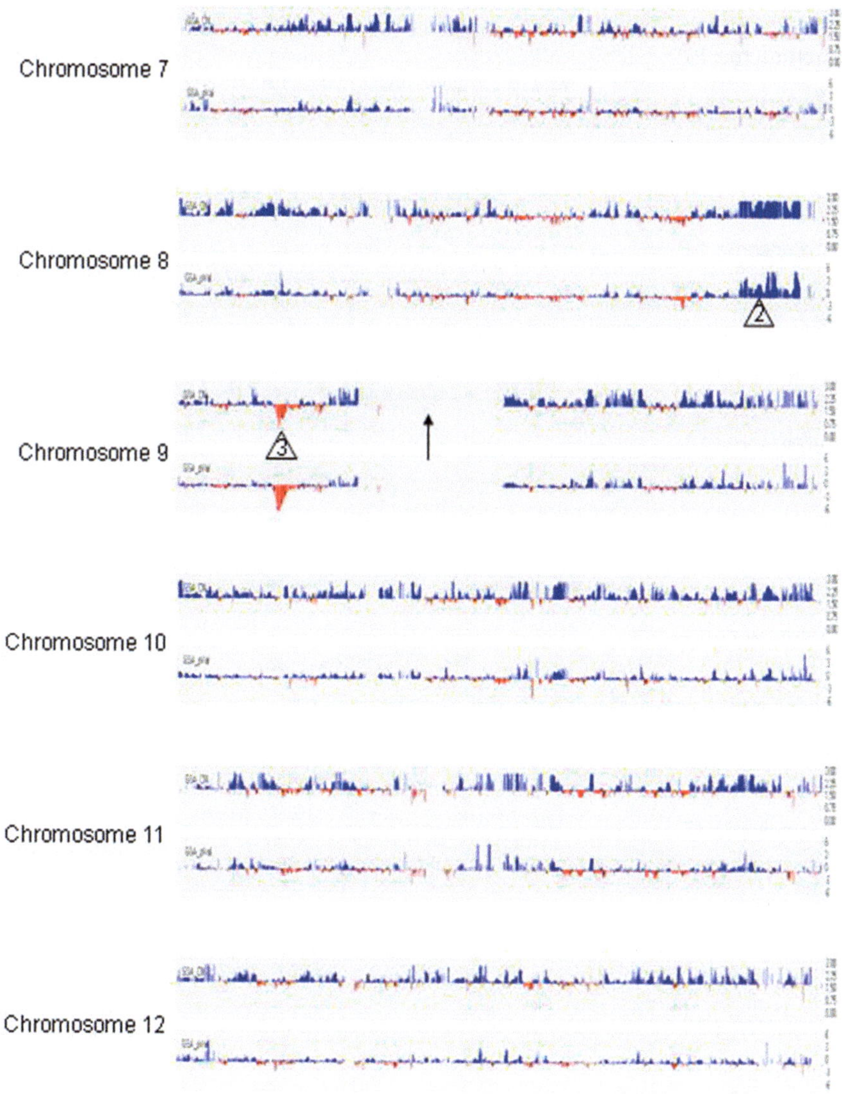

Fig. 3.4 (continued)

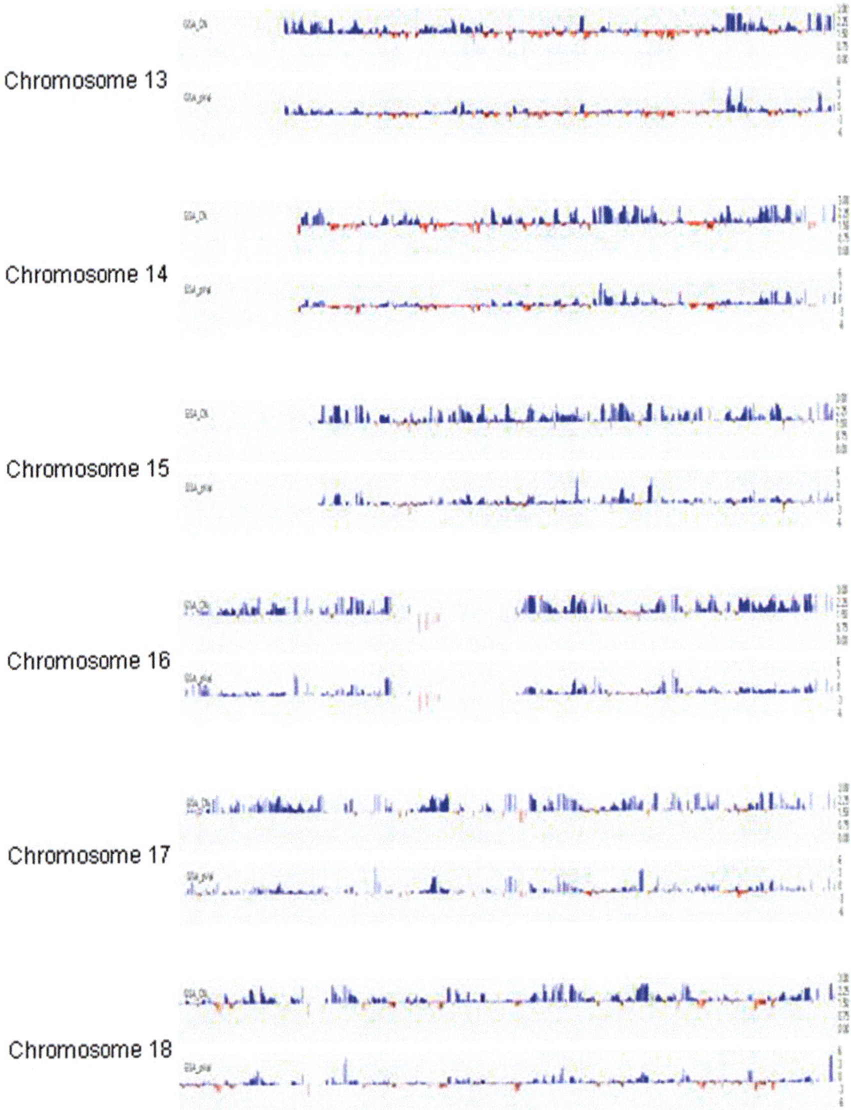

Fig. 3.4 (continued)

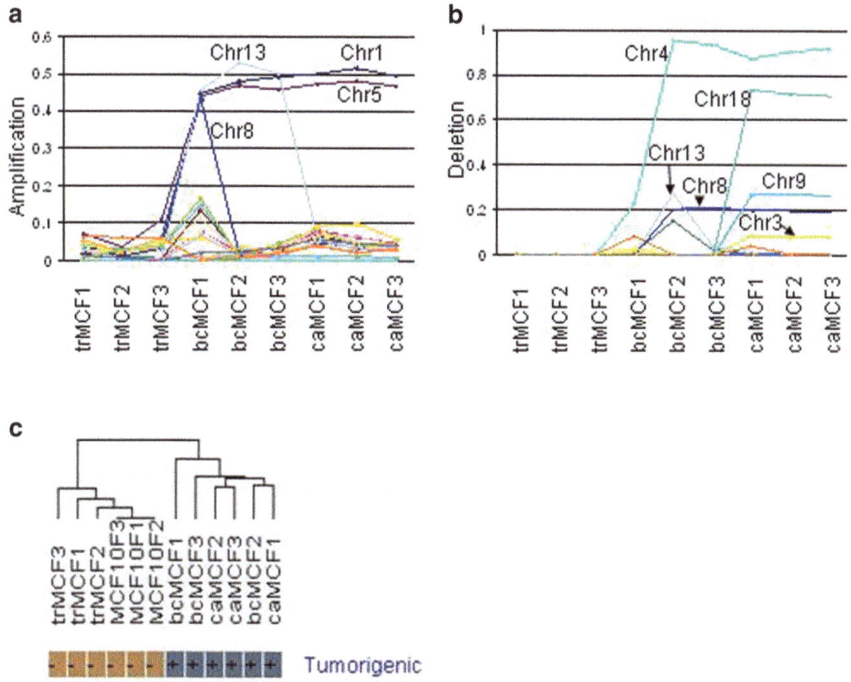

Fig. 3.5 The frequency of copy number (CN) changes and sample clustering. The frequency was calculated as the ratio of the number of amplified SNPs (**a**) or deleted SNPs (**b**) to total number of SNPs in each chromosome. Sample clustering based on the CN profile using dChip software (**c**). Chromosome 8 had a 47 % frequency of amplification in bcMCF1, but not in bcMCF2, bcMCF3 or caMCF (**a**).Chromosome 4 displayed a high frequency of deletion in caMCF and bcMCF (88 %~96 %), except in the subclone bcMCF1 where the frequency was only 23 % (**b**). The 12 samples were correctly clustered into non-tumorigenic (–) and tumorigenic (+) groups demonstrated by their growth in SCID mice (**c**)

cells did not show any CN deletion (Fig. 3.3B). This might imply that certain epigenetic modifications, instead of CN loss, are related to the observed decrease in the number of genes expressed on chromosome 11 of trMCF cells.

Gene ontology (GO) analysis identified several biological processes enriched with dysregulated genes in trMCF, bcMCF and caMCF cells, suggesting that changes in these processes were involved in the *in vitro* transformation phenotype (Table 3.2). Two additional processes, DNA replication and nucleosome assembly, were found only in the trMCF cells. The DNA replication process in trMCF cells contained six genes. Each of these genes was upregulated, including: *replication factor C*, 2.1 fold; *SET translocation*, 1.8 fold; *DNA directed polymerase epsilon 2*, 2.7 fold; *ribonucleotide reductase M2 polypeptide*, 1.7 fold; *minichromosome maintenance deficient 4*, 1.7 fold, and *topoisomerase (DNA) II alpha*, 1.8 fold. Moreover, 11 biological processes were found uniquely in the tumorigenic cells bcMCF and

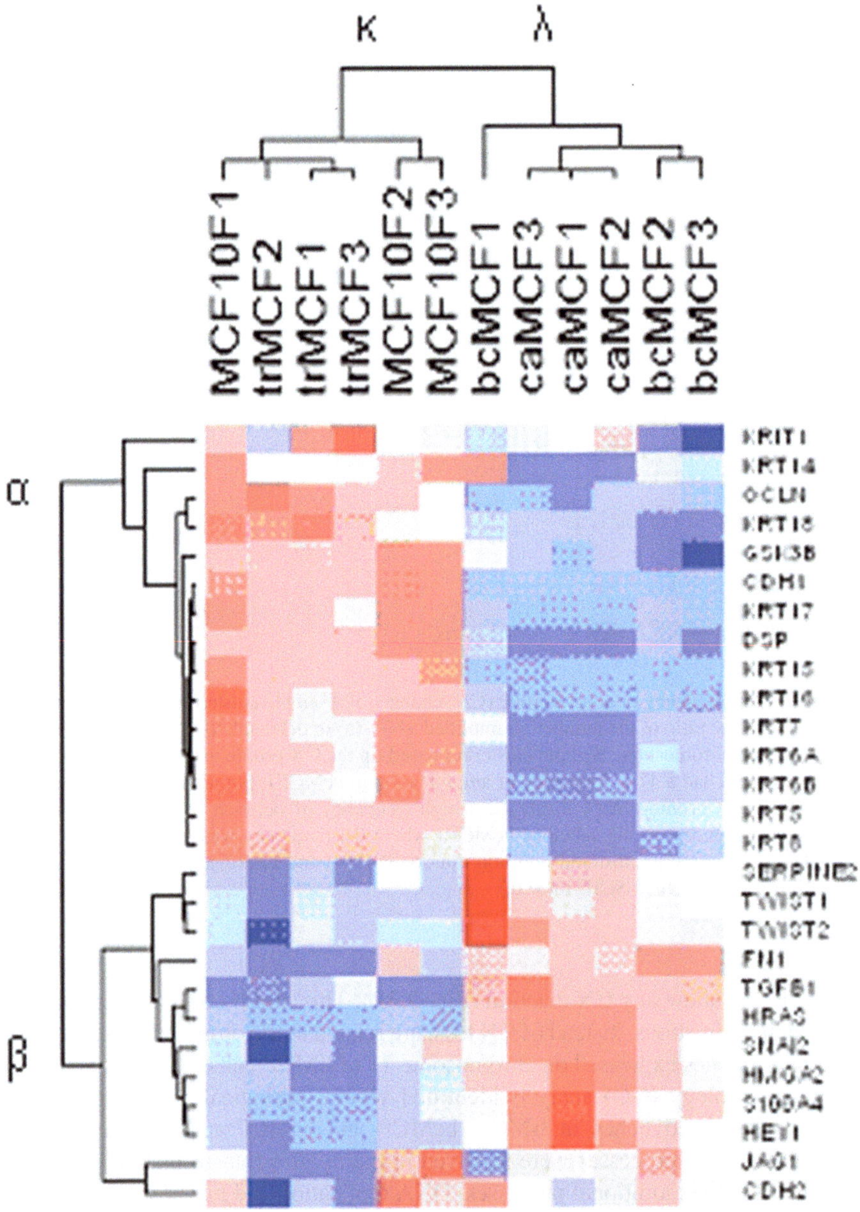

Fig. 3.6 Expression profile of EMT markers and their regulators during malignant cell transformation. A list of EMT markers and promoting genes was generated a priori by literature search (Table 7.6). Hierarchical clustering of cell lines and genes was performed using dChip software. Two sample clusters (k and λ) and two gene clusters (α and β) were identified. *Red*, *white*, and *blue*, level above, at, and below mean expression, respectively

Table 3.1 Histological and immunocytochemistry of the bcMCF clones

Clones	Tumor type	AE1	CAM.2	EMA	Vimentin
bcMCF-1 bcMCF-4	Invasive poorly differentiated spindle cell type	–	–	–	+++
bcMCF-2 bcMCF-6 bcMCF-7	Invasive poorly differentiated epithelial cell type	+/–	+/–	+/–	++
bcMCF-3 bcMCF-5	Invasive poorly differentiated with mix features of spindle and epithelial type	–	+/–	+/–	++

The mouse monoclonal antibodies anti-human cytokeratin of low molecular weight (AE1, Bigenex, San Ramon CA) cytokeran peptides 7 and 8 (CAM5.2, Ventana, Tucson, AZ), epithelial membrane antigen (EMA) clone E29 and vimentin, clone V9, both from DakoCytomation,Inc. (Fort Collins, CO) were used. Negative (–), weak (+/–), moderate (+++) or strong (+++)

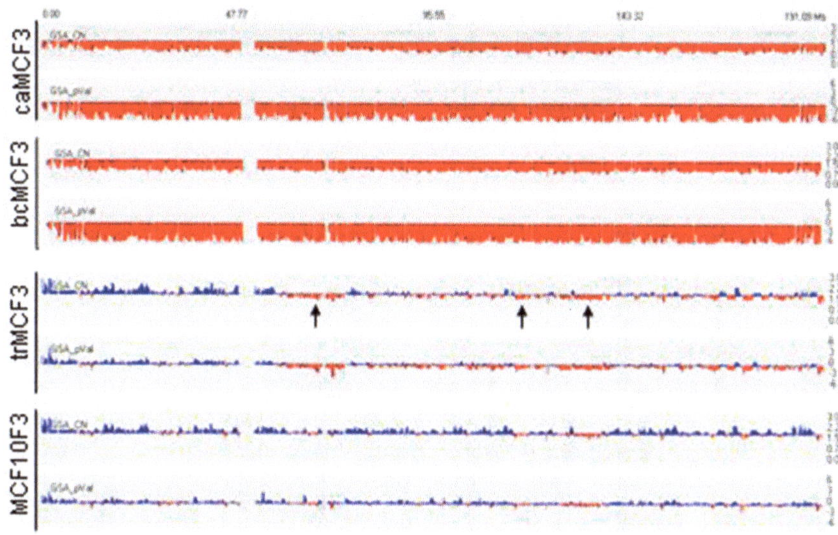

Fig. 3.7 Higher resolution of chromosome 4 copy number is displayed by CNAT v3.0 using genome-smoothing analysis for each cell line. In each panel, the *upper line* shows CN and the *lower line* shows the associated p value. This result was observed consistently in all samples except in bcMCF1 (data not shown).The CNAT default 00+ reference samples were used as diploid control to detect CN changes (Reference: Affymetrix GeneChip Chromosome Copy Number Analysis Tool User Guide Version 3.0, Page 53). Genome-smoothing analysis (GSA) with 0.5 Mbp distance was used to delineate the CN changes

caMCF, suggesting that they were involved in the process of neoplastic transformation (Table 3.2). Most of the genes in the eleven GO categories showed decreased expression relative to MCF-10F cells. For example, the fraction of down-regulated genes was 24/28 in "apoptosis" (Table 3.3), 13/16 in "positive regulation of IKB/NFKB cascade", and 19/21 in "lipid metabolism". Enrichment of dysregu-

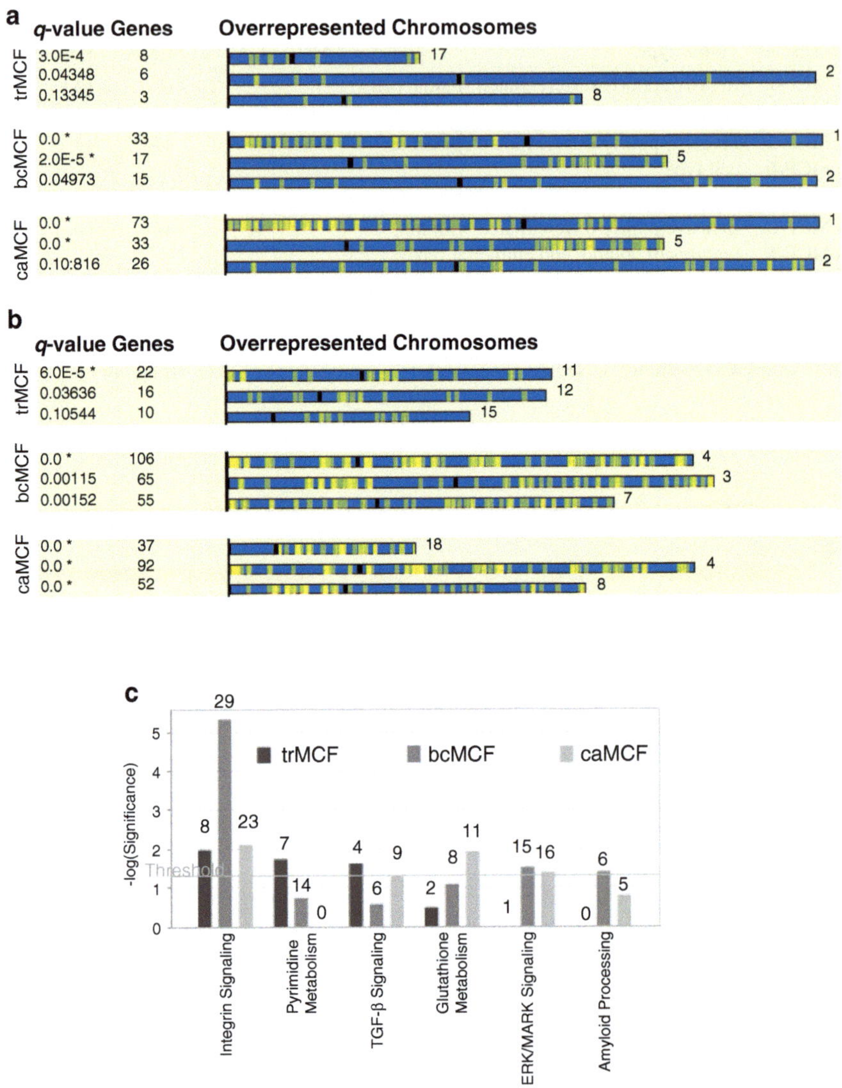

Fig. 3.8 (**a** and **b**) Chromosomes enriched with differentially expressed genes. The top three chromosomes (ranked by *q* value) in trMCF, bcMCF, and caMCF cell lines enriched with up-regulated (**a**) or down-regulated (**b**) genes. *, the chromosomes overrepresented ($q \leq 0.0001$) by up- or down-regulated genes. A *q* value of 0.0 indicates a value smaller than 10^{-5}. (**c**) Canonical pathways altered in progressive malignant cell transformation. The pathways significantly enriched with the dysregulated genes are identified by Ingenuity Pathway Analysis. The number of differentially expressed genes in each pathway is labeled above the corresponding cell. The genes identifying the integrin signaling pathway are displayed in Table 7.4; those in the other five canonical pathways are displayed in Supplementary Table 7.5. *ERK/MAPK,* extracellular signal-regulated kinase/mitogen-activated protein kinase

Table 3.2 Biological processes enriched with differentially expressed genes in tumorigenic and non-tumorigenic cells

		trMCF		bcMCF		caMCF	
GO[a] ID	Function name	Genes	q	Genes	q	Genes	q
In all tumorigenic and non-tumorigenic cell lines							
GO:007155	Cell adhesion	14	6.4E-04	43	3.7E-04		
GO:0007267	Cell-cell signaling	5	5.0E-02	26	5.4E-03	32	1.0E-04
GO:0008544	Epidermis development	7	6.3E-06	20	2.1E-09	19	2.2E-09
GO:0006955	Immune response	11	6.1E-04	26	1.5E-02	30	4.0E-03
GO:0008152	Metabolism	10	8.8E-03	36	6.6E-03	35	1.5E-02
GO:0006508	Proteolysis and peptidolysis	12	2.3E-03	32	4.5E-02	40	3.8E-03
GO:0000074	Regulation of cell cycle	11	3.4E-04	23	2.3E-02	28	2.9E-03
Only in non-tumorigenic cell lines							
GO:0006260	DNA replication	6	1.4E-03	1	NS	6	NS
GO:0006334	Nucleosome assembly	5	5.4E-04	2	NS	3	NS
Only in tumorigenic cell lines							
GO:0006915	Apoptosis	4	NS	28	5.0E-03	27	1.7E-02
GO:0005975	Carbohydrate metabolism	2	NS	20	2.0E-02	24	3.5E-03
GO:0007166	Cell surface receptor linked signal transduction	2	NS	14	4.2E-02	15	3.5E-02
GO:0006935	Chemotaxis	2	NS	11	1.6E-02	11	2.1E-02
GO:0006997	Endocytosis	0	NS	11	2.6E-02	11	2.7E-02
GO:0006629	Lipid metabolism	1	NS	21	1.1E-02	21	1.8E-02
GO:0043123	Positive regulation of IKB/NFKB cascade	1	NS	16	5.0E-04	13	1.5E-02
G0:0019538	Protein metabolism	0	NS	5	1.2E-02	5	1.4E-02
G0:0006986	Response to unfolded protein	1	NS	9	4.7E-03	9	5.8E-03
GO:0016192	Vesicle-mediated transport	1	NS	10	6.6E-04	12	1.2E-03
GO:0007169	Transmembrane receptor	2	NS	12	4.6E-04	12	1.5E-02

Abbreviation: *NS* non-significant GO category

[a]GO categories were identified by Gene Ontology tool, Onto-Express

Table 3.3 Differentially expressed apoptosis genes (GO:00009165) in tumorigenic bcMCF cells

		Fold change		
Symbol	Gene name	trMCF	bcMC	caMCF
AHR	Aryl hydrocarbon receptor	Ns[a]	−2.1	Ns
APP	Amyloid beta (A4) precursor protein	Ns	−1.7	−1.7
BAG1	BCL2-associated athanogene	Ns	−7.6	−9.2
BAG5	BCL2-associated athanogene 5	Ns	−1.8	Ns
BIRC4	Baculoviral IAP repeat-containing 4	Ns	−2.0	−2.7
BNIP3L	BCL2/adenovirus E1B interacting protein 3-like	Ns	−2	−1.9
CASP14	Capase 14,apoptosis-related cysteine peptidase	Ns	−12	−12.3
CASP3	Caspase 3, apoptosis-related cysteine peptidase	Ns	−2.4	Ns
CASP6	Caspase 6, apoptosis-related cysteine peptidase	Ns	−4.1	−4.1
CD14	CD14 antigen	Ns	−4.7	Ns
DAPK1	Death-associated protein kinase 1	Ns	−4.1	−13.4
ELMO3	Engulfment and cell motility 3	Ns	−9.9	−10.8
ELMOD2	ELMO domain containing 2	Ns	−2.5	−2.2
GADD45A	Growth arrest and DNA-damage-inducible alpha	Ns	−2.1	Ns
GADD45B	Growth arrest and DNA-damage-inducible beta	Ns	−2.7	Ns
HIPK2	Homeodomain interacting protein kinase 2	Ns	−3.2	Ns
HIPK3	Homeodomain interacting protein kinase 3	Ns	−3.1	−2.6
MAEA	Macrophage erythroblast attacher	Ns	−2.2	Ns
NGFRAP1	Nerve growth factor receptor associated protein 1	Ns	2.1	2.0
PPP1R13L	Protein phosphatase 1, inhibitory subunit 13 like	Ns	−2.9	−2.8
SGPL1	Sphingosine-1 phosphate lyase 1	Ns	−2.9	−2.7
STK17A	Serine/threonine kinase 17a(apoptosis-inducing)	Ns	−1.7	Ns
STK3	Serine/threonine kinase 3(STE20 homolog, yeast)	Ns	−3.5	Ns
TNFSF7	Tumor necrosis factor superfamily, member 7	−2.0	−7.6	−4.1
TNFSF9	Tumor necrosis factor superfamily, member 9	−2.0	−8.1	−4.5
TP73L	Tumor protein p73-like	Ns	−4.7	11.4
TRAF4	TNF receptor-associated factor 4	Ns	−2.6	−2.5
TTC11	Tetratricopeptide repeat domain 11	Ns	1.9	−1.9

[a]Non-significant gene

Genes were identified by Gene Ontology tool, Onto-Express.

lated genes in the apoptotic process in tumorigenic cells (Table 3.3) and resistance to apoptosis is a hallmark of tumorigenesis. Therefore, suppression of apoptosis in the tumorigenic cells might be a potential mechanism that confers survival onto these cells with disrupted integrin signaling and anchorage-independent growth [6].

There are six canonical pathways significantly dysregulated in one or more groups (Fig. 3.8). The genes associated with each of these pathways, integrin signaling, pyrimidine metabolism, transforming growth factor beta (TGF-β) signaling, glutathione metabolism, ERK/MAPK signaling, and amyloid processing are listed in Tables 3.2 and 3.4. The integrin signaling pathway was the most significantly

Table 3.4 Dysregulated genes involved in integrin signaling pathway

Symbol	Gene name[a]	Fold change		
		trMCF	bcMCF	caMCF
ARF3	ADP-ribosylation factor 3	NS	−1.7	NS
ARF5	ADP-ribosylation factor 5	NS	−2.5	−2.4
ARF6	ADP-ribosylation factor 6	NS	NS	−1.8
ARHGAP5	Rho GTPase activating protein 5	NS	−1.8	NS
CAPN1	Calpain 1(mu/l) large subunit	NS	−2.6	−2.4
CAV1	Caveolin 1	NS	−1.8	NS
CDC42	Cell division cycle 42	NS	1.9	2.5
COL1A1	Collagen type 1, 1	−6.1	−16.8	NS
DDEE1	Develop and differentiation enhancing factor 1	NS	2.2	2.2
FN1	Fibronectin 1	−18.9	4.9	NS
FYN	FYN oncogene related to SRC, FGR, YES	NS	2.2	NS
GSK3B	Glycogen synthesis kinase 3	NS	−2.2	1.9
HRAS	v-Ha-rat Harvey rat sarcoma viral oncogene	NS	6.0	7.4
ITGA6	Integrin $_6$	NS	−1.8	−1.8
ITGAV	Integrin $_v$	−1.8	NS	−2.5
ITGB1	Integrin $_1$	NS	NS	−2.0
ITGB4	Integrin $_4$	−1.9	−3.9	−3.7
ITGB6	Integrin $_6$	−2.6	50.4	−58.5
KRAS	v-Ki–ras2 Kirsten rat sarcoma viral oncogene	NS	−2.7	NS
LAMA3	Laminin 3	−3.5	−100.5	−99.8
LAMA5	Laminin 5	NS	NS	−1.7
LAMB3	Laminin 3	−2.7	−3.6	−4.4
LAMB5	Laminin 2	−3.6	−8.9	−26.2
MYLK	Myosin, light polypeptide kinase	NS	NS	4.4
NCK2	NCK adaptor protein 2	NS	NS	−2.3
RAC1	ras-related C3 botulinum toxin substrate 1	NS	−1.9	NS
RAC2	ras-related c3 botulinum toxin substrate 2	NS	−1.9	NS
RALA	v-related simian leukemia viral oncogene A	NS	−2.0	NS
RALB	v-related simian leukemia viral oncogene B	NS	−1.7	NS
RHOB	ras homologue gene family, member B	NS	2.3	NS
RHOD	ras homologue gene family, member D	NS	−3.2	−4.2
RHOF	ras homologue gene family, member F	NS	−3.8	−4.3
TM4SF6	Tetraspanin 6	NS	−2.5	−1.8
TM4SF7	Tetraspanin 4	NS	1.7	NS
TM4SF8	Tetraspanin 3	NS	NS	−1.9
TM4SF9	Tetraspanin 5	NS	6.5	7.1

Abbreviation: *NS* nonsignificant gene
[a]Genes were identified by Ingenuity Pathway Analysis

altered pathway in the progression of the neoplastic transformation [8]. Integrins function as heterodimeric receptors for extracellular matrix proteins, mediating cell anchorage. The capacity of cells to survive and proliferate in the absence of integrin-mediated adhesion in vitro strongly correlates with tumorigenesis in vivo [9]. The integrin signaling pathway was enriched with dysregulated genes in trMCF, bcMCF, and caMCF, indicating that this pathway was affected in early stages of cell transformation (Fig. 3.4) and is the most significantly altered pathway during the EMT, indicating a continuum change within this pathway throughout the progressive, malignant cell transformation of MCF-10F cells. In this pathway, eight genes were down-regulated in trMCF cells (Table 3.4). Of interest, expression levels of several of these decreased genes, such as *ITGB6*, *LAMA3* and *LAMC,* were inhibited to even a greater extent in the tumorigenic cells. In contrast, *fibronectin 1,* which was completely suppressed (-18.9 fold, 'Absent') in trMCF cells, was strongly induced in bcMCF (4.9 fold) and caMCF (2.8 fold) cells (Fig.. 3.6a, *FN1*). In comparison to the trMCF cells, the number of dysregulated genes associated with integrin signaling was much higher in the bcMCF and caMCF cells, showing both increased and decreased levels of expression (Table 3.4).

Levels of expression of several genes involved in glutathione metabolism were decreased in the bcMCF and caMCF cells (Table 3.5). Multiple glutathione *S*-transferase (*GST*) isoforms were down-regulate including *GST kappa1*, *omega2*, *pi*, and microsomal *GST 2* and *3*. Moreover, glutamate cysteine ligase, the enzyme catalyzing the first rate-limiting reaction in glutathione biosynthesis, was also reduced in the tumorigenic cells. Glutathione is a cellular antioxidant and a substrate for GST, which catalyzes the conjugation reactions with electrophiles. Repressed antioxidant activity is associated with the genomic damage and cancer incidence induced by exposure to the cytotoxic free radical and reactive oxygen species [10].

Analysis of the GO cellular component also identified categories distinct between the non-tumorigenic and tumorigenic cells. The "intermediate filament" component was found in the bcMCF ($q=2.0\times10^{-5}$, 12 genes) and caMCF cells ($q=0.0$, 13 genes), but not in trMCF cells ($q=0.4$, 1 gene) (Table 3.6). Numerous cytokeratins were suppressed or absent, while *vimentin* was strongly induced in bcMCF (7.0 fold) and caMCF (8.1 fold) cells. Because of these findings, we generated a gene list from published literature for epithelial and mesenchymal markers and their regulators (Table 3.6). The 52 genes in the list were filtered by low stringency criteria of combined co-efficient of variation (CV)>0.3 and 'Present calls' in more than 30 % of the samples. The 27 genes passing these criteria were used for sample and gene clustering (Fig. 3.6). Two sample groups and two gene groups were identified. The non-tumorigenic MCF-10F and trMCF cells were grouped into sample cluster k, while the tumorigenic bcMCF and caMCF cells were grouped into cluster λ. On the other side, the genes were grouped into cluster α and β based on their expression pattern. The epithelial markers *E-cadherin, occludin, desmoplakin* and *cytokertins* were decreased, while the mesenchymal markers *fibronectin*, *vimentin* and *N-cadherin* were increased in bcMCF and caMCF cells (Fig. 3.6). By real time RT-PCR, it was confirmed that the expression of FN1, S100A4, SNAI2, HRAS and TGFβ1 were increased, while CDH1 (E-cadherin) was decreased in bcMCF and caMCF cells (Fig. 3.9).

Table 3.5 Canonical pathways enriched with dysregulated genes

Symbol	Gene name	Fold change		
		trMCF	bcMCF	caMCF
TGFβ signaling pathway				
BMP2	Bone morphogenetic protein 2	Ns[a]	Ns	–4.3
BMPR2	Bone morphogenetic protein receptor, type ll	2.0	3.0	2.
HRAS	V-Ha-ras Harvey rat sarcoma viral oncogene homolog	Ns	6.0	7.4
INHBA	Inhibin, beta A (activin A, activin AB alpha polypeptide)	Ns	Ns	−14.9
KRAS	i-Ki-ras2 Kirsten rat sarcoma viral oncogene homolog	Ns	−2.7	Ns
SERPINE1	Serpin peptidase inhibitor, clade E, member 1	–5.1	Ns	Ns
SMAD2	SMAD, mothers against DPP homolog 2	Ns	Ns	−1.8
SMAD5	SMAD, mothers against DPP homolog 5	Ns	Ns	2.1
SMURF2	SMAD specific E3 ubiquitin protein ligase 2	Ns	Ns	2
TGFB1	Transforming growth factor, beta 1	Ns	3.1	3.1
TGFB2	Transforming growth factor, beta 2	–3.6	–9.4	–9.9
TGFBR2	Transforming growth factor, beta receptor II	–2.2	−1.8	Ns
Pyrimidine metabolism pathway				
CANT1	Calcium activated nucleotidase 1	Ns	−1.9	Ns
CDA	Cytidine deaminase	–6.1	Ns	Ns
DCTD	dCMP deaminase	Ns	−2.1	−2.1
DPYSL3	Dihydrophyrimidinase-like 3	Ns	24.1	26.9
ECGF1	Endothelial cell growth factor 1 (platelet-derived)	Ns	−3.2	−3.4
ENTPD4	Ectonucleoside triphosphate diphosphohydrolase 4	Ns	–4.4	−2.9
NME4	Non-metastatic cells 4, protein expressed in	–2	−5.1	–4.7
NME7	Non-metastatic cells 7, protein expressed in	Ns	−2.6	Ns
NT5C2	5′-nucleotidase, cytosolic II	Ns	−2.6	−2.2
NT5E	5′-NUCLEOTIDASE, ECTO (CD73)	–3.3	1.9	2.1
POLB	Polymerase (DNA directed), beta	Ns	−2.2	Ns
POLE2	Polymerase (DNA directed), epsilon 2 ([p89 subunit)	2.7	Ns	Ns
POLR1D	Polymerase (RNA) I polypeptide D, 16 kDa	Ns	−1.8	Ns
POLR2B	Polymerase (RNA) ll (DN directed) polypeptide B	Ns	−1.8	Ns
RFC3	Replication factor C (activator 1) 3, 38 kDa	2.1	Ns	Ns
PRM2	Ribonucleotide reductase M2 polypeptide B	1.7	Ns	Ns
RRM2B	Ribonucleotide reductase M2 B (TP53 inducible)	–2.9	−3.6	−3
UPP1	Uridine phosphorylase 1	Ns	−2.1	−1.8
Amyloid processing pathway				
APP	Amyloid beta (A4) precursor protein	Ns	−1.7	−1.7
CAPN1	Calpain 1(mu/l) large subunit	Ns	−2.6	−2.4
CSNK1E	Casein kinase 1, epsilon	Ns	−3.3	−2.2
GSK3B	Glycogen synthase kinase 3 beta	Ns	−2.2	−1.9

(continued)

Table 3.5 (continued)

Symbol	Gene name	Fold change		
		trMCF	bcMCF	caMCF
PRKAR1A	Protein kinase, cAMP-dependent, regulatory, type I, α	Ns	−1.9	Ns
PSEN1	Presenilin 1 (Alzheimer disease 3)	Ns	−1.8	−1.9
Glutathione metabolism				
GCLC	Glutathione cysteine ligase, catalytic subunit	Ns	−2.9	−2.7
GPX1	Glutathione peroxidase 1	Ns	Ns	−1.7
GSR	Glutathione reductase	2.2	Ns	Ns
GSTK1	Glutathione S-transferase kappa 1	−1.8	−2.3	−2.1
GST02	Glutathione S-transferase omega 2	Ns	−5.6	−6.6
GSTP1	Glutathione S-transferase pi	Ns	−2.0	−2.3
GSTT2	Glutathione S-transferase theta 2	Ns	Ns	2.1
IDH1	Isocitrate dehydrogenase 1 (NADP+), soluble	Ns	−2.2	−2.2
IDH 2	Isocitrate dehydrogenase 1 (NADP+)	Ns	−3.3	−2.3
MGST2	Microsomal glutathione S-transferase 2	Ns	−2.4	−2.4
MGST2	Microsomal glutathione S-transferase 3	Ns	Ns	−2.2
RAB 15	RAB 15,member RA oncogene family	Ns	−2.8	−2.5
ERK/MAPK signaling pathway				
DUSP1	Dual specificity phosphatase 1	3.0	2.7	Ns
DUSP4	Dual specificity phosphatase 4	Ns[a]	−2.2	Ns
DUSP6	Dual specificity phosphatase 6	Ns	2.3	2.4
EIF4E	Eukaryotic translation initiation factor 4E	Ns	2.4	−2.3
EIF4EBP1	(Eu) translation initiation factor 4E binding protein 1	Ns	Ns	−1.9
FOS	v-fos FBJ murine osteosarcoma viral oncogene	Ns	Ns	−11.5
FYN	FYN oncogene related to SRC, FGR, YES	Ns	Ns	2.2
HRAS	V-Ki-ras Harvey rat sarcoma viral oncogene	Ns	6.0	7.4
HSPB1	Heat shock 27kDa protein 1	Ns	−1.9	−2.8
KRAS	v-Ki-ras2 Kristen rat sarcoma viral oncogene	Ns	2.7	Ns
MAP2K1IPI	MAP kinase kinase 1 interacting protein 1	Ns	−3.6	−3.4
MKNK2	MAP kinase interacting serine/threonine Kinase 2	Ns	Ns	−1.9
PPP1CB	Protein phosphatase 1, catalytic subunit, β	Ns	−2.2	−1.9
PPP1R14C	Protein phosphatase 1, inhibitory subunit 1 4C	Ns	20.2	−9.7
PPP2CB	Protein phosphatase 2, catalytic subunit, β	Ns	Ns	−2.0
PPP2R5C	Protein phosphatase 2, regulatory subunit, β, γ	Ns	−2.0	Ns
PRKAR1A	Protein kinase cAMP-dependent regulatory type 1 α	Ns	−1.9	Ns
PRKCI	Protein kinase C iota	Ns	−2.0	−2.6
RAC1	ras-related C3 botulinum toxin substrate1	Ns	−1.9	Ns
RAC2	ras-related C3 botulinum substrate 2	Ns	Ns	Ns
STAT3	Signal transducer and activator of transcription 3	Ns	Ns	−2.1
YWHAZ	Tyrosine 3-monooxygenase zeta polypeptide	Ns	Ns	−1.9

[a]Non-significant gene
Genes were identified by Ingenuity Pathway Analysis

Table 3.6 Dysregulated genes in intermediate filament (GO:0005882)

Symbol	Gene name	Fold change		
		trMCF	bcMCF	caMCF
KRT5	Keratin 5	Ns[a]	−2.2	−17.2
KRT7	Keratin 7	Ns	−2.5	−5.3
KRT8	Keratin 8	Ns	−2.6	−4.4
KRT16	Keratin 16	Ns	−11.7	−50.4
KRT17	Keratin 17	Ns	−37.4	−50.6
KRT18	Keratin 18	Ns	−2.1	−1.9
KRT6B	Keratin 6B	−1.7	−102.9	−114.9
KRTHB1	Keratin, hair, basic, 1	Ns	−6.1	−5.0
DSP	Desmoplakin	Ns	−8.7	−135.7
RSN	Restin	Ns	−2.8	−2.0
SPRR1B	Small proline-rich protein 1B (cornifin)	Ns	−14.2	−15.6
PLEC1	Plectin1	Ns	Ns	4.3
VIM	Vimentin	Ns	7.0	8.1

[a]Non-significant gene

Genes were identified by Gene Ontology tool, Onto-Express for cellular component

Immuno-cytochemical analysis using antibodies against epithelial membrane antigen EMA (also called MUC1) and E-cadherin displayed significant loss of these epithelial markers (Fig. 3.10, *j, k, n, o*) and increased expression of the mesenchymal marker vimentin (Fig. 3.10, *l, p*) in tumorigenic cells. These findings confirmed the EMT phenotype revealed by gene expression profile in Fig. 3.9. Interestingly, while most trMCF cells were E-cadherin positive (+) and vimentin negative (−), a few cells were found to be E-cadherin (−) (Fig. 3.10, *g,* arrow) and vimentin (+) (Fig. 3.10, *h,* arrow).

The cell surface molecules $CD44^{+}/CD24^{-/low}$ phenotype has been identified as marker for tumor initiating cells [11], therefore, the expression pattern of these markers in relation to the tumorigenic capacity of the cells in the model of EMT revealed that *CD44* was slightly decreased in trMCF but increased in bcMCF and caMCF cells, while *CD24* was completely lost in both bcMCF and caMCF cells (Fig. 3.11a, b). The loss of *CD24* in bcMCF and caMCF cells was independently demonstrated by RT-PCR (Fig. 3.11, middle panel). The alternative splicing variants of *CD44* were identified by the size of their PCR products and further validated by DNA sequence analysis. Only the variants of *CD44H* and *CD44E* were expressed (Fig. 3.11, upper panel). *CD44H* was significantly increased in both bcMCF and caMCF cells, while *CD44E* was completely lost in these cells. The ratio of *CD44H/CD44E* was significantly increased in trMCF, bcMCF and caMCF cells (Fig. 3.11d). The CD44 and CD24 gene expression in the different cell lines correlated well with the cell surface protein expression studied by FACS analysis (Fig. 3.12). In addition, the *ESA* (epithelial specific antigen, also termed as *Ep-CAM*, or tumor-associated calcium signal transducer 1), was completely lost in bcMCF and caMCF, but not changed in trMCF cells

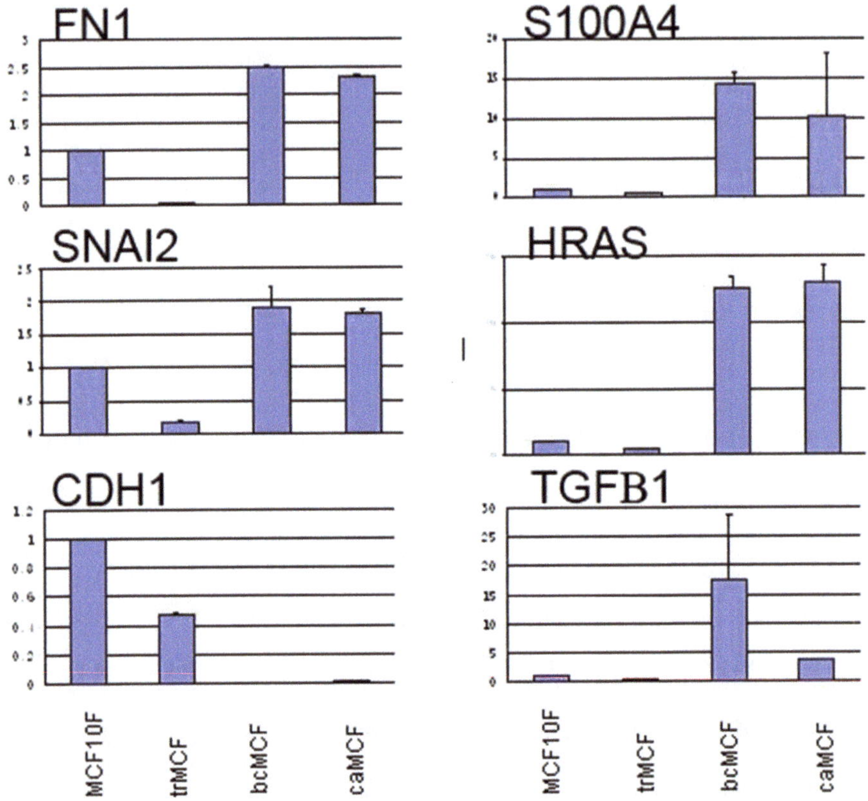

Fig. 3.9 Validation of the expression of epithelial and mesenchymal markers by real time RT-PCR. Dysregulated genes identified by microarray experiments were validated by TaqMan real-time PCR (Applied Biosystems, Foster City, CA). The mesenchymal markers and their regulators Fibronectin 1 (FN1), S100A4, SNAI2, HRAS and TGFβ1 were increased, while the epithelial marker E-cadherin (CDH1) was decreased in the tumorigenic cells bcMCF and caMCF. Briefly, total RNAs was isolated from growing cells at 70–80 % confluence using Trizol (Invitrogen, Carlsbad, CA), treated with DNase I (Invitrogen, Carlsbad, CA) and cleaned using RNeasy kit (Quiagen, Valencia, CA). The TaqMan One Step RT-PCR kit (Applied Biosystems, Foster City, CA) was used for the real time reversetranscriptase PCR and the assays were run using Applied Biosystems 7900 HT instrument. The TATA box-binding protein (TBP) was used as endogenous control. The Taqman primers/probe sets (Applied Biosystems, Foster City, CA) used were: Hs00427620_m1 for TBP (TATA-box binding protein), Hs00171257_m1 for TGFβ1, Hs00610483_m1 for HRAS, Hs00161904_m1 for SNAIL2, Hs00365052_m1 for FN1, Hs00243202_m1 for S100A4 and Hs00170423_m1 for CDH1. The reactions were made in duplicate for each RNA sample and primer/probe and the expression levels were compared to the values in the parental cell line MCF-10F. The Ct is defined as the fractional cycle number at which the fluorescence generated by cleavage of the probe passes a fixed threshold above baseline. The comparative method calculates the relative gene expression using the following equation: Relative quantity = 2 –ΔΔCt

Fig. 3.10 (continued) and vimentin, respectively (×100); (**q** and **r**) invasive ductal carcinoma of the breast as positive control and immunoreacted for EMA and E-cadherin, respectively (×100); (**s**) histologic section of an invasive adenocarcinoma immunoreacted for vimentin (×100)

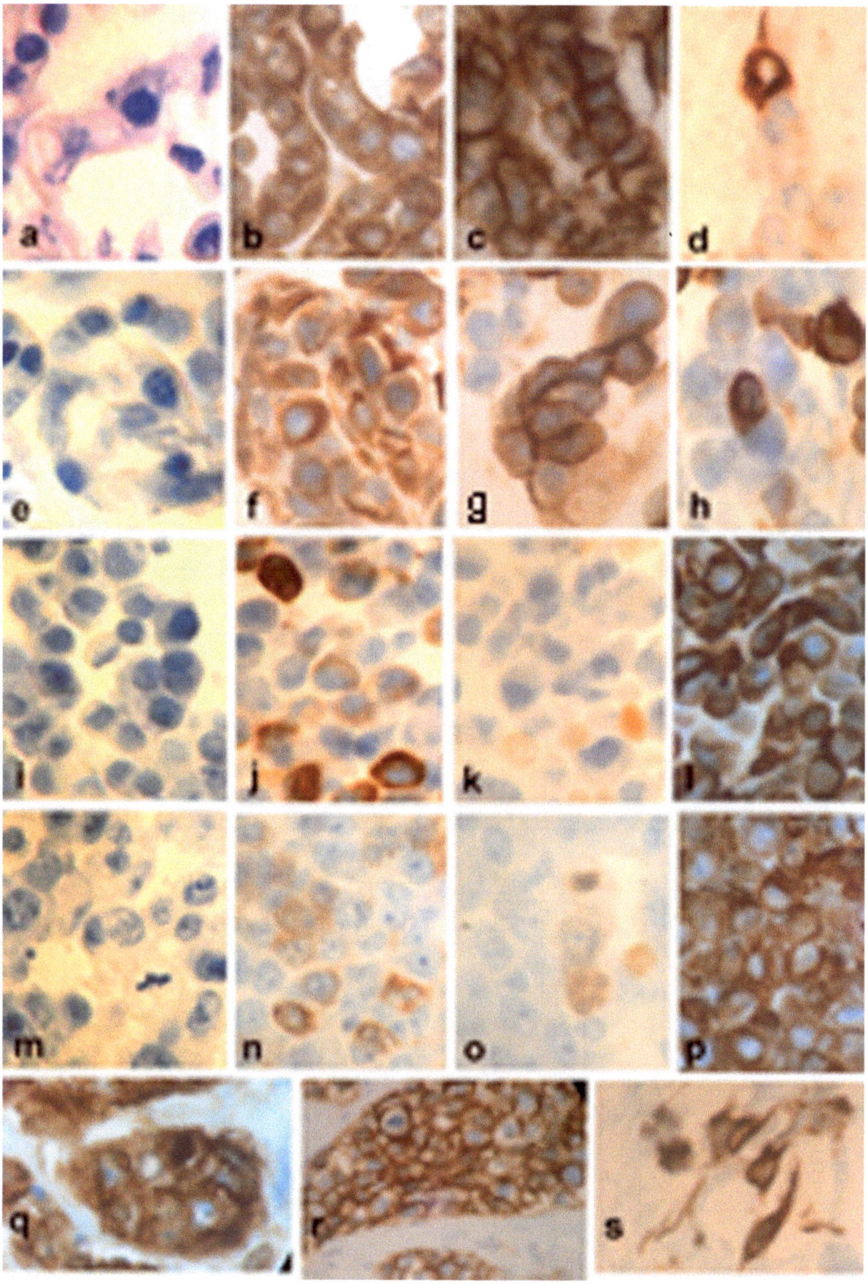

Fig. 3.10 Detection of epithelial and mesenchymal markers by immunocytochemistry. (**a**) Histologic sections of MCF-10F cells, reacted with preimmune mouse serum, were used as the negative control (×100); (**b**) MCF-10F reacted for EMA (×100); (**c**) MCF-10F reacted for E-cadherin (×100); (**d**) MCF-10F reacted for vimentin (×100); (**e**) trMCF cells reacted with preimmune mouse serum used as negative control (×100); (**f**, **g**, and **h**) trMCF cells reacted for EMA, E-cadherin, and vimentin, respectively (×100); (**i**) bsMCF cells reacted with preimmune mouse serum as a negative control (×100); (**j**, **k**, **l**) bsMCF cells reacted for EMA, E-cadherin, and vimentin, respectively (×100); (**m**) caMCF tumor cell line cells reacted with preimmune mouse serum used as negative control (×100); (**n**, **o**, **p**) caMCF tumor cell lines reacted for EMA, E-cadherin,

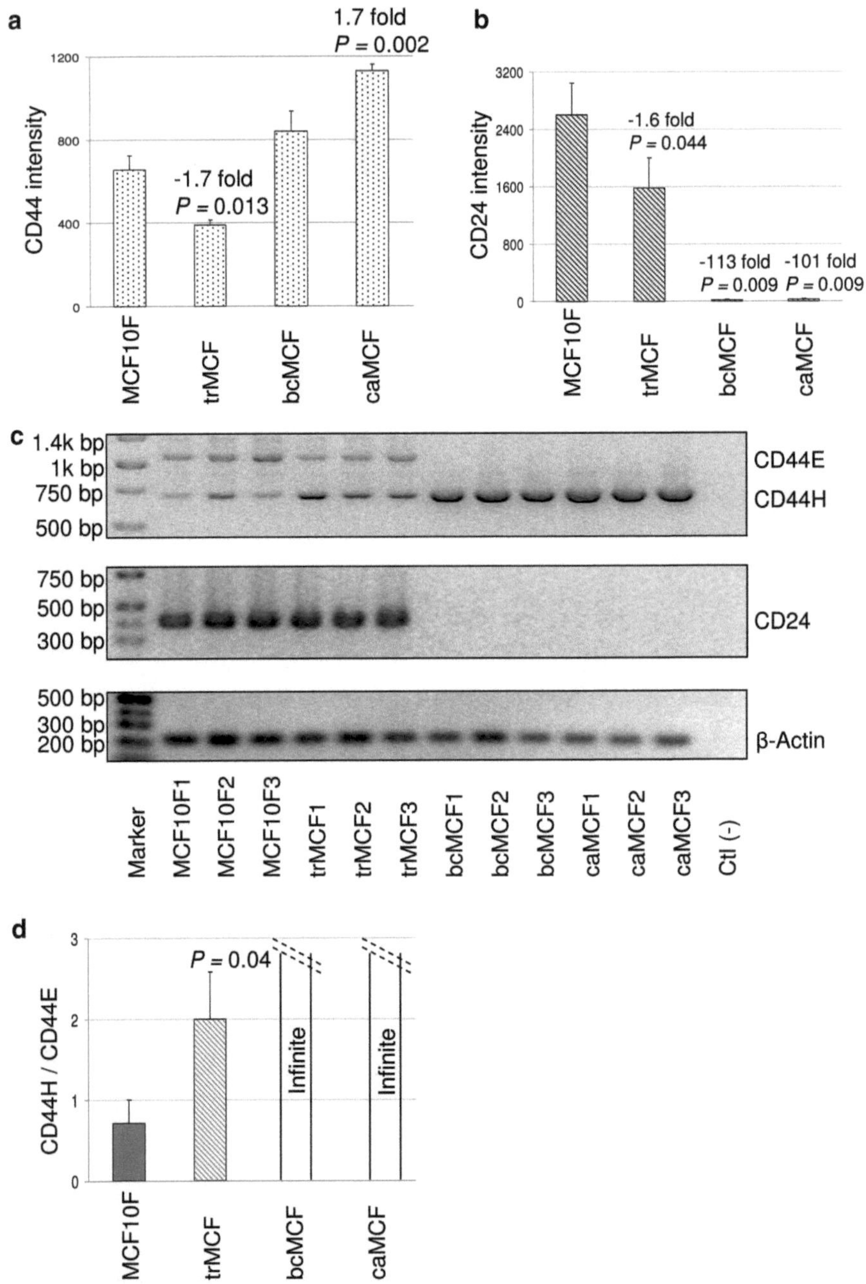

Fig. 3.11 Characterization of the expression of *CD44* and *CD24*. (**a** and **b**) Microarray expression values of *CD44* and *CD24*, respectively. Significance: $P<0.05$ in unpaired t test compared with MCF-10F. (**c**) RT-PCR validation for *CD44* (*top*) and *CD24* (*middle*) expression levels. (**d**) Densitometry ratio of CD44H to CD44E detected in RTPCR

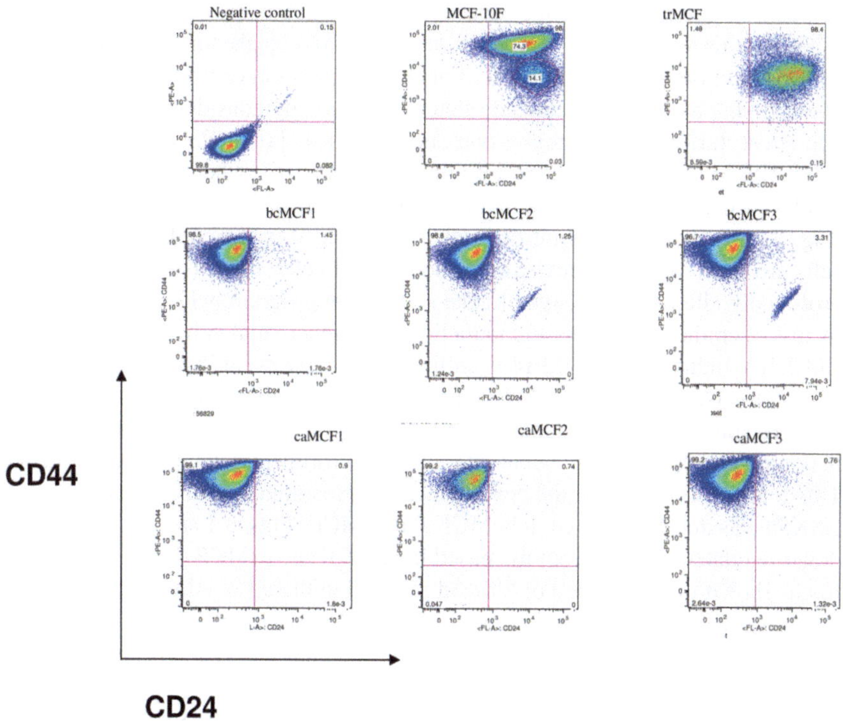

Fig. 3.12 Cell surface proteins *CD44* and *CD24* expression analysis by FACS

It is important to emphasize that the CD44+/CD24-/low is the cell surface marker of tumorigenic breast cancer cells in which the tumorigenic capacity is further increased by additional expression of ESA [11]. These cells are characterized by a 186-gene "invasiveness" gene signature that is associated with risk of death and metastasis in breast cancer [12]. CD24 encodes a small, heavily glycosylated cell-surface adhesion protein. CD44 undergoes extensive alternative splicing within its central region spanning exon 6a to 14, also termed as variable exon v1 to v10. The two variants expressed in the EMT model are mRNA precursor variant 3 with exon 6a to 11 spliced out and variant 4 with exon 6a to 14 spliced out, corresponding to CD44E and CD44H, respectively [13]. The CD44H is mainly expressed on cells of lymphohematopoietic origin; it plays an important role in cell adhesion and its expression promotes tumor cell migration [14]. CD44E is preferentially expressed on epithelial cells and it is involved in the recognition of a common determinant in CD44H and CD44E promoting homotypic cellular aggregation [15]. Microarray and RT-PCR analysis of CD44 and CD24 revealed an expression pattern of CD44Hhigh/CD44E-/CD24-in bcMCF and caMCF. The increased ratio of CD44H/CD44E in trMCF cells might represent an early marker for E2-transformed HBECs. The significant increase of CD44H and complete loss of CD44E might be a novel phenotype associated with the tumorigenic capacity. In addition, the loss of ESA in bcMCF and caMCF indicated that ESA expression is not required for the tumorigenic capacity.

EMT involves dedifferentiation of polarized epithelial cells to a migratory fibroblastoid phenotype, a phenomenon that is increasingly considered to be an important event during cancer progression and metastasis [16, 17]. Cells selected from MCF-10F by Boyden chamber are nontumorigenic [6], whereas all bcMCF cells displayed consistent chromosomal aberrations and formed tumors in SCID mice (Fig. 3.2). Although it cannot be totally ruled out, it is very unlikely that bcMCF cells are derived from preexisting mesenchymal cells. EMT is accompanied by a profoundly altered mesenchymal gene expression program, which is characterized by loss of epithelial keratins and induction of mesenchymal vimentin [18] (Figs. 3.1 and 3.10). Induction of S100A4 is an important early event in the pathway toward EMT [19]. The hallmark of EMT is the loss of expression of the cell adhesion molecule E-cadherin [20]. E-cadherin is a cell-cell adhesion molecule that participates in homotypic, calcium dependent interactions to form epithelial adherens junctions. This function is critical in the development and maintenance of a polar epithelium. GSK3h was down-regulated in bcMCF and caMCF (Figs. 3.1 and 3.9). SNAI2 was down-regulated in trMCF but increased in bcMCF and caMCF (Figs. 3.6 and 3.12). It was shown that inhibition of GSK3h resulted in the up-regulation of SNAI1 and down-regulation of E-cadherin in vivo. SNAI2 (or SLUG) and SNAI1 (or SNAIL) belong to the Snail family of proteins; both contain an NH2-terminal repression domain and a COOH-terminal zinc-finger DNA-binding domain. Snail proteins repress the transcription of E-cadherin [21–24]. E-cadherin loss is believed to contribute to both cancer development and progression [21].

HRAS and TGFh1 were up-regulated in bcMCF and caMCF (Table 3.4). It has been shown that HRAS cooperates with TGFh1 to cause EMT and also it interacts with CD44 directly by increasing its expression [25, 26]. TGFh1 via HMGA2, which was also up-regulated in bcMCF and caMCF, regulates the expression of TWIST, SNAI1, and SNAI2 [27] (Fig. 3.12). EMT is associated with higher tumor grade, high motility index, and ERa (−) status [28]. Therefore, these findings revealed an intrinsic link between EMT and tumorigenic capacity in our model, which, in part, may explain the poor prognosis of ERa (−) human breast carcinomas.

Two recent studies, using a GeneChip containing 1495 SNPs [29] or comparative genomic hybridization [30], have shown that LOH and allelic loss in 4p and 5q occur more frequently in subtypes of breast cancer characterized as ERα (−). SNP mapping reveals that LOH in 4p14-15.3, 5q11-32, and 18q22-23 are significantly associated with a gene expression profile in basal-like subtype [29]. In the experimental model of EMT, the tumor cells caMCF also showed LOH in all these three regions (Fig. 3.4). Moreover, it has been shown that the CD44+/CD24- phenotype is associated with mesenchymal phenotype and invasion in breast cancer cell line, and may define breast cancers of basal/myoepithelial origin rather than luminal origin [31]. These findings indicate a potential correlation of the EMT model with the basal-like subtype. However, the bcMCF and caMCF cells displayed low or absent expression of both luminal and basal cytokeratins in microarray analysis and ERBB2 (−) in immuno-cytochemical staining. Therefore, they cannot simply be classified into basal-like, ERBB2 (+), luminal A or B, or normal breast-like subtype [3]. Carey et al. [32] has identified an "unclassified" subtype using immunohistochemical markers for the classification of breast cancer tissue. This subtype characterized as ERa (–), ERBB2 (–), and progesterone receptor (–), which, for these markers, is the same phenotype as the basal-like subtype. Unlike the basal-like subtype, the unclassified subtype is cytokeratin 5 (–). It also displays a histologic grade and survival prognosis most close to that of basal-like subtype [32]. The experimental model support the concept that E2-induced breast cancer is a polygenic disease having a large range of genomic instabilities. E2 and/or its metabolites can directly cause genomic aberrations without the mediation of ERa. The genomic aberrations lead to changes in gene expression, which result in disrupted integrin signaling and apoptotic pathways, and epithelial to mesenchymal transition. These functional changes lead to colony formation in agar-methocel, loss of ductulogenesis in collagen matrix, invasiveness in vitro, and tumor growth in SCID mice in vivo [6]. However, MCF-10F is a spontaneously immortalized cell line harvested from a woman's breast that was free of malignancies, but had a diagnosis of benign fibrocystic disease [33]. Hence, we cannot rule out its inherited susceptibility to estrogen and the disposition to tumorigenesis.

3.4 Other Factors Involved in the EMT

There is evidence that the invasive and metastatic behavior there are key macrophage-derived signal that are likely to be conveyed by EGF [4]. Studies of mouse breast cancer cells both in vivo and in vitro have shown that the activation of this receptor by EGF causes them to acquire both motility and invasiveness and to secrete CSF-l, the attractant and stimulant of macrophages [34]. Macrophages respond to CSF-l by proliferating and releasing EGF, which activates the cancer cells. Macrophages actually play critical roles in two stages of the invasion-metastasis cascade-initial invasiveness and intravasation. Other stromal signals, such as those carried by EGF and HGF, may also help to elicit many of the changes that we associate with cancer cell invasiveness and the EMT [34]. The most important effectors of these complex changes are the matrix metalloproteinases (MMPs) (Tables 3.6 and 3.7). In carcinomas, the bulk of these proteases are secreted by recruited stromal cells, notably macrophages, mast cells, and fibroblasts, rather than by the neoplastic epithelial cells. By dissolving the dense thickets of ECM molecules that surround and confine individual cells within tissues, these secreted MMPs create spaces for these cells to move. Included among the ECM components that are cleaved by MMPs are fibronectin, tenascin, laminin, collagens, and proteoglycans. During the course of degrading ECM components, MMPs also mobilize and activate certain growth factors that have been tethered in inactive form to the ECM or to the surfaces of cells.

The actions of extracellular proteases, notably the MMPs (Table 3.6), explain at the biochemical level how paths are cleared for the advance of invasive cancer cells through the extracellular matrix and thus through tissues. The motile behavior of cells

Table 3.7 Epithelia-mesenchymal transition markers and their reguators (See reference 8)

Symbol	Gene title
JAG1	Jagged 1(Alagille syndrome)
HEY1	Hairy/enhanced-of-split related with YRPW motif 1
S100A4	S100 calcium binding protein A4
SIP1	Survival of motor meuron protein interacting protein 1
HMGA2	High mobility group AT-hook 2
TGFB1	Transforming growth factor, beta 1
HRAS	v-Ha-ras Harvey rat sarcoma viral oncogene
SERPINE2	Serpin peptidase inhibitor, clade E
TWIST1	Twist homolog 1
TWIST2	Twist homolog 2 (Drosophila)
SNAI2	Snail homolog 2 (Drosophila)
SNAI1	Snail homolog 1(Drosophila)
GSK3B	Glycogen synthase linase 3 beta
CDH1	Cadherin 1, type 1, E-cadherin (epithelial)
FN1	Fibronectin 1
CDH2	Cadherin 2, type 1, N-cadherin (neuronal)
OCLN	Occluding
SOX10	SRY (sex determining region Y(-box 10
FOXC2	Forkhead box C2 (MFH-1, mesenchyme forkhead 1)
GSC	Goosecoid
MMP2	Matrix metallopeptidase 2
MMP3	Matrix metallopeptidase 3
MMP9	Matrix metallopeptidase 9
DSP	desmoplakin
KRIT1	KRIT1, ankyrin repeat containing
KRT10	Keratin 10
KRT12	Keratin 12 (messmann conreal dystrophy)
KRT13	Keratin 13
KRT14	Keratin 14
KRT15	Keratin 15
KRT16	Keratin 16
KRT17	Keratin 17
KRT18	Keratin 18
KRT19	Keratin 19
KRT1B	Keratin 1B
KRT20	Keratin 20
KRT23	Keratin 23 (histone deacetylase inducible)
KRT24	Keratin 24
KRT25A	Keratin 25A
KRT25C	Keratin 25C

(continued)

Table 3.7 (continued)

Symbol	Gene title
KRT2A	Keratin 2A (epidermal ichthyosis bullosa of Siemens)
KRT2B	Cytokeratin 2
KRT3	Keratin 3
KRT4	Keratin 4
KRT5	Keratin 5
KRT6A	Keratin 6A
KRT6IRS	Keratin6 irs
KRT6L	Keratin 6L
KRT7	Keratin 7
KRT8	Keratin 8///keratin 8
KRT9	Keratin 9 (epidermolytic palmoplantar keratoderma)

has been studied extensively with cultured cells, and it is presumed that their crawling on solid substrates *in vitro* reflects the *in vivo* behavior of cancer cells as they invade nearby cell layers and intravasate. Such motility is also presumed to be important for cancer cells' escape from blood vessels or lymph ducts-the process of extravasation. Originally, differentiated epithelial cells grow linked together and move as epithelial sheets [34]. Alternatively, as occurs during embryogenesis, epithelial cells can dissociate and migrate as individual cells. One property of such cells is the ability to detach from the epithelium and express locomotory ability. Migrating cells no longer express cytokeratin filaments and express vimentin filaments associated to a loss of cell-to-cell adhesion [37]. Analyzing the morphology and the immune-histochemical pattern in MCF-10F in collagen gel culture, Tiezzi et al. [2] was able to demonstrate that during the neoplastic transformation the mammary epithelial cell is able to change its phenotype, acquiring a mesenchymal property infiltrating the surrounding extra-cellular matrix. TGFß and HRAS are described as EMT trigger processes [17, 26, 38]. In the EMT model was observed that TGBß1 and TGFß2 are down modulated in transformed and tumorigenic cell lines. On the other hand, the C5-T8 cell line has a significant increase in HRAS transduction (Fig. 3.13). Such an observation supports the hypothesis that the EMT process is also triggered by the HRAS pathway. Oncogene Ras activate the MAPK pathway that is essential for EMT and metastasis processes in an *in vivo* system [17]. The *Smad* group of intracellular proteins transduces signal by members of the TGF-β family from the cell surface to the nucleus. Binding to receptors for the TGF-β family members leads to the phosphorylation and activation of the receptor regulated *Smads,* which include *Smads 1, 2, 3, 5, and 8* [39–41]. The Smad-dependent signaling pathway appears to be necessary for TGFß-induced EMT and Smad5 is activated by members of the bone morphogenetic protein (BMP) branch of the TGF-β superfamily [35, 41, 42]. SMAD5 increased expression was observed in colorectal cancer by immunohistochemistry [43, 44].

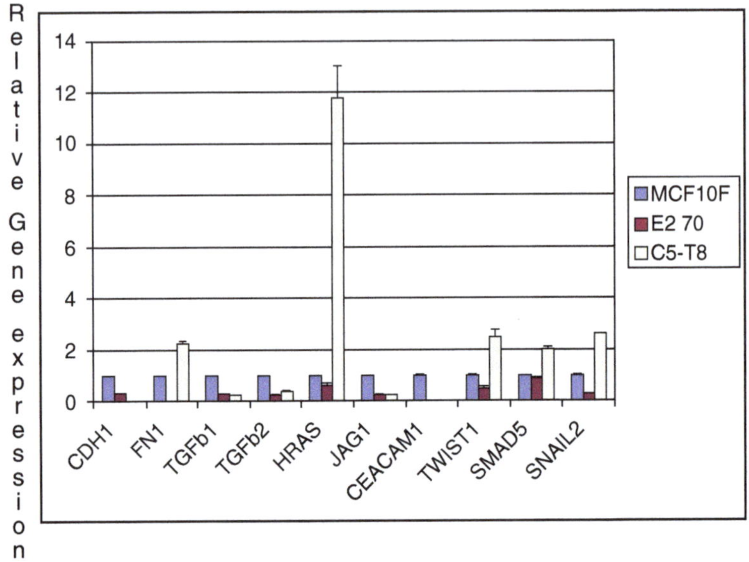

Fig. 3.13 Comparative expression of genes related to EMT process during the neoplastic transformation of breast epithelial cells by estrogen

TWIST was recently reported as an E-cadherin protein repressor and is able to induce EMT [45] and, it has been found to be correlated with metastasis in various cancers including breast, prostate and hepatic cancers [36, 46]. Knocking down Twist by RNAi prevents metastasis and overexpression of Twist in two human epithelial cell lines which causes both complete EMT and E-cadherin repression [45].

SNAIL2 (Slug) is a zinc finger transcriptional repressor closely related to the Snail family. It has been found in vertebrate and is involved in the control of gastrulation [47, 48]. SNAIL2 can down regulate E-cadherin expression. Its expression is related to undifferentiated breast carcinomas and all ductal invasive carcinomas with lymph node positive to SNAIL2 expression [49].

JAG1 and CEACAM1 that are related to the process of duct morphogenesis [35, 50] are also expressed during the EMT. The loss of expression of TGFß1, TGFß2, CEACAM1 and JAG1 is related to the loss of duct morphogenesis, and that the over expression of h-RAS with loss of E-cadherin expression and up modulation of FN1, TWIST1, SNAIL2 and SMAD5 expressions are involved in the EMT modulation.

3.5 Conclusions

It is important to emphasize that the epithelial to mesenchymal transition (EMT) leads to exacerbation of motility and invasiveness in many cell types and is often considered a prerequisite for tumor infiltration and metastasis [5]. The data discussed above clearly shows the complexity of this process and give us a hint of how the breast epithelial cells acquire the invasive properties that are the prerequisite for the metastatic process.

References

1. Christiansen JJ, Rajasekaran AK (2006) Reassessing epithelial to mesenchymal transition as a prerequisite for carcinoma invasion and metastasis. Cancer Res 66:8319–8326
2. Tiezzi DG, Fernandez SV, Russo J (2007) Epithelial to mesenchymal transition during breast cancer progression. Int J Oncol 31:823–827
3. Sorlie T, Tibshirani R, Parker J et al (2003) Repeated observation of breast tumor subtypes in independent gene expression data sets. Proc Natl Acad Sci U S A 100:8418–8423
4. van de Rijn M, Perou CM, Tibshirani R et al (2002) Expression of cytokeratins 17 and 5 identifies a group of breast carcinomas with poor clinical outcome. Am J Pathol 161:1991–1996
5. Russo J, Lareef MH, Tahin Q et al (2002) 17h-Estradiol is carcinogenic in human breast epithelial cells. J Steroid Biochem Mol Biol 80:149–162
6. Russo J, Fernandez SV, Russo PA et al (2006) 17-h-Estradiol induces transformation and tumorigenesis in human breast epithelial cells. FASEB J 20:1622–1634
7. Sorlie T, Perou CM, Tibshirani R et al (2001) Gene expression patterns of breast carcinomas distinguish tumor subclasses with clinical implications. Proc Natl Acad Sci U S A 98:10869–10874
8. Huang Y, Fernandez S, Goodwin S, Russo PA, Russo IH, Sutter T, Russo J (2007) Epithelial to mesenchymal transition in human breast epithelial cells transformed by 17- beta-Estradiol. Cancer Res 67:11147–11157
9. Danen EH, Yamada KM (2001) Fibronectin, integrins, and growth control. J Cell Physiol 189:1–13
10. Waris G, Ahsan H (2006) Reactive oxygen species: role in the development of cancer and various chronic conditions. J Carcinog 5:14
11. Al-Hajj M, Wicha MS, Benito-Hernandez A, Morrison SJ, Clarke MF (2003) Prospective identification of tumorigenic breast cancer cells. Proc Natl Acad Sci U S A 100:3983–3988
12. Liu R, Wang X, Chen GY et al (2007) The prognostic role of a gene signature from tumorigenic breast-cancer cells. N Engl J Med 356:217–226
13. Roca X, Mate JL, Ariza A et al (1998) CD44 isoform expression follows two alternative splicing pathways in breast tissue. Am J Pathol 153:183–190
14. Thomas L, Byers HR, Vink J, Stamenkovic I (1992) CD44H regulates tumor cell migration on hyaluronate-coated substrate. J Cell Biol 118:971–977
15. Droll A, Dougherty ST, Chiu RK et al (1995) Adhesive interactions between alternatively spliced CD44 isoforms. J Biol Chem 270:11567–11573
16. Fuchs IB, Lichtenegger W, Buehler H et al (2002) The prognostic significance of epithelial-mesenchymal transition in breast cancer. Anticancer Res 22:3415–3419
17. Thiery JP (2002) Epithelial-mesenchymal transitions in tumour progression. Nat Rev Cancer 2:442–454
18. Sommers CL, Heckford SE, Skerker JM et al (1992) Loss of epithelial markers and acquisition of vimentin expression in Adriamycin- and vinblastine-resistant human breast cancer cell lines. Cancer Res 52:5190–5197
19. Okada H, Danoff TM, Kalluri R, Neilson EG (1997) Early role of Fsp1 in epithelial-mesenchymal transformation. Am J Phys 273:F563–F574
20. Kang Y, Massague J (2004) Epithelial-mesenchymal transitions: twist in development and metastasis. Cell
21. Nollet F, Berx G, van Roy F (1999) The role of the E-cadherin/catenin adhesion complex in the development and progression of cancer. Mol Cell Biol Res Commun 2:77–85

22. Cano A, Perez-Moreno MA, Rodrigo I et al (2000) The transcription factor snail controls epithelial-mesenchymal transitions by repressing E-cadherin expression. Nat Cell Biol 2:76–83
23. Hajra KM, Chen DY, Fearon ER (2002) The SLUG zinc-finger protein represses E-cadherin in breast cancer. Cancer Res 62:1613–1618
24. Zhou BP, Deng J, Xia W et al (2004) Dual regulation of Snail by GSK-3h-mediated phosphorylation in control of epithelial-mesenchymal transition. Nat Cell Biol 6:931–940
25. Risse-Hackl G, Adamkiewicz J, Wimmel A, Schuermann M (1998) Transition from SCLC to NSCLC phenotype is accompanied by an increased TRE-binding activity and recruitment of specific AP-1 proteins. Oncogene 16:3057–3068
26. Janda E, Lehmann K, Killisch I et al (2002) Ras and TGF[h] cooperatively regulate epithelial cell plasticity and metastasis: dissection of Ras signaling pathways. J Cell Biol 156:299–313
27. Thuault S, Valcourt U, Petersen M et al (2006) Transforming growth factor-h employs HMGA2 to elicit epithelial-mesenchymal transition. J Cell Biol 174:175–183
28. Willipinski-Stapelfeldt B, Riethdorf S, Assmann V et al (2005) Changes in cytoskeletal protein composition indicative of an epithelial-mesenchymal transition in human micrometastatic and primary breast carcinoma cells. Clin Cancer Res 11:8006–8014
29. Wang ZC, Lin M, Wei LJ et al (2004) Loss of heterozygosity and its correlation with expression profiles in subclasses of invasive breast cancers. Cancer Res 64:64–71
30. Loo LW, Grove DI, Williams EM et al (2004) Array comparative genomic hybridization analysis of genomic alterations in breast cancer subtypes. Cancer Res 64:8541–8549
31. Sheridan C, Kishimoto H, Fuchs RK et al (2006) CD44+/CD24 breast cancer cells exhibit enhanced invasive properties: an early step necessary for metastasis. Breast Cancer Res 8:R59
32. Carey LA, Perou CM, Livasy CA et al (2006) Race, breast cancer subtypes, and survival in the Carolina Breast Cancer Study. JAMA 295:2492–2502
33. Soule HD, Maloney TM, Wolman SR et al (1990) Isolation and characterization of a spontaneously immortalized human breast epithelial cell line, MCF-10. Cancer Res 50:6075–6086
34. Weinberg RA (2014) The biology of cancer, 2nd edn. Garland Science, Taylor and Francis Group, New York
35. Zavadil J, Cermak L, Soto-Nieves N, Bottinger EP (2004) Integration of TGF-beta/Smad and Jagged1/Notch signalling in epithelial-to-mesenchymal transition. Embo J 23:1155–1165
36. Lee TK, Poon RT, Yuen AP et al (2006) Twist overexpression correlates with hepatocellular carcinoma metastasis through induction of epithelial-mesenchymal transition. Clin Cancer Res 12:5369–5376
37. Thiery JP, Sleeman JP (2006) Complex networks orchestrate epithelial-mesenchymal transitions. Nat Rev Mol Cell Biol 7:131–142
38. Oft M, Peli J, Rudaz C, Schwarz H, Beug H, Reichmann E (1996) TGF-beta1 and Ha-Ras collaborate in modulating the phenotypic plasticity and invasiveness of epithelial tumor cells. Genes Dev 10:2462–2477
39. Miettinen PJ, Ebner R, Lopez AR, Derynck R (1994) TGF-beta induced transdifferentiation of mammary epithelial cells to mesenchymal cells: involvement of type I receptors. J Cell Biol 127:2021–2036
40. Massague J, Wotton D (2000) Transcriptional control by the TGF-beta/Smad signaling system. Embo J 19:1745–1754
41. Itoh S, Itoh F, Goumans MJ, Ten Dijke P (2000) Signaling of transforming growth factor-beta family members through Smad proteins. Eur J Biochem 267:6954–6967
42. Flanders KC (2004) Smad3 as a mediator of the fibrotic response. Int J Exp Pathol 85:47–64
43. Korchynskyi O, Landstrom M, Stoika R, Funa K, Heldin CH, ten Dijke P, Souchelnytskyi S (1999) Expression of Smad proteins in human colorectal cancer. Int J Cancer 82:197–202
44. Van Beijnum JR, Moerkerk PT, Gerbers AJ, De Bruine AP, Arends JW, Hoogenboom HR, Hufton SE (2002) Target validation for genomics using peptide-specific phage antibodies: a study of five gene products overexpressed in colorectal cancer. Int J Cancer 101:118–127
45. Yang J, Mani SA, Donaher JL et al (2004) Twist, a master regulator of morphogenesis, plays an essential role in tumor metastasis. Cell 117:927–939

46. Kwok WK, Ling MT, Lee TW et al (2005) Up-regulation of TWIST in prostate cancer and its implication as a therapeutic target. Cancer Res 65:5153–5162
47. Leptin M (1991) twist and snail as positive and negative regulators during Drosophila mesoderm development. Genes Dev 5:1568–1576
48. Nieto MA (2002) The snail superfamily of zinc-finger transcription factors. Nat Rev Mol Cell Biol 3:155–166
49. Blanco MJ, Moreno-Bueno G, Sarrio D, Locascio A, Cano A, Palacios J, Nieto MA (2002) Correlation of Snail expression with histological grade and lymph node status in breast carcinomas. Oncogene 21:3241–3246
50. Huang J, Hardy JD, Sun Y, Shively JE (1999) Essential role of biliary glycoprotein (CD66a) in morphogenesis of the human mammary epithelial cell line MCF10F. J Cell Sci 112(Pt 23):4193–4205

Chapter 4
The Invasive Breast Cancer Types

4.1 Introduction

Several morphologic and endocrinologic criteria are commonly used to classify breast tumors; these include histologic pattern, degree of nuclear pleomorphism, number of mitoses, presence of inflammatory response, blood vessel invasion, lymph node involvement and the presence of hormone receptors [1–25]. In this chapter we will describe the classical histological type for evaluation of primary breast cancer. This knowledge cannot be replaced by those obtained with the immuno-cytochemical or genomic classification, but by the contrary will help the readers to understand the heterogeneity of breast cancer.

4.2 The Invasive Cancer Subtypes

Invasive breast cancer includes all the tumors in which stromal invasion are detectable including the microinvasive carcinoma. The vast majority of breast carcinomas are described as invasive ductal carcinomas based on architectural patterns and cytological features (Table 4.1). This invasive ductal carcinoma also called no otherwise specified (NOS) comprise 70 to 80 of the primary breast cancer. The remainder of the invasive carcinomas are classified as lobular, tubular, medullary, etc. as depicted in Table 4.1.

J. Russo, *The Pathobiology of Breast Cancer*,
DOI 10.1007/978-3-319-40815-6_4

Table 4.1 Types of invasive carcinoma of the breast

Invasive ductal
Invasive ductal with an extensive intraductal component
Invasive lobular
Mucinous
Medullary
Papillary
Tubular
Adenoid cystic
Secretory
Apocrine
Cribriborm
Paget's disease of the nipple
With invasive carcinoma
Without invasive carcinoma
Carcinoma with metaplasia
Squamous type
Spindle cell type
Cartilaginous and osseous type
Mixed type
Inflammatory

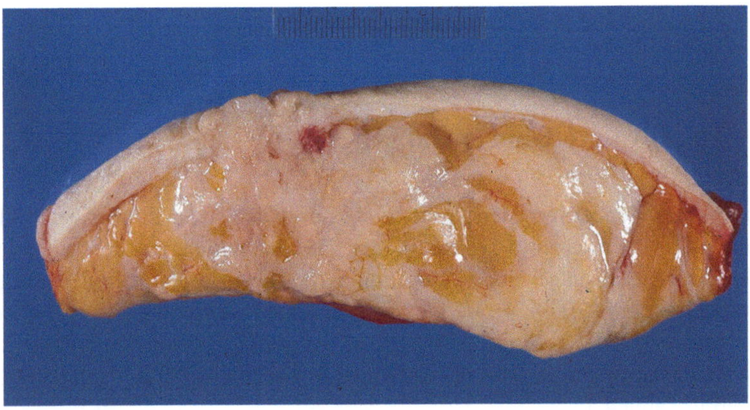

Fig. 4.1 Gross appearance of invasive ductal carcinoma showing the irregular shape of the tumor with white fibrous appearance, and chalky streaks

4.2.1 Invasive Ductal Carcinoma No Otherwise Specified (NOS)

The gross appearance is of this type of tumor (Fig. 4.1) is very variable in size and shape. In general they are firm and poorly circumscribed. Necrosis, hemorrhage, and cystic degeneration may be found. Microscopically, the cancer cells can grow in diffuse sheets (Fig. 4.2), well-defined nests (Figs. 4.3 and 4.4), cords (Fig. 4.5),

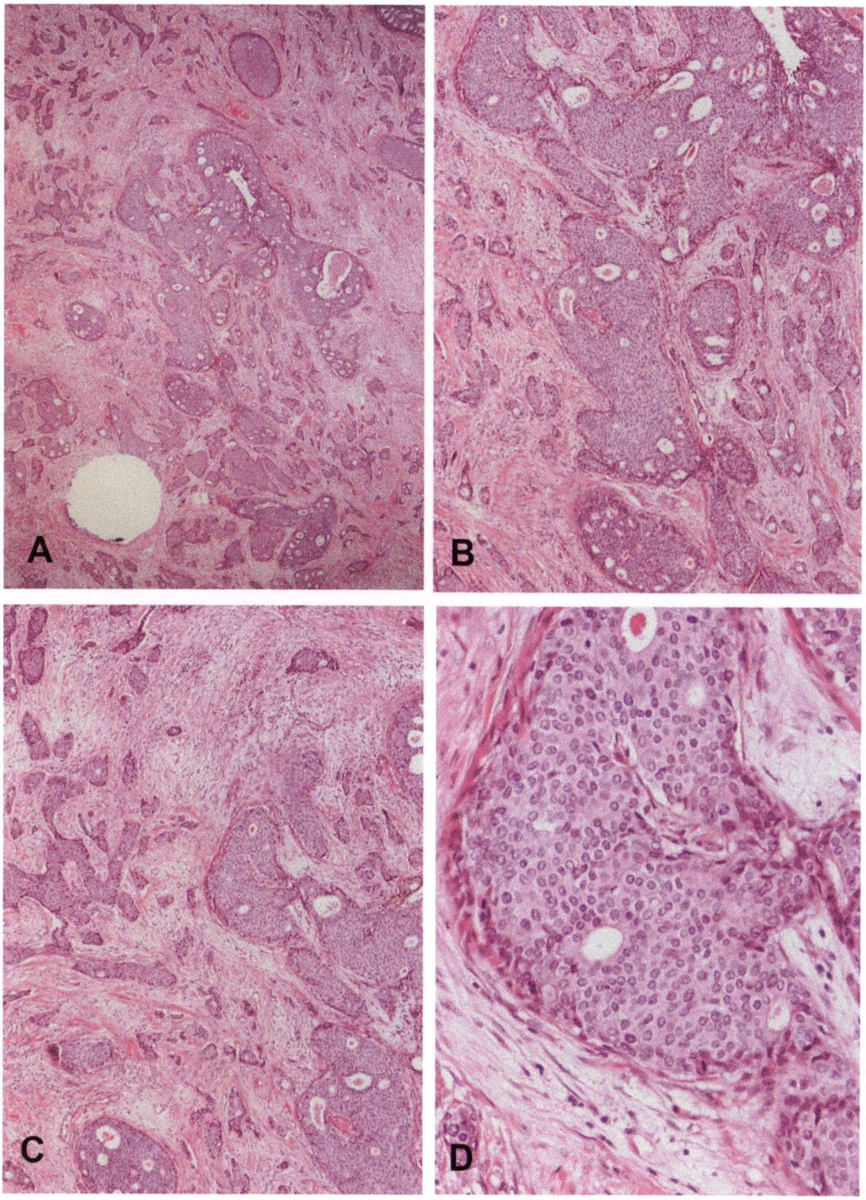

Fig. 4.2 Invasive cribriform carcinoma. (**a**) Low magnification showing the nest of neoplastic cells invading the stroma containing the cribriform pattern, ×4. (**b** and **c**) Detail of Fig. 4.2a, ×10. (**d**) The cribriform pattern is characterized by the formation of luminal spaces, ×40

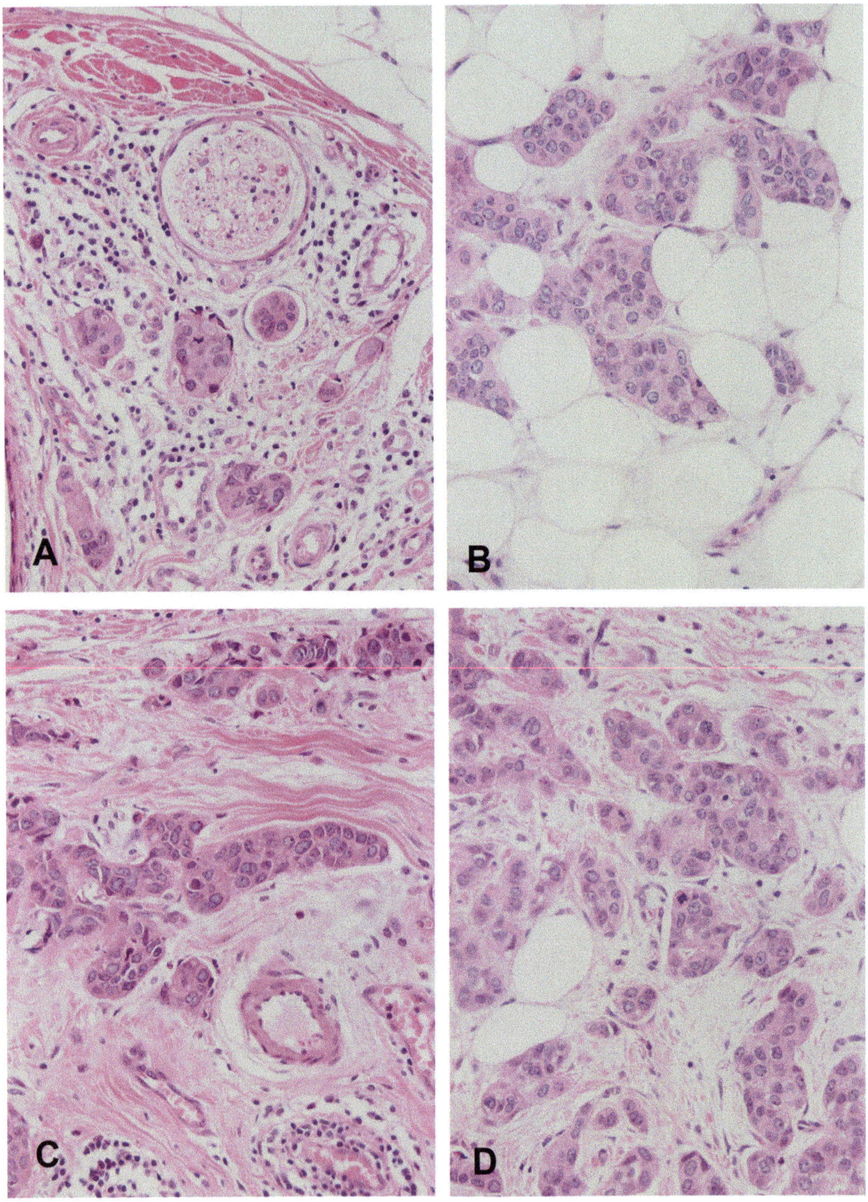

Fig. 4.3 Invasive ductal carcinoma. (**a**) Nests and single cells surrounding a nerve and intermingle with inflammatory cells. (**b**) Cells invading the fat. (**c**) Nest and single cells surrounded by collagen fibers and inflammatory cells. (**d**) Desmoplastic reaction elicited by the cancer cells in the fat stroma, ×40

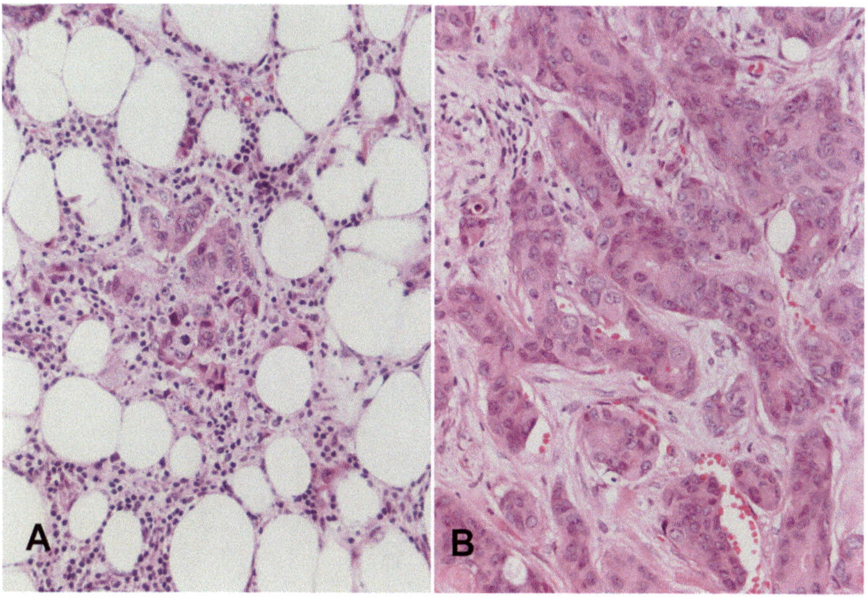

Fig. 4.4 Invasive ductal carcinoma. (**a**) Nests and single cells invading the fat and surrounded by inflammatory cells. (**b**) Nest and single cells surrounded by collagen fibers and few lymphocytes, ×40

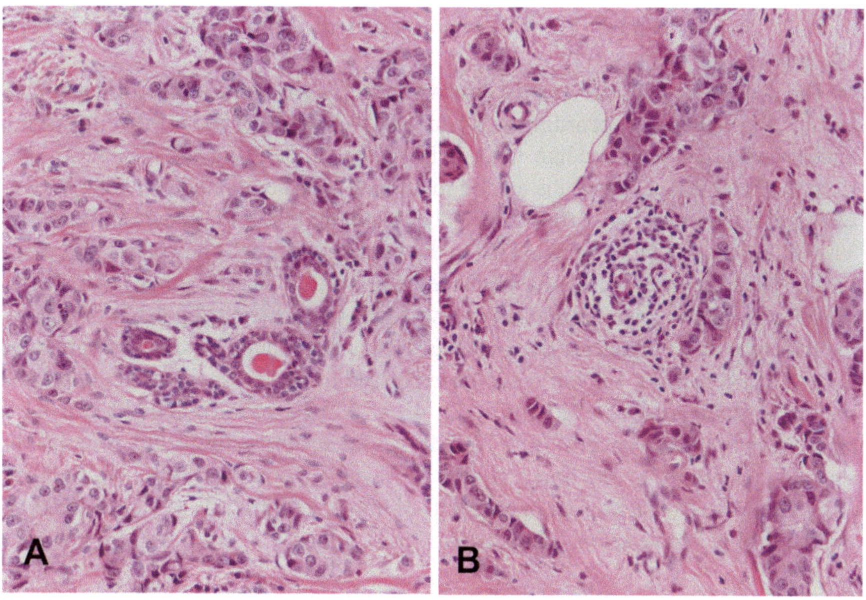

Fig. 4.5 Invasive ductal carcinoma. (**a**) Nests and single cancer cells surmounting a normal ducts and (**b**) Desmoplastic reaction elicited by the cancer cells in the stroma, ×40

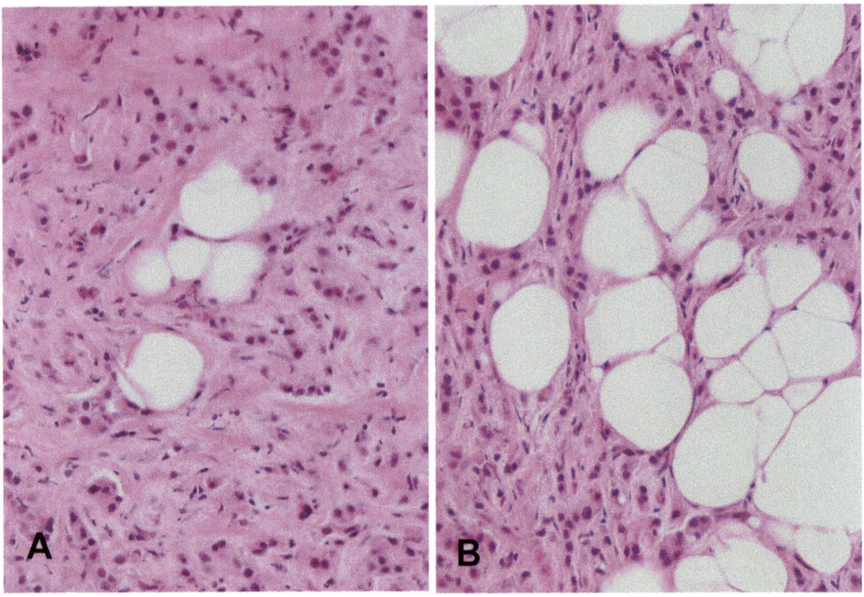

Fig. 4.6 Invasive ductal carcinoma. (**a**) Strands of cancer cells surrounded by desmoplastic reaction in the stroma. (**b**) Cells invading the fat stroma, ×40

or as individual cells (Fig. 4.6). In some cases the neoplastic invading sheets of cells form a cribriform pattern (Fig. 4.2) or presenting areas of necrosis forming a comedo pattern (Fig. 4.7).

The neoplastic cells are mostly positive for different types of keratin type 8, 18 and 19 (Fig. 4.8a) and according to Rosai [26] 70 % of cases are positive for lactalbumin and other markers like CEA (Fig. 4.8b), B72.3, and BCA-225 [27–31]. Epithelial membrane antigen is a good marker of breast epithelial cell (Fig. 4.9), as well as E-cadherin (Figs. 4.10 and 4.11), beta catenin (Fig. 4.12) and P53 (Fig. 4.13). Vimentin may also be expressed [32] in the breast cancer cells but also is expressed in the stromal cells (Fig. 4.14). We have reported [33] as well other [34, 35] that breast carcinomas can be immune-reactive for S-100 protein (Fig. 4.15). The basement membrane components laminin (Fig. 4.16) and collagen IV show a fragmented pattern of expression. Vimentin can be expressed in certain areas of the tumor and may be an indication of EMT [36] (Fig. 4.14). Marker of stem cells like CD44 is also shown in breast cancer (Fig. 4.17). Marker for desmin is frequently found in the stromal component of the tumor (Fig. 4.18), whereas markers for smooth muscle antigen (SMA) can be present in remnants of myoepithelial cells as well as decorating the wall of blood vessels (Fig. 4.19).

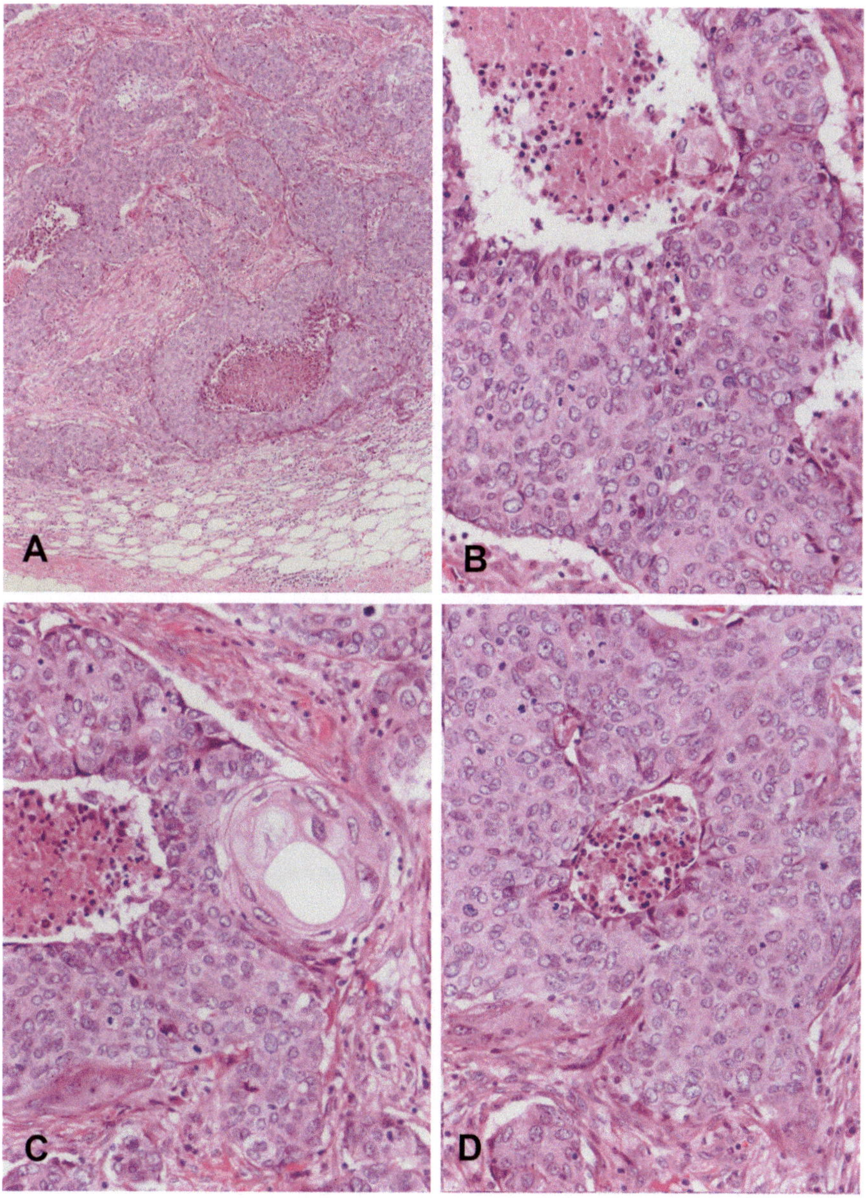

Fig. 4.7 Invasive comedo carcinoma. (**a**) Low magnification showing the nest of neoplastic cells invading the stroma containing the comedo pattern, 4x. (**b**, **c** and **d**) The comedo pattern is characterized by nuclear pleomorphism abundant mitosis and necrotic material, ×40

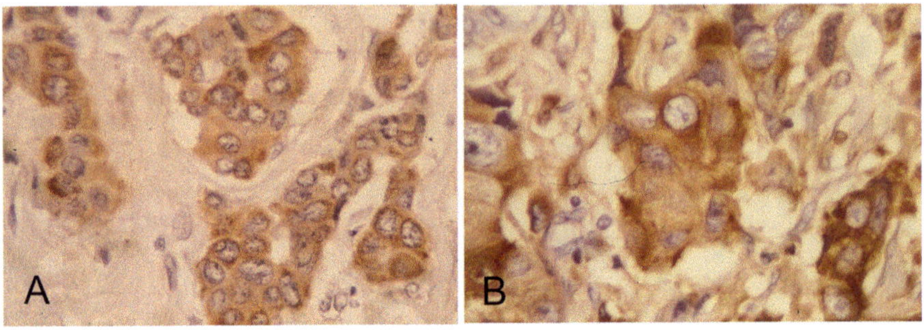

Fig. 4.8 Invasive ductal carcinoma. (**a**) The histological section has been reacted with an antibody that recognizes the keratin 18 (K18). This antibody against K18 reacts in the luminal cells (LC) and is not reactive in the basal and myoepithelial cells. *We have used the monoclonal mouse anti-human cytokeratin 18, from Biogenex, San Ramon, Ca.* (**b**) *The section of the tumor has been reacted using a monoclonal antibody against* CEA or Carcinoembryonic antigen that is an oncofetal glycoprotein. *We have used the monoclonal mouse anti-human CEA* clone II-7, *from DAKO,* Denmark

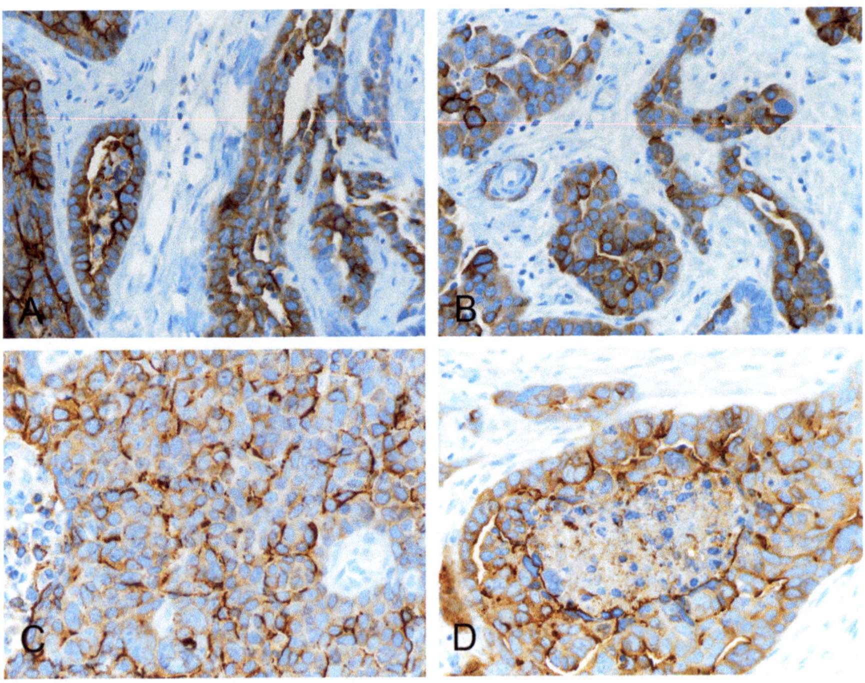

Fig. 4.9 (**a**, **b**, **c** and **d**) Invasive ductal carcinoma of the breast reacting with an antibody against the epithelial membrane antigen (EMA). The protein is located in the cell membrane. *We have used a mouse monoclonal* antibody at a dilution 1:200 from abcam, ×40

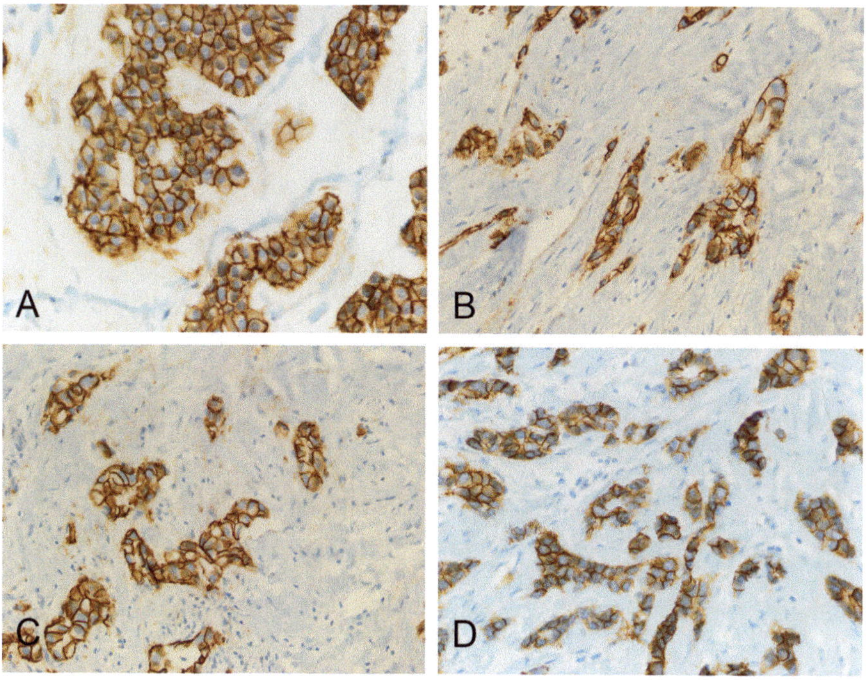

Fig. 4.10 (**a**, **b**, **c** and **d**) Invasive ductal carcinoma of the breast reacting with an antibody against the E cadherin. The protein is located in the cell membrane. *We have used a mouse monoclonal* antibody at a dilution 1:100 from Biogenex, CA, ×40

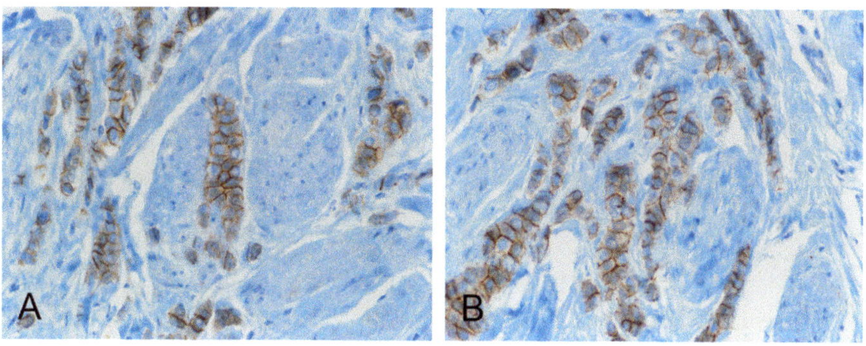

Fig. 4.11 (**a** and **b**) Invasive ductal carcinoma of the breast reacting with an antibody against the E cadherin. The protein is located in the cell membrane and allow to recognize epithelial cancer cells and in this case invading the peri-neural space. *We have used a mouse monoclonal* antibody at a dilution 1:100 from Biogenex, CA, ×40

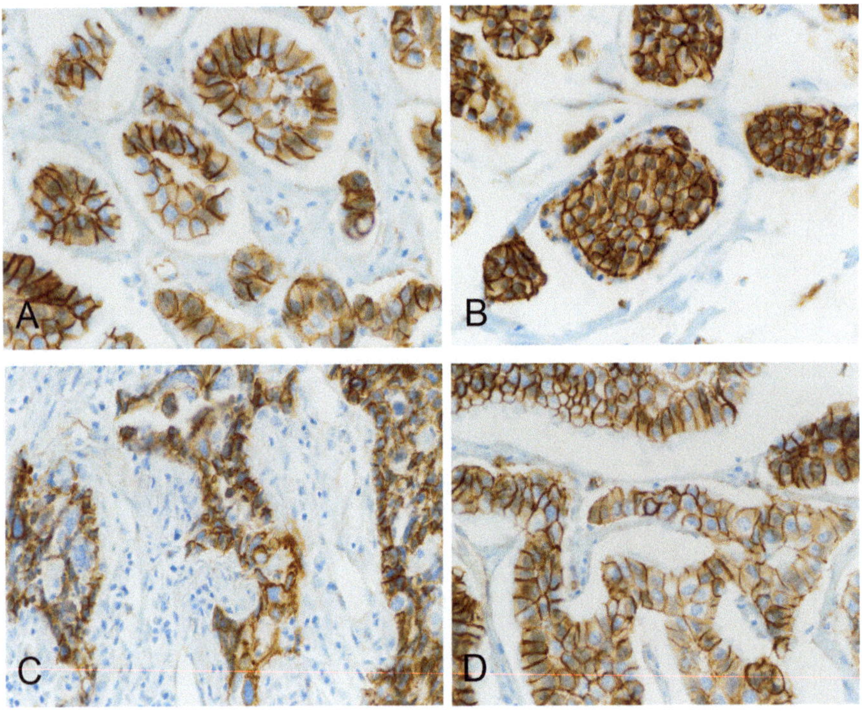

Fig. 4.12 (**a**, **b**, **c** and **d**) Invasive ductal carcinoma of the breast reacting with an antibody against the beta catenin. The protein is located in the cell membrane. *We have used a rabbit monoclonal* antibody at a dilution 1:200 from abcam, ×40

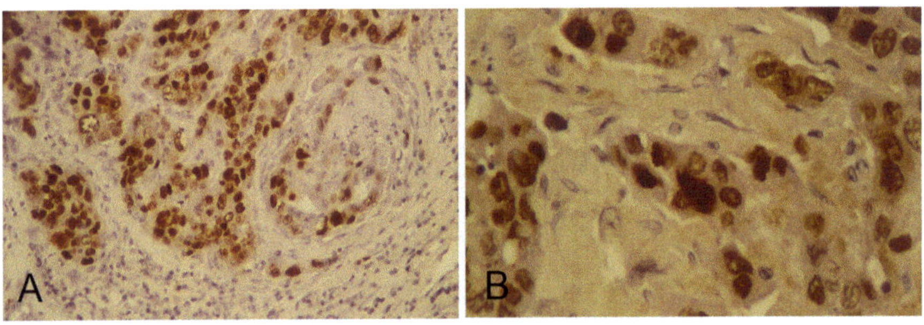

Fig. 4.13 (**a** and **b**) Invasive ductal carcinoma of the breast reacting with an antibody against p53. The protein is located in the nucleus of cancer cells that are overexpressing this protein, ×20 and ×40 respectively. *We have used a mouse monoclonal* antibody at a dilution 1:100 [(DO-1): sc-126] from Santa Cruz Biotechnology, CA

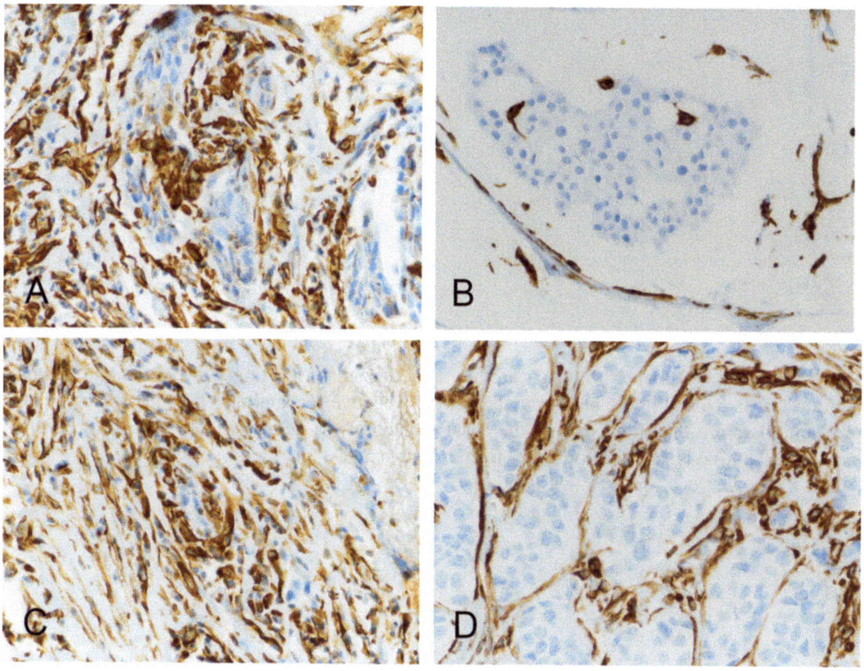

Fig. 4.14 (**a**, **b**, **c** and **d**) Invasive ductal carcinoma of the breast reacting with an antibody against Vimentin. The protein is located in the cytoplasm of stromal cells or in those epithelial cells undergoing epithelial mesenchymal transition like those in figures (**b**) and (**c**), ×40. We have used a mouse monoclonal antibody at a dilution 1:200 from Biogenex, CA

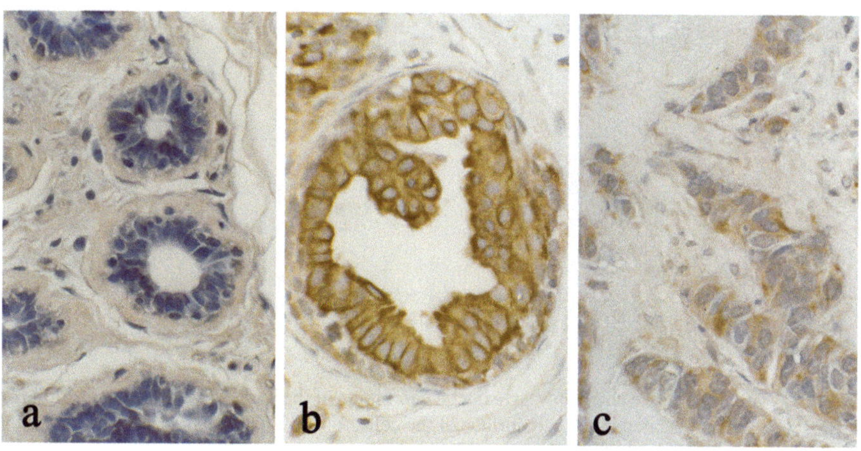

Fig. 4.15 Immunocytochemical localization of S100P in parafin section of fomalin fixed tissue. (**a**) Normal breast, (**b**) Carcinoma in situ, (**c**) Invasive ductal carcinoma, ×40. *We have used a mouse monoclonal* antibody, clone 6 at a dilution 1:50 from Transduction Laboratories, Lexington, KY

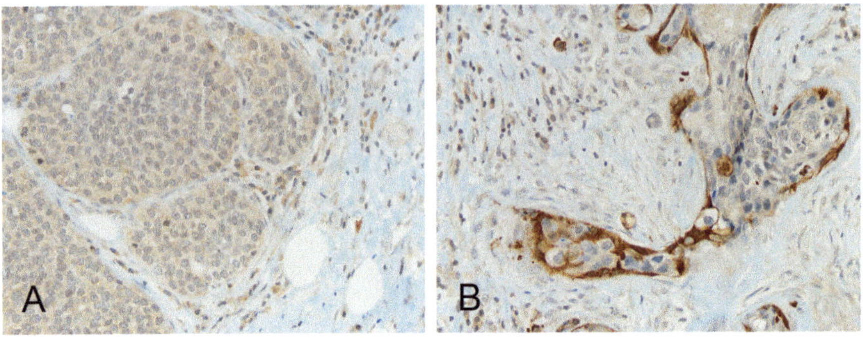

Fig. 4.16 (**a** and **b**) Invasive ductal carcinoma of the breast reacting with an antibody against laminin that is located in the basal membrane. (**a**) Negative for laminin. (**b**) The cancer cells are producing laminin that is decorating the basal surface of the cells, ×40. *We have used a rabbit monoclonal* antibody at a dilution 1:100 from Biogenex, CA

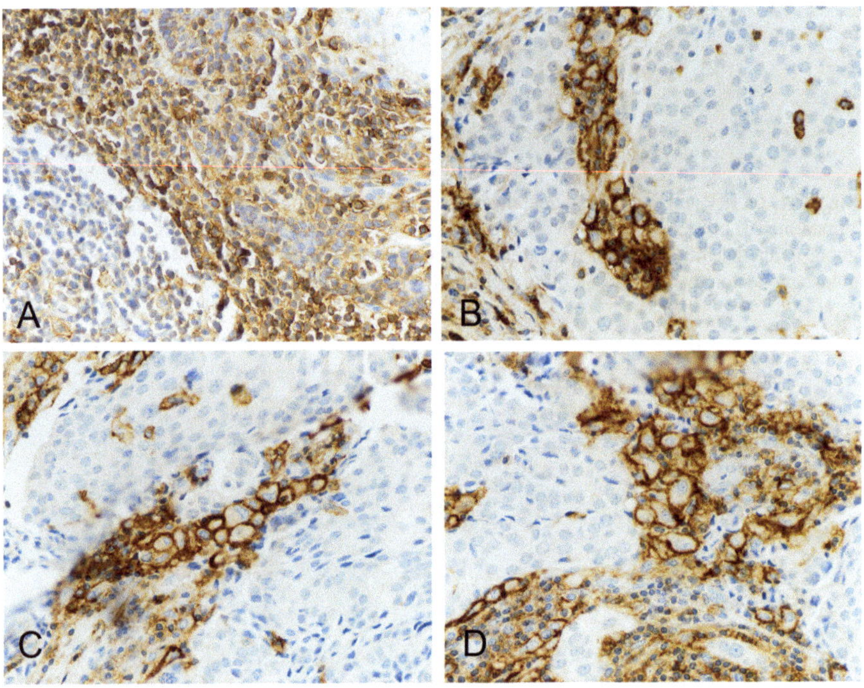

Fig. 4.17 (**a**) Tonsil, (**b**, **c** and **d**) invasive ductal carcinoma of the breast reacting with an antibody against cd44 that is located in the cell surface of the cells, ×40. *We have used a mouse monoclonal* antibody at a dilution 1:100 from Biogenex, CA

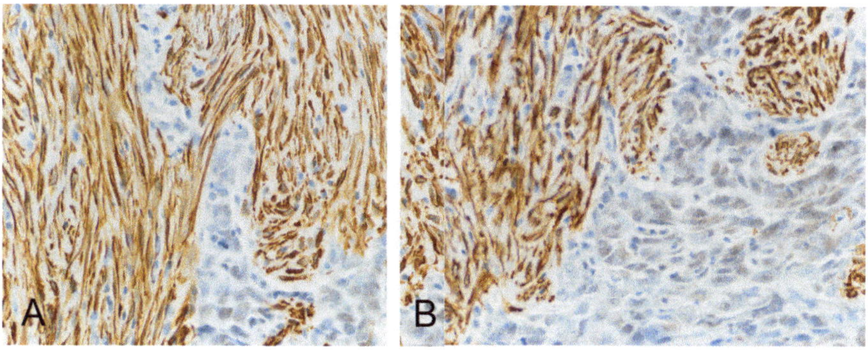

Fig. 4.18 (**a** and **b**) Invasive ductal carcinoma of the breast reacting negatively with an antibody against desmin that is positive in the stromal cells, ×40. *We have used a mouse monoclonal* antibody at a dilution 1:100 from Santa Cruz Biotechnology, CA

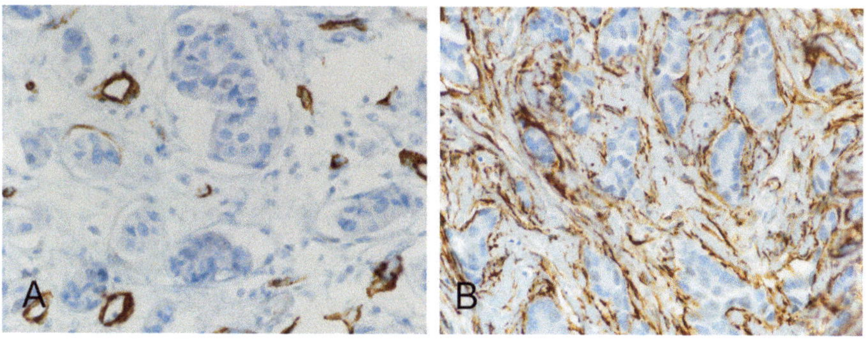

Fig. 4.19 (**a** and **b**) Invasive ductal carcinoma of the breast reacting negatively with an antibody against SMA or smooth muscle antigen. The reactivity is observed in the wall of small blood vessel in A or in myoepithelial cells in B, ×40. *We have used a rabbit polyclonal* antibody at a dilution 1:100 from abcam

Under the electron microscope the pattern invasive ductal carcinoma shows cells with pleomorphic nuclei and prominent nucleoli (Fig. 4.20). In general they do not produce basement membrane and myoepithelial cells are not observed [37]. Whereas the light microscopy and immunocytochemistry solve most of the cases of breast cancer, there are undifferentiated tumors in which the electron microscope is helpful in identify luminal spaces border by microvilli that indicate the glandular origin (Figs. 4.21, 4.22 and 4.23). The presence of desmosomes or secretory granules are also found in the undifferentiated breast cancer cells and only visible under the electron microscope (Figs. 4.24 and 4.25). For more details on the use of the electron microscope in tumor diagnosis see reference [37].

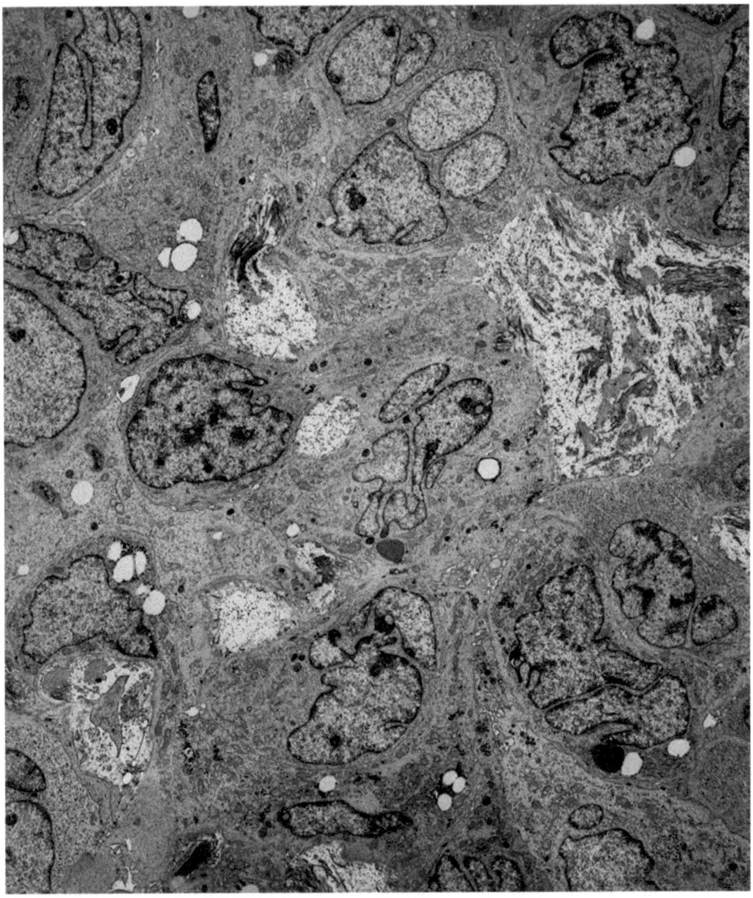

Fig. 4.20 Ultrastructure image of an Invasive ductal carcinoma showing cells with pleomorphic nuclei and prominent nucleoli, ×3000

4.2.2 Invasive Cribriform Carcinoma

The tumor has a cribriform appearance similar to that seen in ductal carcinoma in situ but exhibiting stromal invasion (Fig. 4.2). This type of tumor is not frequent and has an excellent prognosis [38, 39].

4.2.3 Mucinous Carcinoma

This type of tumors also called colloid, or gelatinous carcinoma (Fig. 4.26). Under the microscope small clusters of tumor cells are surrounded by mucin that is extracellularly located. Most of the mucin is of acid or neutral type [40]. It has been

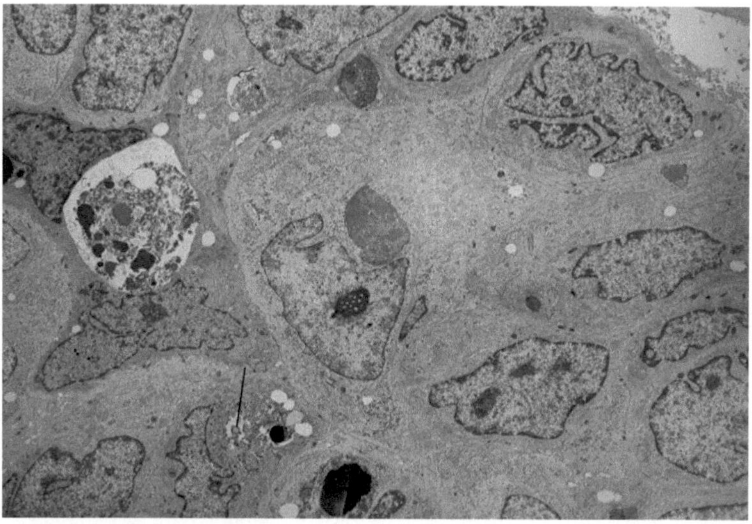

Fig. 4.21 Poorly differentiated invasive ductal carcinoma showing a luminal space (*arrow*) bordered by microvilli that indicate the glandular origin, ×3000

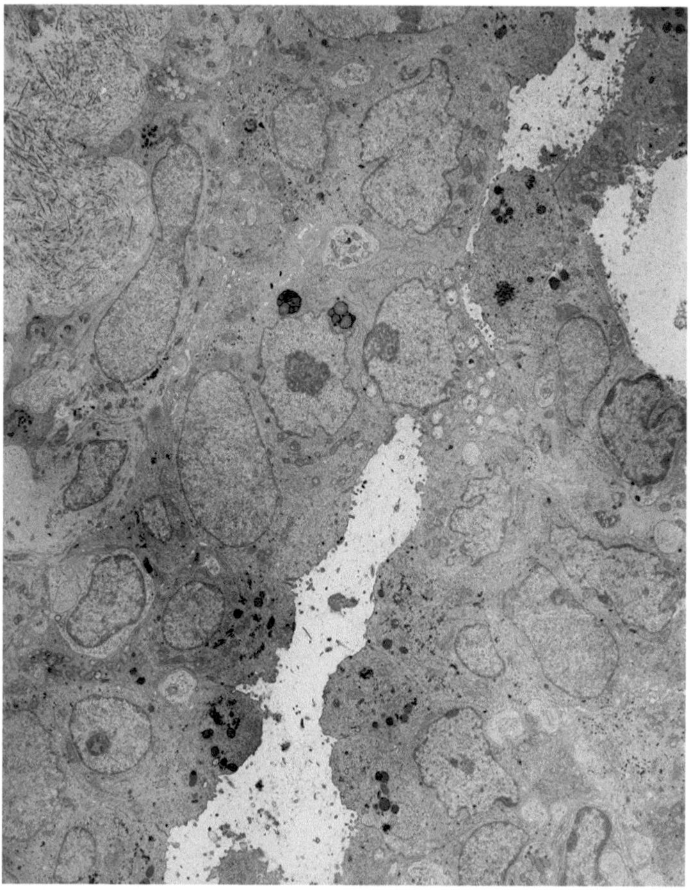

Fig. 4.22 Invasive ductal carcinoma showing the fine structure of the luminal cells. The apical portion of the cells show the small microvilli and secretory material, ×3000

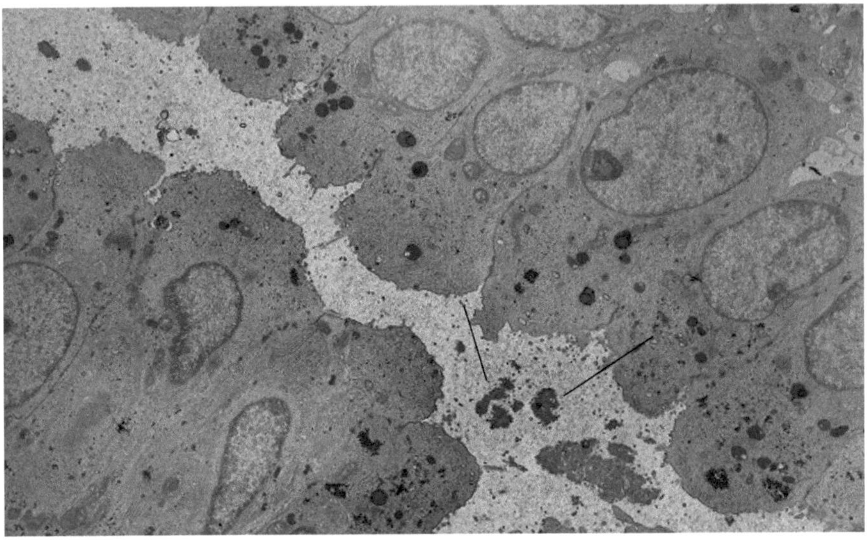

Fig. 4.23 Invasive ductal carcinoma showing the apical portion of the cells containing the electron dense secretory material that is typical of the breast epithelia and the same material in the luminal space (*arrows*), ×3000

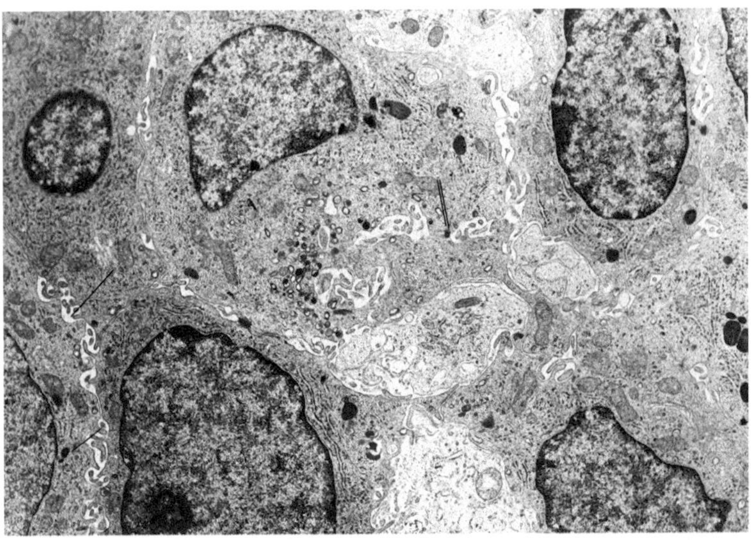

Fig. 4.24 Poorly differentiated invasive ductal carcinoma showing the small microvilli protruding toward the intercellular spaces (*arrows*) and the desmosome joining the cells (*two arrows*), ×3000

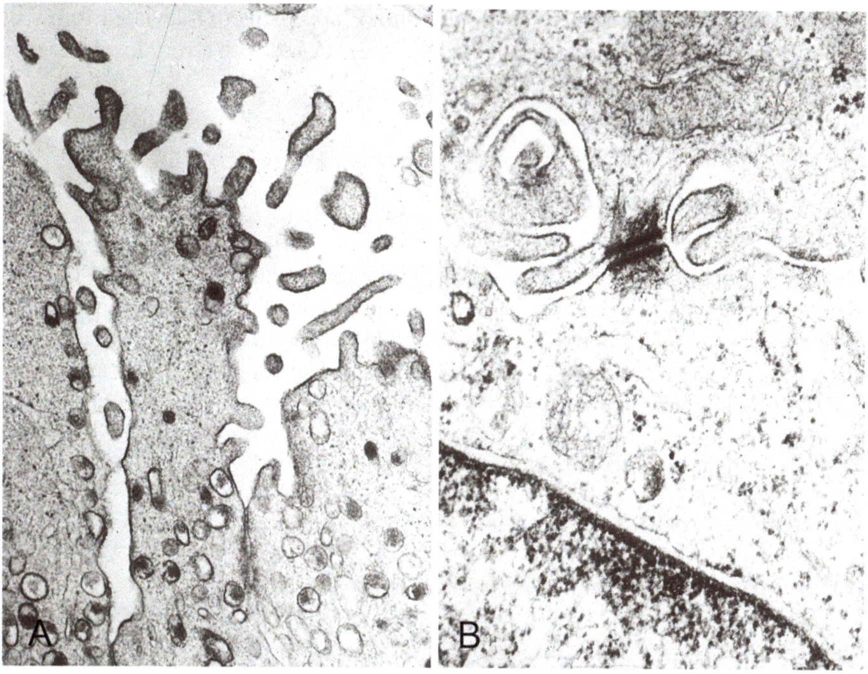

Fig. 4.25 The presence of secretory granules (**a**) and desmosomes (**b**) are markers that can be used in the diagnosis of breast cancer, ×36,000

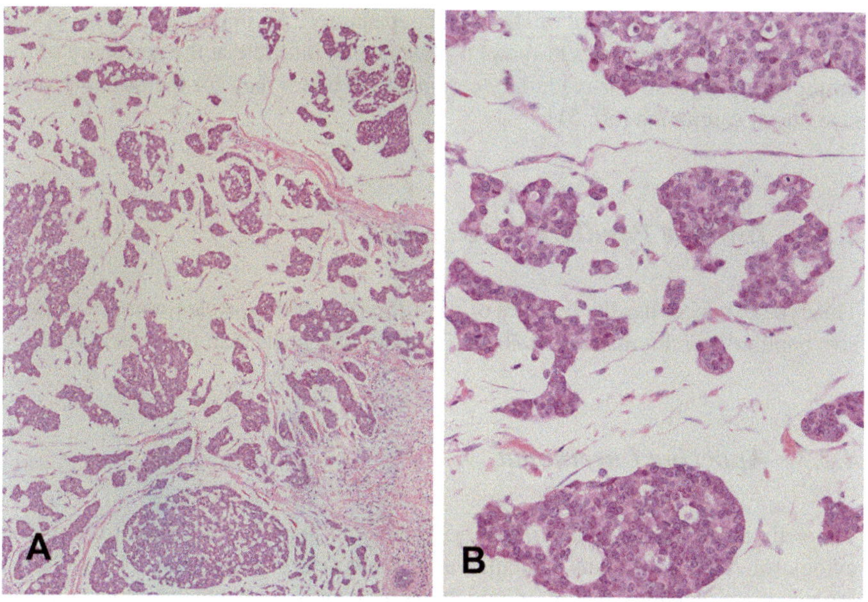

Fig. 4.26 (**a**) Mucinous carcinoma of the breast, ×4. (**b**) Clusters of well differentiated tumor cells surrounded by mucin, ×40

reported that the mucins secreted by this tumor are distinct O-acylated forms of sialomucins [41]. Pure mucinous carcinoma is associated with a very low incidence (2 % to 4 %) of nodal metastases [42–44].

4.2.4 Tubular Carcinoma

Tubular carcinomas are in general small, with a mean diameter of about 1 cm [45, 46]. This type of tumor is characterized by the prominent tubular arrangement of the cancer cells (Figs. 4.27 and 4.28) surrounded by prominent stroma of collagen that is some areas are hyaline and positive for amyloid. The glandular spaces are irregular and often angulated and they lack myoepithelial cell component and basement membrane. Metastases to axillary nodes occur at very low frequency (10% of cases) [45, 47, 48] and the prognosis is excellent [49].

4.2.5 Medullary Carcinoma

This tumor is a well circumscribed and the borders are always of the "pushing" type. The pattern of growth is diffuse; with minimal or no glandular pattern (Fig. 4.29a). The tumor cells are large and pleomorphic, with large nuclei and prominent nucleoli and numerous mitoses (Fig. 4.29b). The cell borders are indistinct, giving the tumor a syncytial or sheet-like appearance. An important feature of this type of tumor is the prominent lympho-plasmacytic infiltrate at the periphery of the tumor. The prognosis for medullary carcinoma is better than for the ordinary invasive ductal carcinoma [50, 51].

4.2.6 Invasive Papillary Carcinoma

This type of cancer has the features of an ordinary ductal-type carcinoma. The invasive papillary carcinoma (Fig. 4.30) has a better prognosis than the typical NOS.

4.2.7 Apocrine Carcinoma

The cells in the apocrine carcinoma [52] have an abundant acidophilic, granular cytoplasm. The nuclei are vesicular and nucleoli are prominent. The cells are arranged in a glandular pattern in which the luminal portion has the typical characteristic bulbous expansion or "apocrine snout".

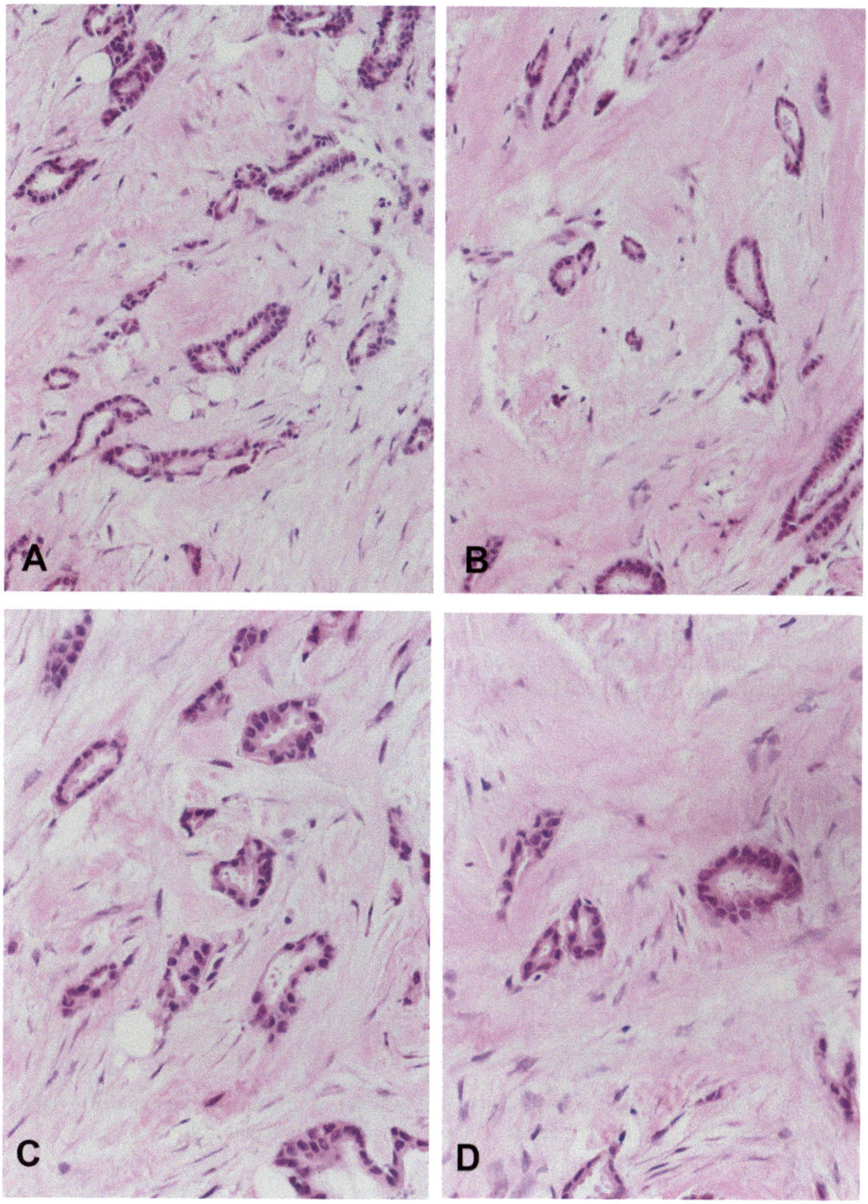

Fig. 4.27 (**a**) Tubular carcinoma of the breast. (**b**, **c** and **d**) show the angulated shape of the glandular structures produced by the cancer cells, ×20

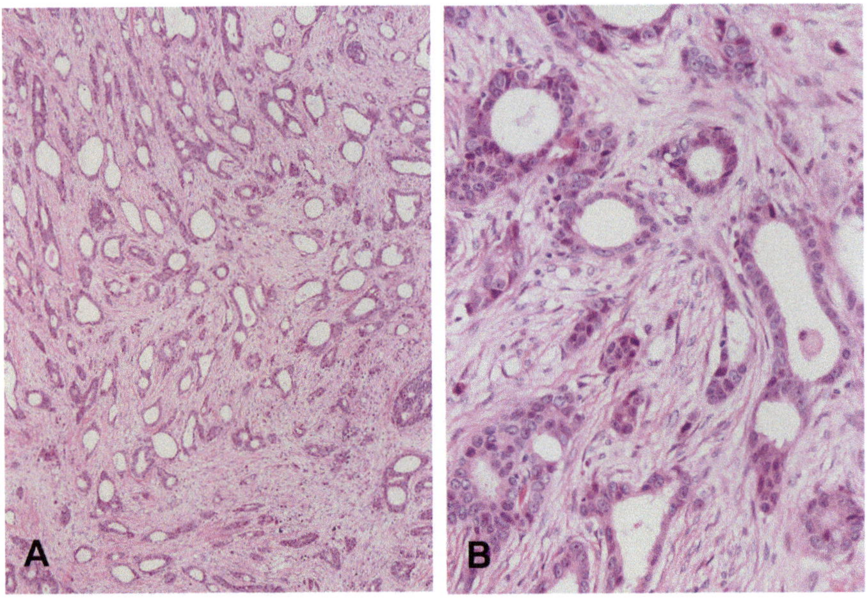

Fig. 4.28 (**a**) Low magnification of the tubular carcinoma of breast, ×4. (**b**) Detail of the angulated shape of the glandular structures produced by the cancer cells. The glandular spaces are lacking myoepithelial cell component and basement membrane, ×40

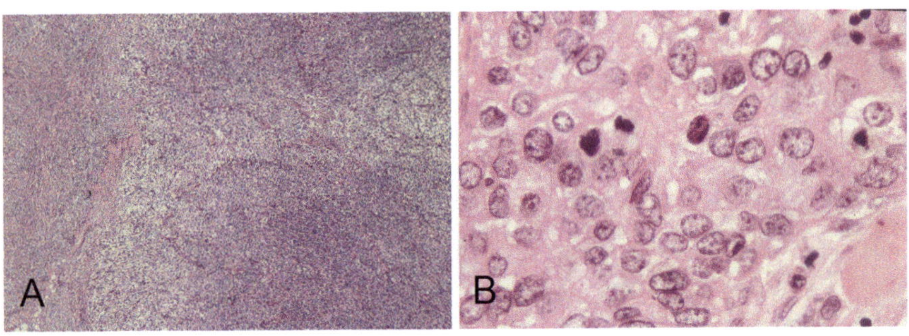

Fig. 4.29 Medullary carcinoma. (**a**) Pushing and well demarcated border with marked lymphoplasmacytic infiltrate, ×4. (**b**) The tumor cells are large and pleomorphic, ×40

4.2.8 Juvenile (Secretory) Carcinoma

This is a well circumscribed and usually small tumor. Under the microscope there are tubule-alveolar and focally papillary formations lined by cells with a vacuolated cytoplasm forming lumina filled by an eosinophilic PAS-positive secretion [53–55]. The overall prognosis is excellent, most series quoting a 5-year survival rate close to 100 % [53].

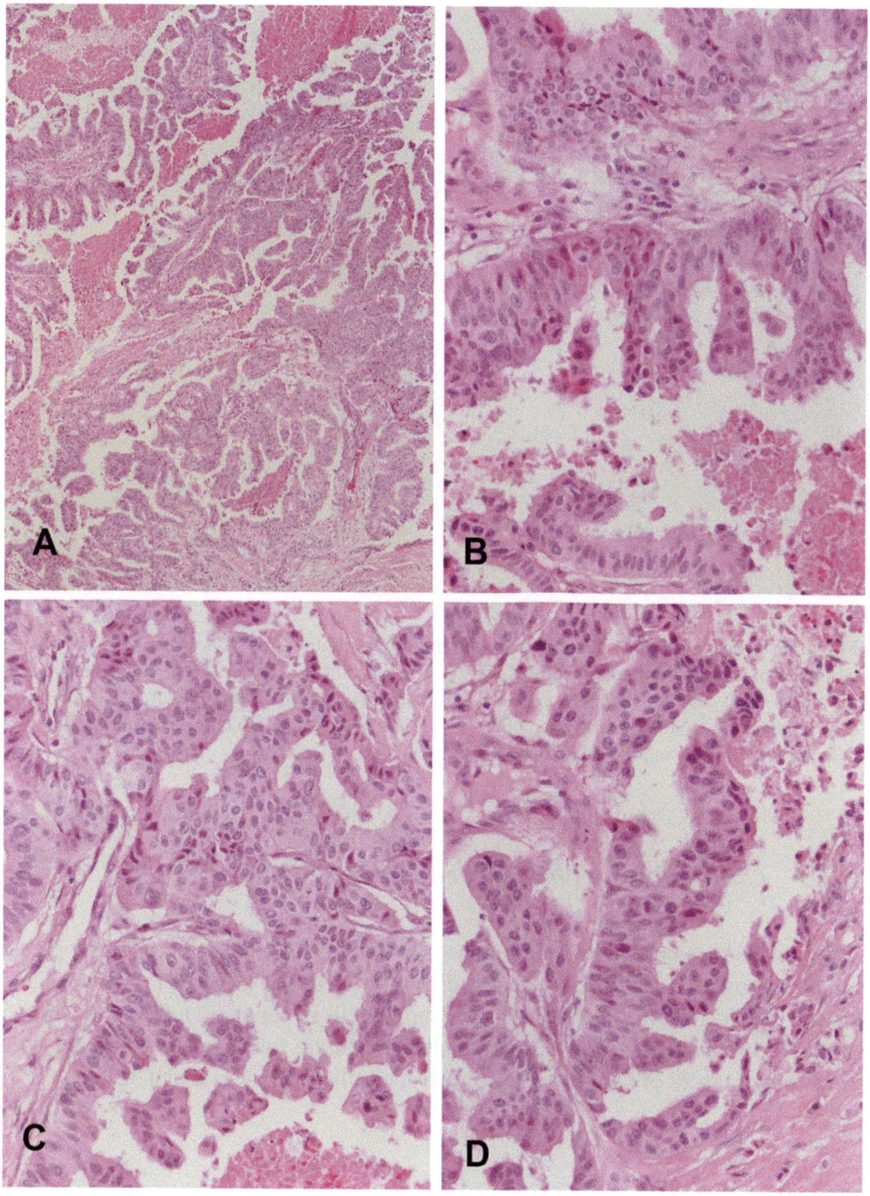

Fig. 4.30 (**a**) Papillary carcinoma, ×4. (**b**, **c** and **d**) Details showing the papillae, ×40

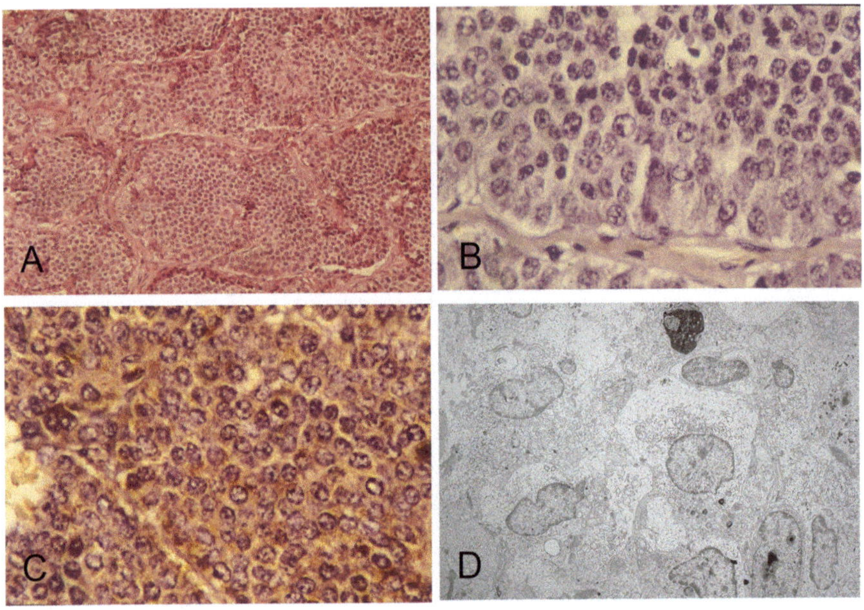

Fig. 4.31 Breast carcinoma with neuroendocrine differentiation. (**a**) This tumor is characterized by their small cells, arranged in solid nests separated by fibrous tissue, ×4. (**b**) The cells are uniform with scanty cytoplasm, ×40. (**c**) The cells are reactive against chromogranin A, ×40; We have used a rabbit monoclonal antibody at a dilution 1:100 from abcam. (**d**) Ultrastructural appearance with some cells showing the typical neurosecretory granules, ×2000

4.2.9 Carcinomas with Neuroendocrine Features

This tumor does not have specific gross appearance and under the microscope are small, arranged in solid nests separated by fibrous tissue (Fig. 4.31). Ribbons and rosette-like formations may be seen. The tumor cells of carcinoid tumor of the breast are argyrophilic (Fig. 4.31) but not argentaffin and are found to contain dense-core secretory granules of various types ultrastructurally (Figs. 4.31 and 4.32).

4.2.10 Metaplastic Carcinoma

The metaplastic carcinoma according to some authors [26] is a generic term for breast carcinoma of ductal type in which the predominant component of the neoplasm has an appearance other than epithelial and glandular and more in keeping with another cell type (Fig. 4.33). Pattern like *sarcomatoid carcinoma,* carcinoma with sarcoma-like stroma, and carcinosarcoma; *Spindle cell carcinoma; Carcinoma with osteoclast-like giant cells; Squamous cell carcinoma and the pleomorphic*

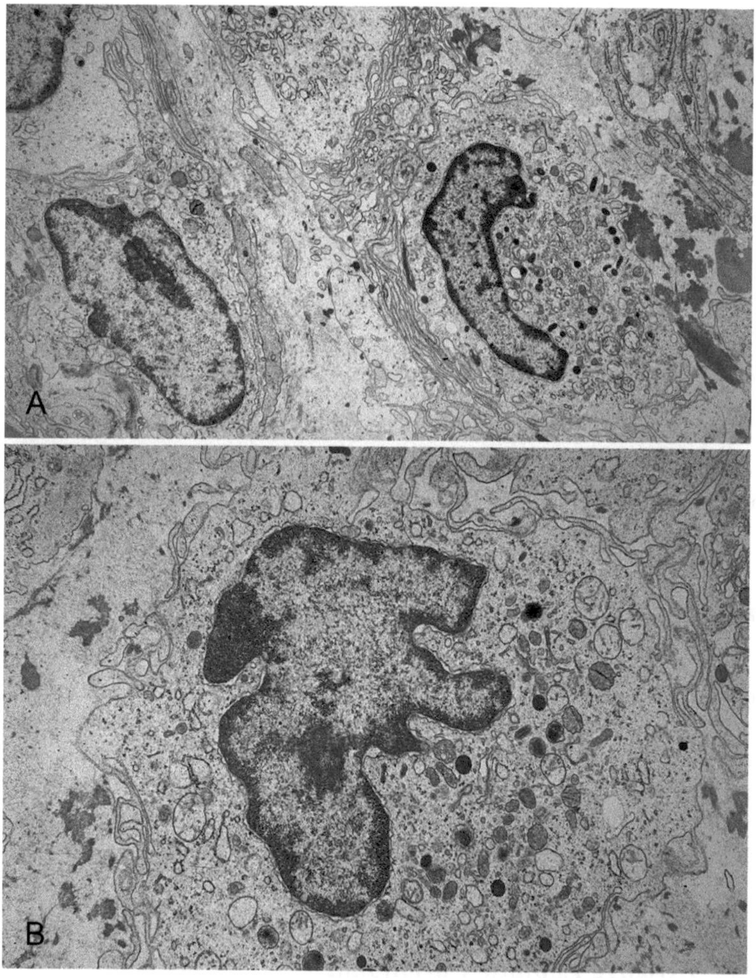

Fig. 4.32 (**a**) Electron microscopic appearance of breast carcinoma with neuroendocrine differentiation with the typical dense core neurosecretory granules, ×6000. (**b**) Higher magnification showing the cells containing the neurosecretory granules ranging in size from 140 to 225 nm, ×8500

carcinoma . These tumors are rare and in general are more aggressive than that of ordinary invasive ductal-type carcinoma [56, 57].

4.2.11 Inflammatory Carcinoma

The term inflammatory carcinoma is a clinical term for a type of breast carcinoma in which the entire breast was reddened and warm, with widespread edema of the skin, thus simulating the appearance of mastitis. Pathologic studies in some of those

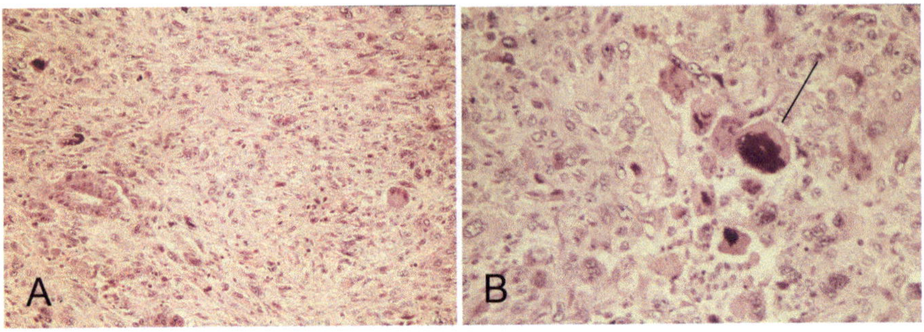

Fig. 4.33 Metaplastic carcinoma. (**a**) Presence of glandular epithelium (*arrow*) and its surrounding sarcoma-like components, ×10; (**b**) Osteclastic like cells (*arrow*), ×40

cases revealed the lesion to be an undifferentiated carcinoma with widespread carcinomatosis of the dermal lymphatic vessels.

4.2.12 *Paget's Disease*

Paget's disease is the name given to an eczema-like lesions centered in the nipple caused by breast carcinoma (Fig. 4.34) [58]. It is most of the time accompanied by an underlying breast carcinoma of in situ ductal type, with or without associated stromal invasion. Under the microscope the tumor is characterized by large clear cells with atypical nuclei are seen within the epidermis, usually concentrated along the basal layer but also permeating the malpighian layer (Fig. 4.35).

4.2.13 *Invasive Lobular Carcinoma*

Grossly the invasive lobular carcinoma (ILC) can be very small or large like the one depicted in Fig. 4.36. Histologically is characterized by the presence of small and relatively uniform tumor cells growing singly, in Indian file, and in a concentric ("targetoid") fashion around lobules involved by in situ lobular neoplasia (Figs. 4.37 and 4.38). The stroma is formed by dense fibrous type, and contains foci of periductal and perivenous elastosis in virtually every case. A lymphocytic infiltrate may be present. These tumors react positively to HMW keratin, lack of accumulation of p53, and-most importantly-decrease or absence of E-cadherin [59–62]. An important differential diagnosis must be done with carcinoma with neuroendocrine features and malignant lymphoma.

A variety of lobular invasive carcinoma is the pleomorphic lobular carcinoma that has the pattern of growth of a classical breast carcinoma but exhibits a marked degree of nuclear pleomorphism and abundant cytoplasm [63]. It also frequently

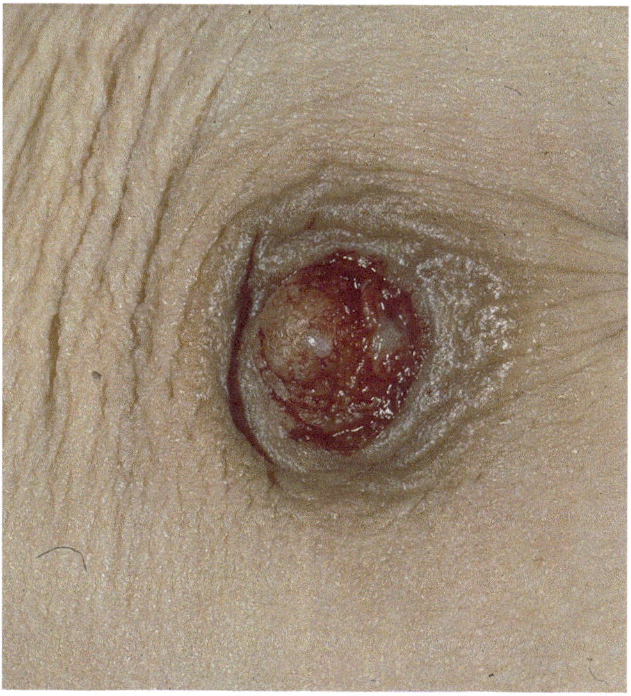

Fig. 4.34 Eczema-like clinical appearance of Paget's disease

shows apocrine differentiation, focal signet ring morphology, lack of hormone receptors, higher expression of p53 and *HER2 / neu*, occasional expression of chromogranin, and lack of E-cadherin staining [64–67].

Another variant of lobular invasive carcinoma I the tubule-lobular carcinoma that is characterized by the admixture of small tubular formations having a minute or undetectable lumen ("closed" or "almost closed" tubules) with cords of tumor cells growing in a lobular configuration similar to that of invasive lobular carcinoma [68].

4.2.14 Mixed Ductal and Lobular Carcinoma

Biphasic carcinomas composed in part of a component with definite features of invasive ductal carcinoma and in part of a component with definite features of invasive lobular carcinoma do occur.

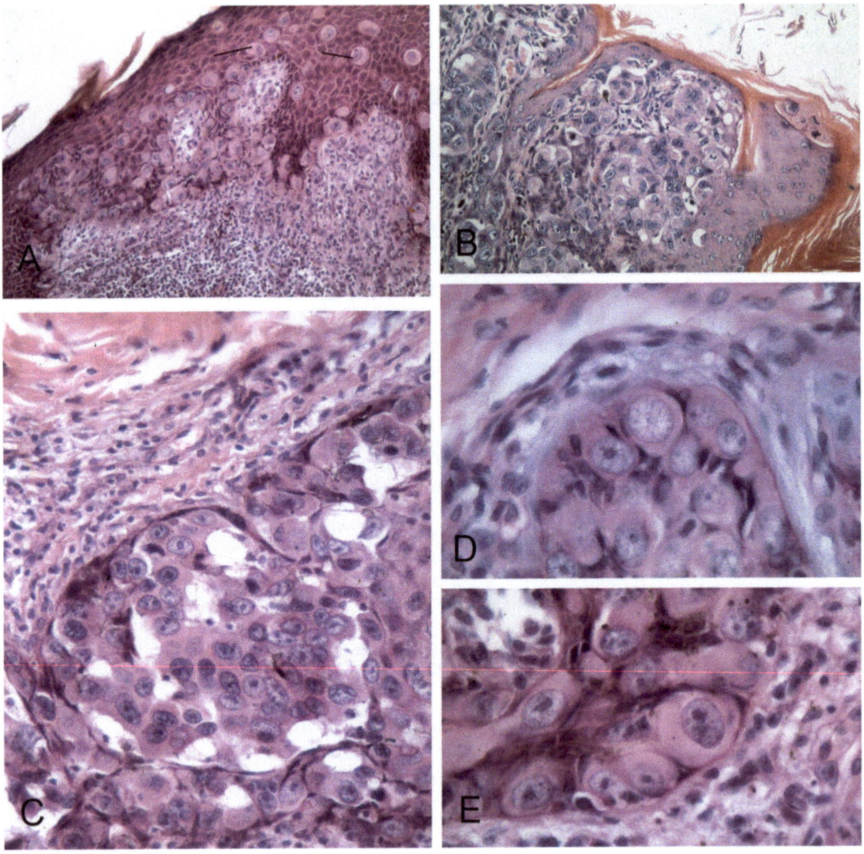

Fig. 4.35 (**a**) Low magnification of the histological appearance of Paget's disease, ×4. (**b**) (×10); (**c**) (×40), (**d**) and (**e**) (×100) —Large clear cells with atypical nuclei concentrated along the basal layer of the epidermis but also permeating the malpighian layer

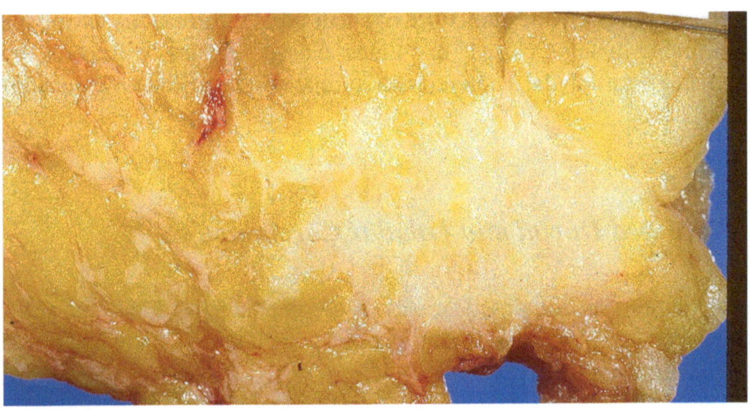

Fig. 4.36 Gross appearance of a lobular invasive carcinoma showing irregular infiltrative margins with white discoloration due to the intense desmoplastic reaction

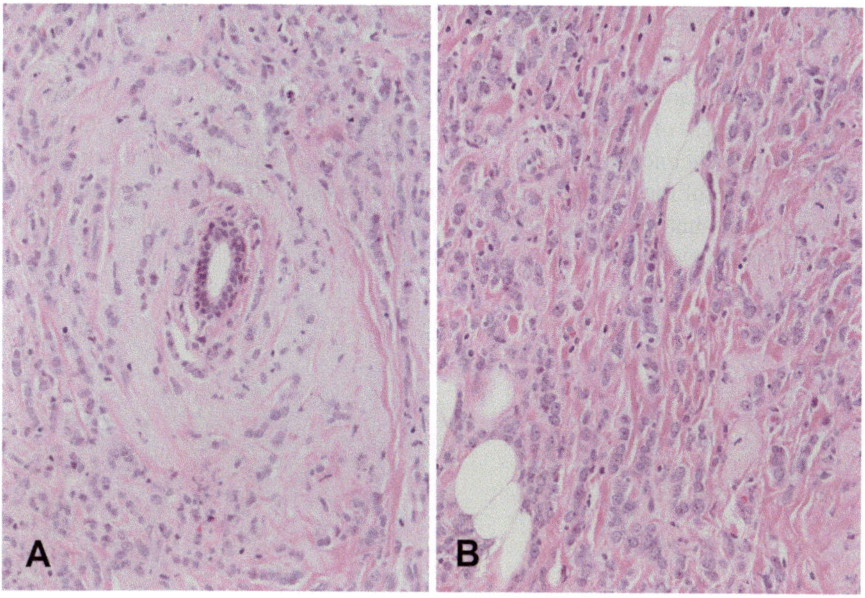

Fig. 4.37 (**a** and **b**) Invasive lobular carcinoma composed of small and uniform cells with round nuclei growing in an Indian file pattern, ×40

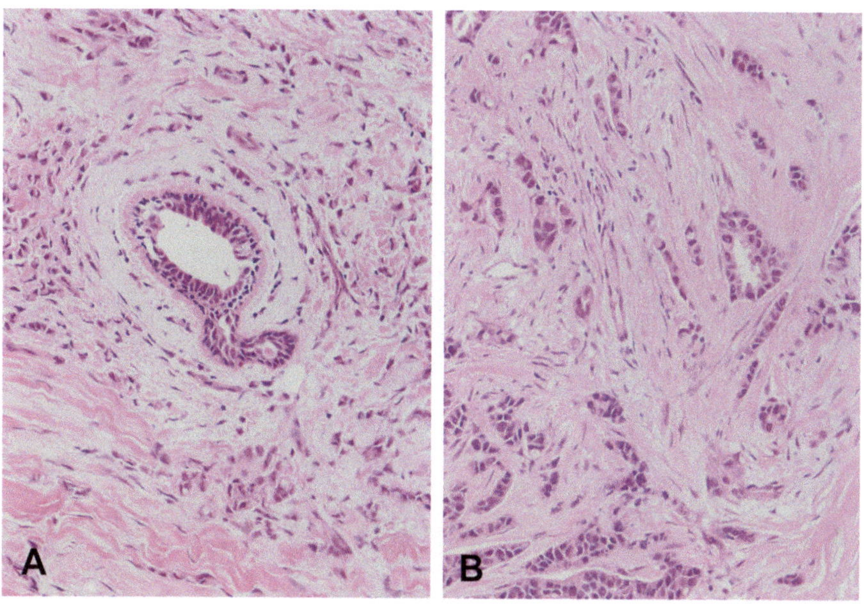

Fig. 4.38 (**a**) Invasive lobular carcinoma with the typical target-like growth around an uninvolved duct. (**b**) The invasive lobular carcinoma elicit a severe desmoplastic reaction in the stroma, ×40

4.3 Microinvasive Breast Carcinoma

Micro-invasive carcinoma is defined as any carcinoma in situ of the breast showing one or more areas of stromal invasion (Fig. 4.39) not surpassing 1 mm in thickness. Immuno-histochemical evaluation with myoepithelial and basement membrane markers is useful for a confirmation of the diagnosis. Overall patients with micro-invasive carcinoma are at risk for nodal metastases but their survival rate is better than for patients with T1 invasive carcinoma [69].

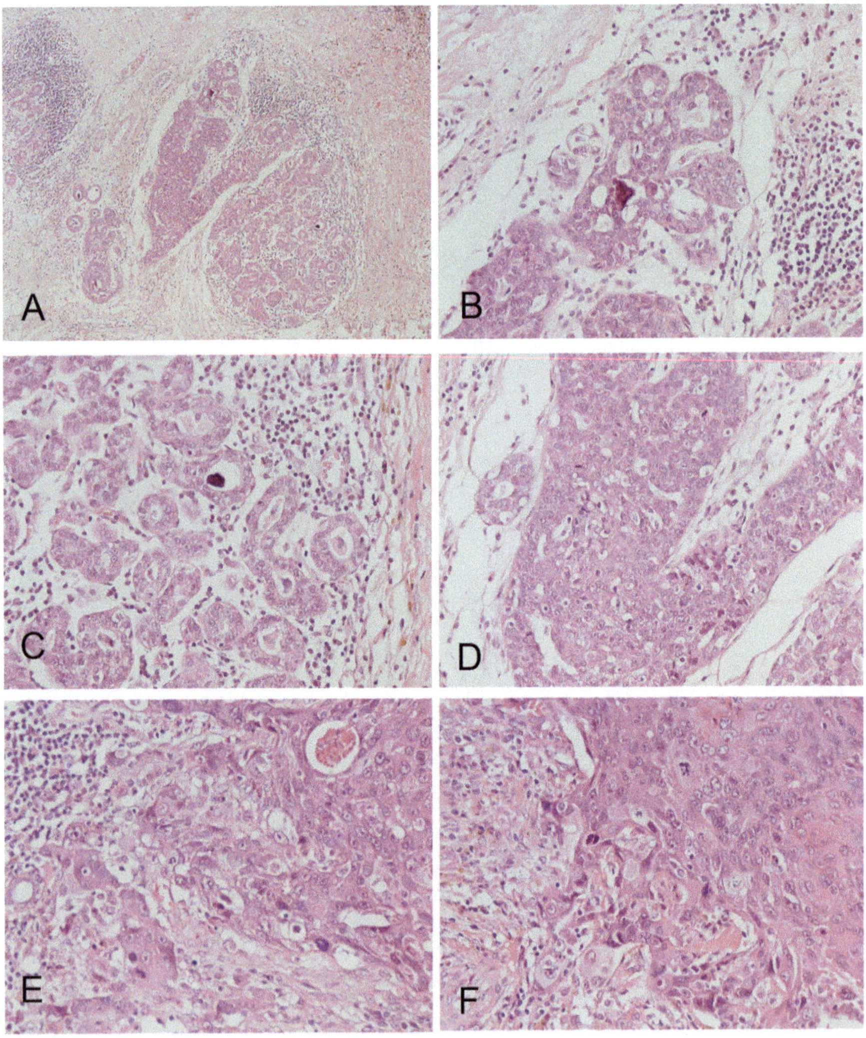

Fig. 4.39 (**a**, **b**, **c** and **d**) Carcinoma in situ involving the terminal ducts of the lobule type 1. (**e** and **f**) Details of the micro invasive areas of the ductal carcinoma in situ, ×40

References

1. Aamdal S, Bormer O, Jorgensen O et al (1984) Estrogen receptor and long-term prognosis in breast cancer. Cancer 53:2525–2529
2. Antomiades K, Spector H (1979) Correlation of estrogen receptor levels with histology and cytomorphology in human mammary cancer. Am J Clin Pathol 71:497–503
3. Bishop HM, Blarney RW, Elston CW, Hoybittle JL, Nicholson RI, Griffiths K (1979) Relationship of estrogen-receptor status to survival in breast cancer. Lancet 1:283–284
4. Black MM, Opler SR, Speer FD (1955) Survival in breast cancer cases in relation to the structure of the primary tumor and regional lymph nodes. Surg Gynecol Obstet 100:543–551
5. Black MM, Speer FD, Opler SR (1956) Structural representations of tumor-host relationships in mammary carcinoma. Biologic prognostic significance. Am J Clin Pathol 26:250–265
6. Black MM, Speer FD (1957) Nuclear' structure in cancer tissues. Surg Gynecol Obstet 105:97–102
7. Bloom HJG, Richardson WW (1957) Histologic grading and prognosis in breast cancer: a study of 1409 cases of which 359 have been followed for 15 years. Br J Cancer 11:359–377
8. Cutler SJ, Black MM, Goldenberg IS (1963) Prognostic factor in cancer of the female breast. Cancer 16:1589–1597
9. Cutler SJ, Black MM, Friedell GH, Vidone RA, Goldenberg IS (1966) Prognostic factors in cancer of the female breast. II, Reproductibility of histopathologic classification. Cancer 19:75–82
10. Fisher B, Slack NH (1970) Number of lymph nodes examined and the prognosis of breast cancer. Surg Gynecol Obstet 131:79–88
11. Fisher ER, Gregorio RM, Fisher B (1975) The pathology of invasive breast cancer-a syllabus derivcd from findings of the National Surgical Adjuvant Breast Project (Protocol N. 4). Cancer 36:1–84
12. Fisher ER, Palekar AS, Redmond C, Barton N, Fisher B (1980) Pathologic findings from the National Surgical Adjuvant Breast Project. (Protocol N. 4). VI. Invasive papillary cancer. Am J Clin Pathol 73:313–322
13. Fisher ER, Redmond CK, Lieu H, Rockette H, Fisher B (1980) Collaborating NSABP Investigators: correlation of estrogen receptor and pathologic characteristics of invasive breast cancer. Cancer 45:349–353
14. Foote FW Jr (1959) Surgical pathology of cancer of the breast. In: Parsons WH (ed) Cancer of the breast. Charles C Thomas, Springfield, pp 37–38
15. Hahnel R, Wooding T, Vivian AB (1979) Prognostic value of estrogen receptor in primary breast cancer. Cancer 44:671–675
16. Knight WA, Livingston RB, Bregory EJ, McGuire WL (1977) Estrogen receptor as an independent prognostic factor for early recurrence in breast cancer. Cancer Res 37:4669–4671
17. Lane N, Gokiel H, Salerno RA, Haagensen CD (1961) Clinocopathologic analysis of the surgical curability of breast cancers: a minimum ten year study of a personal series. Am Surg 153:483–498
18. Masters JRW, Sangster K, Hawkins RA, Shivas AA (1976) Elastosis and estrogen receptors in human breast cancer. Br J Cancer 33:342–343
19. Maynard PV, Blaney RN, Elston CW, Haybittle JL, Griffiths K (1978) Oestrogen receptor assay in primary breast cancer and early recurrence of the disease. Cancer Res 38:4292–4295
20. McDivitt RW, Stewart FW, Berg JW (1968) Tumors of the breast, Atlas of tumor pathology, second series. Armed Forces Institute of Patho!ogy, Washington, DC
21. McGuire WL (1975) Current status of estrogen receptors in human breast cancer. Cancer 36:638–644
22. Patchefsky A, Shaber GS, Schwartz GF, Feig SA, Norlinger RE (1977) The pathology of breast cancer detected by mass population screening. Cancer 40:1659–1670
23. Rich MA, Furmanski P, Brooks SC (1978) The Breast Cancer Prognostic Study Surgery and Pathology Associates: prognostic value of estrogen receptor. Cancer Res 38:4296–4298

24. Rosen PP (1979) The pathological classification of human mammary carcinoma, Past, present and future. Ann Clin Lab Sci 144:156
25. Sanfilippo O, Daidone MG, DiFronzo G, Silvestrini R (1979) Short term tissue culture of human breast cancer. Cancer 43:2365–2369
26. Rosai J (2004) Breast. Chapter 20. In: Rosai and Ackerman's surgical pathology. Mosby, New York, pp 1802–1818
27. Kuhajda FP, Offutt LE, Mendelsohn G (1983) The distribution of carcinoembryonic antigen in breast carcinoma. Diagnostic and prognostic implications. Cancer 52:1257–1264
28. Loy TS, Chapman RK, Diaz-Arias AA, Bulatao IS, Bickel JT (1991) Distribution of BCA -225 in adenocarcinomas. An immunohistochemical study of 446 cases. Am J Clin Pathol 96:326–329
29. Prey MU, Bedrossian CW, Masood S (1991) The value of monoclonal antibody B72.3 for the diagnosis of breast carcinoma. Experience with the first commercially available source. Hum Pathol 22:598–602
30. Robertson JF, Ellis IO, Bell J, Todd JH, Robins A, Elston CW, Blarney RW (1989) Carcinoembryonic antigen immunocytochemistry in primary breast cancer. Cancer 64:1638–1645
31. Tavassoli FA, Jones MW, Majeste RM, Bratthauer GL, O'Leary TJ (1990) Immunohistochemical staining with monoclonal Ab B72.3 in benign and malignant breast disease. Am J Surg Pathol 14:128–133
32. Domagala W, Wozniak L, Lasota I, Weber K, Osborn M (1990) Vimentin is preferentially expressed in high-grade ductal and medullary but not in lobular breast carcinomas. Am J Pathol 137:J059–1064
33. Guerreiro Da Silva ID, Hu YF, Russo IH, Ao X, Salicioni AM, Yang X, Russo J (2000) S100P Ca^{+2}-binding protein overexpression is associated with immortalization and neoplastic transformation of human breast epithelial cells *in vitro* and tumor progression *in vivo*. Int J Oncol 16:231–240
34. Dwarakanath S, Lee AKC, DeLellis RA, Silverman ML, Frasca L, Wolfe HI (1987) S-100 protein positivity in breast carcinomas. A potential pitfall in diagnostic immunohistochemistry. Hum Pathol 18:1144–1148
35. Lunde S, Nesland JM, Holm R, Johannessen JV (1987) Breast carcinomas with protein 5-100 immunoreactivity. An immunocytochemical and ultrastructural study. Pathol Res Pract 182:627–631
36. Huang Y, Fernandez S, Goodwin S, Russo PA, Russo IH, Sutter T, Russo J (2007) Epithelial to mesenchymal transition in human breast epithelial cells transformed by 17- beta- estradiol. Cancer Res 67:11147–11157
37. Russo J, Tait L, Russo IH (1986) Guidelines in the ultrastructural diagnosis of tumors. In: Russo J, Somers SC (eds) Tumor diagnosis by electron microscopy. Field, Rich and Associates, New York, pp 15–126
38. Page DL, Dixon JM, Anderson TJ, Lee D, Stewart HJ (1983) Invasive cribriform carcinoma of the breast. Histopathology 7:525–536
39. Venable JG, Schwartz AM, Silverberg SG (1990) Infiltrating cribriform carcinoma of the breast. A distinctive clinicopathologic entity. Hum Pathol 21:333–338
40. Walker RA (1982) Mucoid carcinomas of the breast. A study using mucin histochemistry and peanut lectin. Histopathology 6:571–579
41. Sáez C, Japón MA, Poveda MA, Segura DI (2001) Mucinous (colloid) adenocarcinomas secrete distinct O-acylated forms of sialomucins: a histochemical study of gastric, colorectal and breast adenocarcinomas. Histopathology 39:554–560
42. O'Connell JT, Shao ZM, Drori E, Basbaum CB, Barsky SH (1998) Altered mucin expression is a field change that accompanies mucinous (colloid) breast carcinoma histogenesis. Hum Pathol 29:1517–1523
43. Norris HI, Taylor HE (1965) Prognosis of mucinous (gelatinous) carcinoma of the breast. Cancer 18:879–885

44. Rasmussen BB, Rose C, Christensen IB (1987) Prognostic factors in primary mucinous breast carcinoma. Am J Clin Pathol 87:155–160
45. McDivitt RW, Boyce W, Gersell D (1982) Tubular carcinoma of the breast. Clinical and pathological observations concerning 135 cases. Am J Surg Pathol 6:401–411
46. Peters GN, Wolff YI, Haagensen CD (1981) Tubular carcinoma of the breast. Clinical pathologic correlations based on 100 cases. Ann Surg 193:138–149
47. Carstens PHB, Huvos AG, Foote FIN Jr, Ashikari R (1972) Tubular carcinoma of the breast. A clinicopathologic study of 35 cases. Am J Clin Pathol 58:231–238
48. Deos PH, Noris HJ (1982) Well-differentiated (tubular) carcinoma of the breast. A clinicopathologic study of 145 pure and mixed cases. Am J Clin Pathol 78:1–7
49. Diab SG, Clark GM, Osborne CK, Libby A, Allred DC, Elledge RM (1999) Tumor characteristics and clinical outcome of tubular and mucinous breast carcinomas. J Clin Oncol 17:1442–1448
50. Bloom HJG, Richardson WW, Fields JR (1970) Host resistance and survival in carcinoma of breasts. A study of 104 cases of medullary carcinoma in a series of 1,411 cases of breast cancer followed for 20 years. Br Med J 3:181–188
51. Richardson WW (1956) Medullary carcinoma of the breast. A distinctive tumour type with a relatively good prognosis following radical mastectomy. Br J Cancer 10:415–423
52. Abati AD, Kimmel M, Rosen PP (1990) Apocrine mammary carcinoma. A clinicopathologic study of 72 cases. Am J Clin Pathol 94:371–377
53. McDivitt RW, Stewart FW (1966) Breast carcinoma in children. JAMA 195:388–390
54. Rosen PP, Cranor ML (1991) Secretory carcinoma of the breast. Arch Pathol Lab Med 115:141–144
55. Tavassoli FA, Norris HJ (1980) Secretory carcinoma of the breast. Cancer 45:2404–2413
56. Bussolati G, Papotti M, Sapino A, Gugliotta P, Ghiringhello B, Azzopard JG (1987) Endocrine markers in argyrophilic carcinomas of the breast. Am J Surg Pathol 11:248–256
57. Oberman HA (1987) Metaplastic carcinoma of the breast. A clinicopathologic study of 29 patients. Am J Surg Pathol 11:918–929
58. Paget J (1874) On disease of the mammary areola preceding cancer of the mammary gland. St Barth Hosp Rep 10:87–89
59. Acs G, Lawton TJ, Rebbeck TR, Li Volsi VA, Zhang PJ (2001) Differential expression of E-cadherin in lobular ductal neoplasms of the breast and its biologic and diagnostic implications. Am J Clin Pathol 115:85–98
60. Domagala W, Harezga B, Szadowska A, Markiewski M, Weber K, Osborn M (1993) Nuclear protein accumulates preferentially in medullary and highcgrade ductal but in lobular breast carcinomas. Am J Pathol 142:669–674
61. Goldstein NS, Bassi D, Watts JC, Layfield LJ, Yaziji H, Gown AM (2001) E-cadherin reactivity of 95 non-invasive ductal and lobular lesions of the breast implications for the interpretation of problematic lesions. Am J Clin Pathol 115:534–542
62. Lehr HA, Folpe A, Yaziji H, Kommoss F, Gown AM (2000) Cytokeratin 8 immunostaining pattern and E-cadherin expression distinguish lobular from ductal breast carcinoma. Am J Clin Pathol 114:190–196
63. Weidner N, Semple JP (1992) Pleomorphic variant of invasive lobular carcinoma of the breast. Hum Pathol 23:1167–1171
64. Frolik D, Caduff R, Varga Z (2001) Pleomorphic lobular carcinoma of the breast: its cell kinetics, of oncogenes and tumour suppressor genes compared with invasive ductal carcinomas and classification infiltrating lobular carcinomas. Histopathology 39:503–513
65. Middleton LP, Palacios DM, Bryant BR, Krebs P, Otis CN, Merino MJ (2000) Pleomorphic lobular carcinoma: morphology, immunohistochemistry, and molecular analysis. Am J Surg Pathol 24:1650–1656
66. Radhi JM (2000) Immunohistochemical analysis of pleomorphic lobular carcinoma: Higher expression of p53 and chromogranin and lower expression of ER and PgR. Histopathology 36:156–160

67. Wahed A, Connelly J, Reese T (2002) E-cadherin expression in pleomorphic lobular carcinoma: an aid to differentiation from ductal carcinoma. Ann Diagn Pathol 6:349–351
68. Fisher ER, Gregorio RM, Redmond CI, Fisher B (1977) Tubulolobular invasive breast cancer. A variant of lobular invasive cancer. Hum Pathol 8:679–683
69. Padmore RF, Fowble B, Hoffman J, Rosser C, Hanlon A, Patchefsky AS (2000) Microinvasive breast carcinoma; clinicopathologic analysis of a single institution experience. Cancer 88:1403–1409

Chapter 5
The Molecular Basis of Breast Cancer Subtypes

5.1 Introduction

The name of "breast cancer" is the general term for carcinomas of the breast tissue. While this seems fairly concise, breast cancer extends to a diverse group of maladies. The reason is that breast carcinomas are heterogeneous and vary greatly in physical appearance, or their phenotype, on both a macroscopic and molecular scale (see Chap. 3). This phenotypic diversity corresponds to diversity in gene expression producing classifications for breast cancer that more accurately predict tumor behavior. The rationale for a genomic classification are basically two; one is that knowing the gene expression profile of a tumor lead us to understand cancer behavior and second rat the clinical level can help us to identify genes that are associated with specific cancer phenotype. Knowing the genomic signature of a given breast cancer provides the molecular pathways that can help better targeting of available therapy.

5.2 Initial Genomic Classification

The first distinguishing classification came from the study performed by Perou and Sorlie [1]. In this study they acquired samples of breast tissue from forty-two individuals. Forty of these were breast cancers (mostly invasive ductal carcinomas), one was a fibro-adenoma and the last sample normal breast tissue. In addition, contained in this study twenty-two pairs of tumor sample from which 20 were paired before and after a chemotherapy regimen, and two pairs from primary tumors paired with their respective lymph node metastasis. Using these samples, they isolated the RNA and performed cDNA microarray. From these microarrays 8,102 genes were

J. Russo, *The Pathobiology of Breast Cancer*,
DOI 10.1007/978-3-319-40815-6_5

initially identified and a subset of these genes was selected based on the variation in their expression using at least plus-or-minus four-times the median level of expression. Using these criteria they finally selected 1,753 for hierarchical cluster for their final classification. They selected genes based on gene expressions that were similar in any sample taken from the sample tumor, but varied between different tumors and for that purpose they utilized the 22 paired tumor samples and identified 496 genes from the 1,753 identified earlier based on the variation in their gene expression, with the added distinction of having greater variation between different tumors than from the same tumor. They called this new cluster the "Intrinsic Gene Subset". From this new subset, they were not only able to determine the expression levels for each sample, but were also able to group them based off their expression within the two layers found in mammary gland structures (the lobules and ducts): the inner luminal epithelial cells and the surrounding basal myoepithelial cells. From these groupings, they were able to further distinguish each group into gene clusters. They identified one cluster for the luminal epithelial cells, the Luminal epithelial/estrogen cluster. They also identified 3 clusters for the basal epithelial cells: the ERBB2 overexpression cluster, the Basal epithelial cell-associated cluster, containing keratins 5 and 17, and the Basal epithelial-cell-enriched gene cluster [1].

The clusters identified were later refined into subtypes [2]. The Luminal Cluster was separated into Luminal Subtype A, Luminal Subtype B, and Luminal Subtype C. The Basal Clusters were redefined: the ERBB2 over expression cluster became simply the ERBB2+ subtype, the Basal epithelial cell-associated cluster became the Basal-like subtype, and the Basal epithelial-cell-enriched gene cluster became the Normal breast-like subtype. From these produced subtypes, they also looked at the clinical features indicated by each subtype. To accomplish this, they utilized data acquired from 49 breast cancer patients showing all 5 subtypes, whom only had diseases local to the breast and with little-to-no metastasis present [2]. They specifically looked at the overall survival (survival months) and relapse-free survival (RFS) probabilities for each subtype over a 4-year period, in comparison to the other subtypes. In addition, they also looked at the outcomes when Luminal Subtype C and B were grouped with the other subtypes Analysis showed that the Basal-like and ERBB2+ subtypes had the both the lowest RFS and overall survival. Additionally, the Luminal Subtype C was shown to have the worst overall survival of all the Luminal Subtypes, and Subtypes ERBB2+ and Luminal B were shown to share certain genes associated with a poor prognosis.

5.3 Extended Classifications: Molecular Subtypes

Since the defining of the Intrinsic Subtypes, many studies have gone on to further refine and expand upon the initial breast cancer classifications of Perou and Sorlie [1, 2] redefining them as the Molecular Subtypes of Breast Cancer. The modern, accepted subtypes are the Luminal A, Luminal B, Basal-like, ERBB2+/HER2+, and the Normal Breast-like subtypes. There are several accepted means for distinguishing between the five subtypes. The primary method is the presence of absence of three different cellular receptors in the breast cancer tumors. The three receptors are the Estrogen Receptor (ER), the Progesterone Receptor (PR), and the Human Growth Factor Receptor 2 (HER2). Over expressions of these receptors has been observed in breast cancers, but have often only been looked at individually. In addition, different breast cancer tumors have been shown to have different expression levels of these receptors. Thus, by looking at all the receptors together and identifying which are over- expressed (or absent) in the tumor cells, there can be a clear classification used to distinguish between breast cancers. The protein Ki-67, a known prognostic factor associated with proliferation, is utilized in a similar manner, looking at the low or high levels for subtype distinction. The grade of the tumor is another factor incorporated into identifying molecular subtypes, which looks at how the tumors appear in comparison to normal, well-differentiated breast tissue. High grades are described as "poor," or not well-differentiated, while Low grades are described as "good," or well-differentiated. A summary of these features in the molecular subtypes can be seen in Table 5.1 [3, 4].

Additional classifications have set out to further distinguish between these accepted subtypes. This includes the further subtypes for Luminal B: the HER2+ and HER- subtypes. The distinguishing factor between them is that Ki-67 levels are generally high in HER2+ Luminal B breast cancers [3]. Another distinction is made between Triple-Negative breast cancers (TNBC) and Basal-like breast cancers. Though TNBC (named for being negative for all three receptors) have traditionally been grouped with the Basal-like subtype, they are not synonymous, and there is at most an 80 % overlap between the two [4]. TNBCs have thus been separated into two further subtypes: Basal-like and Non-Basal-like. The major distinction is the expression of Cytokeratins 5 and 6 (CK5/6), as well as Epidermal Growth Factor Receptor (EGFR), for the Basal-like subtype [5].

Other studies have set out to describe additional molecular subtypes, distinct from the accepted five. One proposed subtype is the claudin-low subtype. They are characterized with low expression of the claudin proteins (found within cellular junctions), and are associated with mammary stem-cells [6]. Although similar to the Basal-like subtype in being triple-negative, the claudin-low subtype clinically shows to have a better prognosis that the basal-like subtype.

Table 5.1 Summary of the standard features for each of the 5 molecular subtypes [3, 4]

Molecular subtype	Estrogen (ER)	Progesterone (PR)	HER2	Ki-67	Tumor grade
Luminal A	+ and/or	+	–	Low	Low
Luminal B	+ and/or	+	+/–	High (HER2+)	High
ERBB2+/HER2+	–	–	+	High	High
Basal-like ("Triple-Negative")	–	–	–	High	High
Normal breast-like	Normal	Normal	Normal	Normal	Low

5.4 The Molecular Taxonomic Breast Cancer International Consortium (METABRIC) Classification

Despite the usefulness of the molecular subtype classifications of breast cancer, there are several limitations. One of the major limitations is the apparent lack of understanding of the variation in response to therapies specific to the subtype. Such variation has limited value in a clinical setting, where proper treatment is crucial to patient survival. Because of these limitations, several groups have sought out newer, more reliable studies for means of classifying breast cancers. Among them are the studies done by the Molecular Taxonomic Breast Cancer International Consortium (METABRIC) [7]. The methodology used by this consortium used sequencing technologies that identified the mutational patterns and genomic instabilities characteristic of different breast cancers. This new classification also set out to integrate the classical classifications of breast cancer, describing features such as receptors and tumor grade, as well as direct comparisons with the molecular classifications. In the study, about 2000 breast tumors were analyzed, to acquire both their genomic and transcriptomic sequences, identifying where gene alterations had occurred. This included inherited variation to the genome, specifically Single-Nucleotide Polymorphisms (SNPs) and Copy Number Variants (CNVs), but also looked at acquired variation via Single Nucleotide Variants (SNVs, aka mutations) and Copy Number Aberrations (CNAs) [7]. From this data, they were able to produce ten novel subtypes of breast cancer, which they designated Integrative Clusters (IntClust 1-10). Each cluster was primarily distinguished by the CNAs, which were identified to have the greatest variation, but also found to have overall differing gene expressions. The extent to which the clusters associated with the accepted intrinsic subtypes was analyzed for each cluster, as was the expression of the accepted prognostic receptors of estrogen, progesterone, and HER2. Further analysis identified the clinical implications for each cluster, such as the genomic instabilities, and distinguishing somatic mutations, but also more specific characteristics for each cluster including age of diagnosis and survivability probabilities.

5.5 Genomic Classification Based on the Normal Cell Subtype

Despite the benefits of the newer genomic classifications of breast cancer, alternative means of classifications still arise to confront new or unaddressed issues. Such issues were addressed in a study by Santagata et al. [8]. In their study, they set out to produce a Normal Cell Subtype-based classification system, where they focused on utilizing normal cell types found in normal breast tissue as references for breast cancer classifications. This method, they argue, has successfully been used before to characterize hematopoietic tumors (lymphomas, leukemias, etc.) by other research groups [9], but have rarely been emulated due to a poor understanding of cell-type diversity among tissues. They argue that their new classification, unlike previously produced ones, forms actual disease taxonomy for breast cancers. That is, previous classification systems have heavily relied on differing clinical results (based on different molecular platforms for analysis) to form categories based only on overall prognosis. These categories also vary greatly, no new classification system is truly agreed upon in clinical settings, seeing them as unreliable for patient prognosis and treatments. Their new classification system aimed to provide such clinical reliability. In this classification, identified that breast cancers, being heterogeneous, can vary depending on their cellular origin: in the luminal epithelial layers or the myoepithelial layers [8]. Thus, they analyzed about 15,000 normal breast cells for cellular markers distinguishing between the two layers. They focused on identifying bimodal expression markers (which produced a clear negative/positive distinction), and utilizing these markers to distinguish between varying differentiation states of the cell populations [8]. Three of the major markers identified were hormone receptors of the luminal epithelia: the Vitamin D receptor (VDR), the Androgen receptor (AR), and the Estrogen Receptor (ER). Additional markers included different keratins, claudins, cluster of differentiation (CD) markers, and even Ki-67. Identifying the different expression of these markers in the different cell populations allowed for the formation of eleven luminal layer subtypes (L1-11) and 2 myoepithelial layer subtypes (My1 and My2). Following these classified layers, the study focused on actually classifying human breast tumors based on normal cell types. Four unique subtypes, called "Hormone Receptor Subtypes," were identified: HR0, HR1, HR2, and HR3. Each subtype is based on the expression of the three major hormone receptors (VRD, AR, and ER) and how many were expressed (0 to 3). The previously characterized luminal subtypes were then distinguished based on these novel subtypes. Next, the study looked to identify if breast tumors maintain the same expression patterns characteristic of the normal cell-type, specifically the differentiation-state-specific patterns, this involved identifying the gene expression patterns among the luminal and basal markers (including the three major markers), as well as the specific marker of K5/K14 (found to be a reliable distinguisher between luminal layers. The expression of these markers were identified in ER+, HER2+, and Triple-Negative Breast Cancer (TNBC) tumors, and compared with the expressions found in normal breast tissues with the same distinguished expressions. An example of this comparison can best be seen for the ER+ tumors, where they identified the ER+ tumors to co-express VDR in 93 % of the tumors and AR in 59 % of the tumors, and the K5/K14 were found to be negative in these tumors. When compared to the counterpart normal cells there was found to be a near identical expression pattern: they both co-express VDR and AR to the same levels, and both rarely expressed K5 or K14. Such identical expressions were seen in the HER2+ and TNBC tumors and their counterparts, verifying their Normal Cell Subtype Classification method [8]. Finally, the study aimed to identify the clinical significance behind the new Hormone Receptor subtypes. This involved acquiring tumor data from patients in a separately performed study by the Nurse' Health Study, which had previously been classified as ER+, HER2+, and TNBC based on the presence of the classical receptors (ER, PR, and HER2) within their tumors. These classically assigned tumors were compared with the new HR subtypes, which showed that the HR subtypes provided more distinguished groups of tumors than the previous classification. In addition, the HR subtypes were clinically identified from each other by overall survival and Relapse-free survival, which identified that the HR0 subtype had the worst prognosis and the HR3 subtype had the best prognosis [8].

References

1. Perou CM, Sorlie T et al (2000) Molecular portraits of human breast tumours. Nature 406(6797):747–752
2. Sorlie T, Perou M et al (2001) Gene expression patterns of breast carcinomas distinguish tumor subclasses with clinical implications. Proc Natl Acad Sci U S A 98(5):10869–10874
3. Nishimura R et al (2010) Ki-67 as a prognostic marker according to breast cancer subtype and a predictor of recurrence time in primary breast cancer. Exp Ther Med 1(5):747–754
4. Goldhirsch A et al (2013) Strategies for subtypes—dealing with the diversity of breast cancer: highlights of the St Gallen International Expert Consensus on the Primary Therapy of Early Breast Cancer 2011. Ann Oncol 22(8):1736–1747
5. Nielsen TO, Hsu FD, Jensen K, Cheang M, Karaca G, Hu Z et al (2004) Immunohistochemical and clinical characterization of the basal-like subtype of invasive breast carcinoma. Clin Cancer Res 10(16):5367–5374
6. Prat A, Perou CM (2011) Deconstructing the molecular portraits of breast cancer. Mol Oncol 5(1):5–23
7. Dawson SJ, Rueda OM et al (2013) A new genome-driven integrated classification of breast cancer and its implications. EMBO J 32(5):617–628
8. Santagata S, Thakkar A et al (2014) Taxonomy of breast cancer based on normal cell phenotype predicts outcome. J Clin Invest 124(2):859–870
9. Swerdlow SH (2013) Lymphoma classification and the tools of our trade: an introduction to the 2012 USCAP Long Course. Mod Pathol 26 Suppl 1:S1-S14

Chapter 6
Stem Cells in Breast Cancer

6.1 Introduction

In the mammary gland, DeOme et al. [1] demonstrated that fragments of different parenchymal portions were able to generate fully functional mammary outgrowths in mice, forming ductal and lobulo-alveolar structures composed by epithelial and myoepithelial cells. This seminal work demonstrated that stem cells in adult structures have the ability for self-renewal and for generating a differentiated progeny. The progeny from a single cell might comprise the epithelial population of a fully developed lactating mammary outgrowth in mice was further demonstrated by Kordon and Smith [2]. Russo and coworkers [3–5], demonstrated that cancer, started in terminal end buds (TEBs) present in the mammary gland of young virgin rats. The analysis of these structures by electron microscopy allowed them to characterize their cellular composition based upon cell and nuclear size, nuclear-cytoplasmic ratio, amount of chromatin condensation, electron density of the cytoplasm, number and distribution of organelles, and presence or absence of Mg^{++} and $Na^{+}K^{+}$-dependent ATPases. Based upon these criteria, in addition to myoepithelial cells, three types of epithelial cells were identified: Light, intermediate and dark cells [4, 5]. Dark cells were found to be the predominant type in TEBs, intermediate and myoepithelial cells were present in significantly lower percentages and light cells were only occasionally seen, therefore their percentage was combined with that of intermediate cells. The analysis of the DNA labeling index revealed that all the cell types proliferated, although at different rates, depending upon the type of cells and of their location within the mammary gland tree. Cell proliferation was maximal in intermediate cells located in TEBs, being significantly lower in dark and myoepithelial cells found in the same location. High cell proliferation was associated with greater incorporation of H^3-DMBA, and a progressive dominance of intermediate cells in DMBA-induced intraductal proliferations (IDPs) and in ductal carcinomas [5, 6]. These results indicated that intermediate cells were not only the targets of the carcinogen but also the stem cells of mammary carcinomas. Bennett et al. [7]

J. Russo, *The Pathobiology of Breast Cancer*,
DOI 10.1007/978-3-319-40815-6_6

demonstrated that intermediate cells isolated from DMBA-induced mammary tumors originated two cell types in culture, the dark cell, representing a terminally differentiated cell or a class in transition to differentiation, and intermediate cells, which could represent an undifferentiated, or stem cell, a progenitor of dark and myoepithelial cells. Rudland and coworkers [8] isolated and characterized from the normal rat mammary gland and from DMBA-induced mammary adenocarcinomas epithelial cells that were cuboidal and gave rise to a mixture of cuboidal and spindle-shaped cells resembling fibroblasts. In confluent cultures, cuboidal cells acquired the morphology of a third type of cells, which were dark, polygonal and with many small vacuoles, resembling the dark cells ultrastructurally described by Russo et al. [5]. Chepko and Smith [9] differentiated three division-competent cell populations in the murine mammary epithelium, a subset of "large light cells" structurally and functionally compatible with early stages of secretory differentiation, "small light cells" that were the least differentiated, suggesting that the large light cells were a direct precursor to terminally differentiated cells, both secretory and myoepithelial, confirming the Russo's work [5].

6.2 Cell Markers for Identifying the Stem Cell in the Mammary Gland

Smith et al. [10] utilizing the expression of keratins 6 and 14 in mouse mammary epithelium defined subsets of morphologically distinct luminal mammary epithelial cells with kinetic properties expected for latent mammogenic stem cells. Keratin 6 was confined to a small number of mammary epithelial cells found in the growing end buds and among the luminal epithelium, whereas keratin 14 was expressed in basally located fusiform cells as the myoepithelial cells. Stingl et al. [11, 12] utilized new molecular markers for selecting subpopulations of cells with distinct differentiation potential. They described bipotent human mammary epithelial progenitor cells based on the expression of epithelial specific antigen (ESA), sialomucin 1 (MUC1), common acute lymphoblast antigen (CALLA/CD10,) and alpha-integrin, in combination with exclusion of rhodamine dye. Hebbard et al. [13] observed that CD44, a member of the family of cell surface proteins that is expressed in breast carcinomas, is also expressed in the normal mammary gland. In rodents, CD44 expression is first detected at puberty, and thereafter it is regulated by the estrous cycle; it disappears during lactation, reappearing during involution, suggesting that the expression of this protein is a marker of a stem cell. Novel studies in mice mammary gland [14] have identified stem cells in TEBs and ducts by pulse labeling HC-11 primary mammary epithelial cells with fluorescent TRITC-cell linker membrane label and BrdU, the cells were then transplanted into cleared juvenile syngenic mammary fat pads, in which they were identified as long-lived, label-retaining mammary epithelial cells (LRCs) in mammary ducts that were actively growing or static. This study demonstrated that LRCs are stem cells and their progeny (transitional cells) are arranged as transitional units (TUs) and that both express Zonula Occludens-1 and alpha-catenin proteins, data that suggest that transitional units retain stem cells.

The study of markers for other stem cells has been useful in the identification of mammary stem/progenitor cells. Sca1 (stem cell antigen 1) was first described in mice as a hematopoietic stem cell antigen [15]. Welm et al. [16] detected in the luminal epithelium of mice a Sca1+ cell population that is enriched for functional stem/progenitor cells. These cells are BrdU label retaining, lack expression of differentiation markers, and are progesterone receptor negative. The Sca1+ population also shows "side population" (SP) properties, a characteristic first defined in bone marrow cells [15], as cells with Hoechst dye-effluxing properties that have phenotypic markers of multipotential hematopoietic stem cells. It has been proposed that the protein responsible for that phenotype is breast cancer resistance protein 1 (BCRP1), suggesting that the expression of this protein could serve as a marker for stem cells from various sources [17]. Mammary epithelial cells with SP properties were also identified in human mammary gland. Alvi et al. [18] showed that 0.2–0.45 % of both human and mouse epithelia were formed by distinct SP cells. These cells generated ductal and lobulo-alveolar structures when transplanted into murine cleared mammary fat pads. The SP cells had a high expression of BCRP1, sca1, telomerase catalytic subunit, and low levels of differentiated markers for luminal (epithelial membrane antigen and cytokeratin 19) and myoepithelial cell types (cytokeratin 14). These cells were detected in all human breast samples studied, but their presence was not correlated with age, parity, contraceptive use or day of menstrual cycle. Further investigations identified new markers which may be specific for the human stem/progenitor cells. Gudjonsson et al. [19] isolated a cell line derived from human mammary cells expressing epithelial specific antigen (ESA) and lacking sialomucin (MUC) expression that could give rise to both luminal epithelial and myoepithelial cells in culture. One single ESA+/MUC- cell had the ability of generating a terminal ductal-lobular unit-like structure in basement membrane gel, similar to that formed when the cell line was implanted in mice. In contrast, an ESA+/MUC+ subpopulation was differentiated, luminal epithelial-restricted without stem cell properties.

Dontu and coworkers [20] developed a system to enrich the population of human mammary progenitor/stem cells by culturing them in suspension where they formed "nonadherent mammospheres". These structures were able to differentiate along all three mammary epithelial lineages and to clonally generate complex functional structures in 3D culture systems. Cytological and immuno-cytochemical analysis of secondary mammospheres revealed that these structures contained cells positive for alpha-6 integrin, cytokeratin 5, which was widely expressed, and CD10; ESA-positive and cytokeratin 14-positive cells were less frequently found. Muc 1, alpha-smooth muscle antigen (ASMA), and cytokeratin 18 were not detected. In addition to cells, mammospheres contained extracellular material (ECM). However, immunostains for fibronectin and collagen IV, the classical components of adult gland ECM material were negative, although ~20 % of the mammospheres stained positive for laminin. In contrast, abundant expression of the embryonic ECM components tenascin and decorin, was detected in mammospheres (see [20]). Moreover, the comparison of the genomic profile of undifferentiated cells from mammospheres to that of differentiated cells cultured on collagen identified genes candidates for stem/progenitor cell markers. Some of these genes were already described as involved in stem/progenitor cell-specific functions or regulation of self renewal and abnormal expression of some of them has been correlated with breast cancer development such as proliferation, cell survival and invasion. Recently, new studies showed that the null mutation of the peroxisome proliferator-activated receptor-binding protein (PBP) resulted in defective mammary gland development and in the inability of the mammary epithelial cells to form mammospheres, suggesting a role of PBP in the survival of mammary stem cells or in mammosphere formation process [21].

6.3 Estrogen Receptor as a Marker of Stem Cells in the Mammary Gland

The importance of the role played by the ER alpha in mammary gland development has been highlighted by the development of the ER alpha KO mouse [22]. At birth, the mammary gland of intact animals consists of a rudimentary ductal tree that develops and fills the stroma of the gland in response to increased ovarian estrogen at puberty. The mammary gland of ER alpha KO females does not grow beyond the rudimentary ducts, illustrating the role of estrogens in ductal elongation. The importance of active ductal growth driven by estrogen has been further emphasized by the higher susceptibility of the breast to be transformed during a "high risk" window in the lifespan of a female encompassed between menarche and a first full term pregnancy [6]. This period is characterized by rapid ductal growth and active proliferative activity of the mammary epithelium of Lob 1. These structures are composed of a rapidly proliferating epithelium that has a high content of ER alpha and progesterone receptor (PR) positive cells. With the progressive maturation of Lob 1 to Lob 2, Lob 3, and Lob 4 there is a progressive decrease in the percentage of proliferating cells, a reduction in the percentage of cells positive for steroid hormone receptors, and a reduction in the susceptibility of the cells to be transformed by chemical carcinogens [23]. These data indicate that the stem cells that originate the mammary tree as well as cancerous lesions are located in a specific compartment of the mammary parenchyma, namely the Lob 1 or TDLU; these are the cells that have been called Stem cell 1 by Russo and Russo [24]. Supporting studies by Petersen et al. [25] have shown that a subset of suprabasal breast luminal epithelial cells that are able to generate themselves as well as differentiated luminal epithelial and myoepithelial cells, and to form terminal ductal lobular unit (TDLU)-like structures are distinguished by cytokeratin 19. The suprabasal population of breast stem cells consists of undifferentiated "intermediate" cells with Hoechst dye-effluxing "side population" (SP) properties. These cells lack expression of myoepithelial and luminal apical membrane markers such as CALLA and MUC1. They are rich for ER alpha-positive cells and express several fold higher levels of the ER alpha, p21 (CIP1) and

Msi1 genes than non-SP cells. They also form branching structures in matrigel which included cells of both luminal and myoepithelial lineages. These data suggest a model where scattered steroid receptor-positive cells are stem cells that self-renew through asymmetric cell division and generate patches of transit amplifying and differentiated cells [26, 27]. ERalpha/PR+ breast cancers exhibit loss of the two key regulators of asymmetric cell division, Musashi-1 and Notch-1 and thus they may arise from symmetric division of the ERalpha/PR+ stem cell [26]. These data are supported by the observations of Russo et al. [23] that epithelial cells of the Lob 1 co-express ERalpha, PR and the proliferation marker Ki67, suggesting that these cells could originate ERalpha positive tumors. However, these cells represent less than 1 % of the total cell population, whereas the majority of ERalpha/PR+ cells do not express the proliferation marker, an indication that the cells that contain the receptors are not capable of proliferating. The findings that proliferating cells are different from those that are ERalpha- and PR-positive support data that indicate that estrogen controls cell proliferation by an indirect mechanism. Further support is the finding that when Lob 1 of normal breast tissue are placed in culture, they lose the ERalpha-positive cells, indicating that only proliferating cells that are also ERalpha-negative can survive, representing a type of stem cell that may originate ER negative tumors [23]. The fact that the majority of proliferating breast epithelial cells do not express ERalpha and PR could explain Clayton and coworkers [28] data that cells characterized as human mammary stem cells, present ESA expression, Hoechst dye exclusion, low levels of MUC-1 and CALLA, and lack detectable expression of ER alpha and beta. Cells expressing that phenotype had high cloning efficiency in culture from a single cell, generating mixed colonies containing luminal and myoepithelial cells.

6.4 MCF10F Cells Behave as a Stem Cell In Vitro

The human breast epithelial cell MCF-10F behaves like a stem cell by growing in collagen matrix forming ductular structures mimicking the breast epithelia *in vivo* [29, 30]. cDNA microarray has shown the differential gene expression profile between the cells growing in monolayer and those growing in a tri-dimensional matrix and determining the role of the stroma in the ductulogenic process (Fig. 6.1). There are 161 genes differentially expressed and up-modulated (log mean of difference >2.0, with $p<0.05$) in the ductular structures in comparison to the MCF-10F cells in monolayer [31]. Those genes are related to several biological functions such as gene transcription or regulation of transcription (such as Myeloid cell nuclear differentiation antigen) [31–47], protein biosynthesis (such as Stromal cell-derived factor 2) [48], amino acid transport and membrane trafficking (collagen, type IV, alpha 5 and RAB4, member RAS oncogene family), DNA repair system (such as ADP-ribosyltransferase (NAD+, poly (ADP ribose) polymerase)-like 2) [49–59] and genes related to regulation of cell transformation such as transforming growth factor, beta receptor III. Two of the genes that are highly relevant are the myeloid nuclear differentiation antigen of

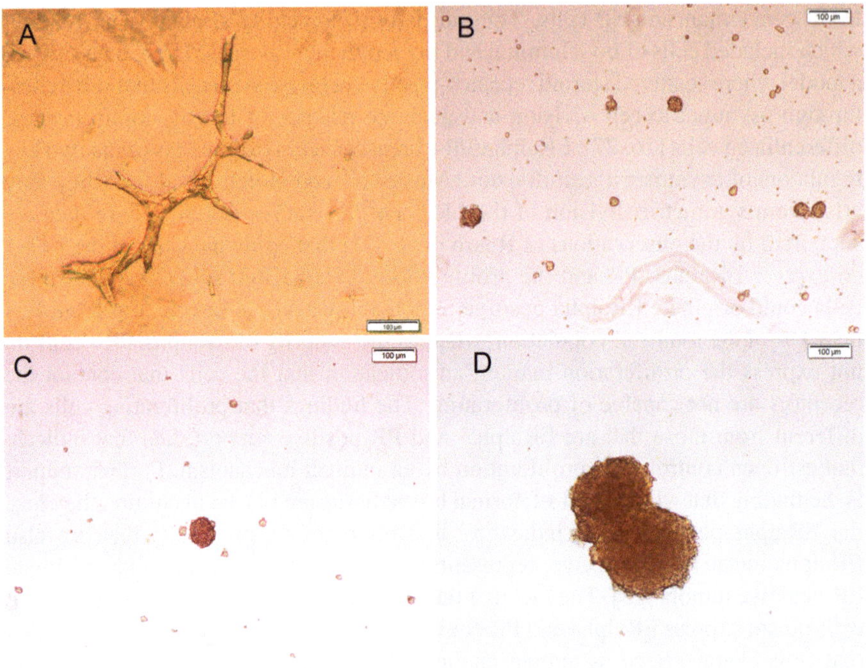

Fig. 6.1 Ducts and mammospheres formed by MCF 10F and cells derived from MCF 10F. (**a**) 3D culture of MCF10F cells in bovine type I collagen. Cells were mixed with collagen and plated onto pre-coated 24-well plate at 1500 cells/well; pictures of structures formed in collagen were acquired after 6 days of culture. One representative duct is shown. (**b**) Mammospheres formed by MCF 10F cells after 1 week of culture. (**c**) Mammospheres formed by trMCF cells after 1 week of culture. (**d**) Mammospheres formed by bsMCF cells after 4 days of culture. Scale bar, 100 µm. Magnification: 40×

MNDA and the ADP-ribosyltransferase (NAD+, poly (ADP ribose) polymerase)-like 2 or PARP. The myeloid nuclear differentiation antigen (MNDA) is expressed in a lineage-specific manner in myeloid cells [44, 46]. MNDA may have an important role in myelomonocytic cell differentiation by exerting an antiproliferative effect on myeloid cell growth. In our specific model the MCF-10F cells express it under the stroma like effect of the collagen matrix and if could be involved in the expression of the organization of the ductal structures. MNDA may be related to the stem cell differentiation process in the human breast epithelial cells.

ADP-ribosyltransferase (NAD+, poly (ADP ribose) polymerase)-like 2 is also upregulated in the ductular structures and is thought to participate in chromatin condensation, maintenance of genomic stability, and the repair of oxidative DNA damage [49]. PARP binds to double and single DNA strand breaks, generated by reactive oxygen species and the DNA-bound repair enzymes during the repair process. Upon binding to strand breaks sites via two zinc fingers in its N-terminal region, PARP's catalytic activity increases 500-fold [50, 51]. PARP catalyzes the transfer of the ADP-ribose moiety of NAD+ onto a host of acceptor proteins such as histones, DNA topoisomerases, p53, DNA-dependent protein kinase, and other DNA binding proteins, including itself, thus forming long branched polymers of

ADP-ribose [52–55]. The high negative charge associated with poly ADP-ribosylation electrostatically repels the modified proteins from DNA and this is thought to clear the damaged site of chromatin and other extraneous proteins and facilitate repair [56, 57]. Although PARP is necessary for the repair of damaged DNA that allows continued cell survival, it is widely recognized that in the face of extensive DNA strand breaks, PARP activation can lead to depletion of NAD+, decreases in intracellular ATP levels, and cell death [58, 59].

6.5 The MCF-10F in Estrogen Induced Carcinogenesis

There is substantial amount of epidemiological, clinical and experimental evidence pointing to estrogens, e.g. 17beta-estradiol (E_2), as being one of the most important etiological factors for the development and progression of breast cancer [36]. To test the transforming ability of estrogens on MCF-10F cells that is ERalpha negative, ERbeta positive and progesterone receptor negative, and for this purpose considered as a Stem cell, were treated twice a week during two weeks with 70 nM 17beta-E_2 [36]. These cell expressed transformation phenotypes such as the formation of colonies in agar methocel, and the loss of the ductulogenic capacity when they grew in a collagen matrix. In order to identify the more aggressive transformed cells capable of forming tumors after E_2 treatment, we have selected the highly invasive populations of MCF-10F cells treated with E_2 in their 9th passage by seeding them in Boyden chambers [29, 30]. Those cells crossing the membrane were collected, expanded, and designated B2, B3, B4, B5, C2, C3, C4 and C5. Four of them, B2, C3, C4 and C5 cells were injected to severe combined immune depressed (SCID) mice. Only C3 and C5 cells were tumorigenic in 2/12 and 9/10 animals injected, respectively [29, 30]. The tumors were poorly differentiated adenocarcinomas, ERalpha and PR negative, and expressed basic keratin of high molecular weight, E-Cadherin, CAM5.2, and vimentin. The genomic profile of C3 and C5 cell analyzed by cDNA microarray, revealed that C5 cells overexpressed more than 5-fold tankyrase (chr 8p23.1), claudin 1 (chr 3q28), homeobox C10 (HOX-C10; chr 12q13.3), and Notch homolog 3 (chr 19p13.12). It also exhibited downregulation of telomeric repeat binding factor 1 (chr 8q21.11) and tumor metastasis suppressor LASS2 (chr 1q21.3) genes. Four tumoral cell lines were obtained from four of the nine tumors derived from C5: C5-A1-T1, C5-A4-T4, C5-A6-T6 and C5-A8-T8 and all of them produced tumors when they were injected to SCID mice. Comparative genomic hybridization (CGH) analysis was performed to identify gains and losses of genetic material in the different cell lines during the tumorigenic process. CGH analyses shown that the cells treated with E_2 have not differences when compared with untreated MCF-10F cells, except a lost in 9p11-13. The four tumors (An1, An 4, An6 and An 8) showed identical pattern of genomic imbalances. CGH analysis showed similar genomic patterns between the four tumors (A1, A4, A6 and A8) and the four cell lines derived from them (C5-A1-T1, C5-A4-T4, C5-A6-T6 and C5-A8-T8); and there was no additional chromosomal alterations after *in vitro* cell culture. All the tumors and derived cell lines showed gains of 1p, 5q15-qter and

8q24.1-qter and losses of chromosome 4, 3p12.3-13, 8p11.1-21, 9p21-pter and 18q24.1. The gain of 8q 24.1 shown in the tumors have also been shown by MCF-10F, the cells treated with E_2 and C5. The parental cell line C5 has shown a tendency for gain of 1p and 5q15-qter and loss of chromosome 4 and it is likely that in the tumors derived from C5, a sub-clone with these changes had a selective advantage *in vivo* and became more distinct. Losses of 3p12.3-13, 8p11.1-21, 9p21-pter and 18q appear to be new changes in the tumor. Interestingly, C5 and the cells treated with E_2 had loss of 9p11-13 while in the tumors the 9p21-pter was lost. The chromosomal alterations that we found *in vitro* are similar to those reported in primary breast cancer. Altogether the data shows both that 17beta-estradiol is able to transform a HBEC and that the MCF-10F has all the properties of a Stem cell that is able to generate a tumor when challenged with the carcinogen agent.

6.6 The Evidence for the Role of Stem Cells in the Pregnancy Preventive Effect During Carcinogenesis

Epidemiological studies in humans and experimental carcinogenesis models have provided wide evidence of the protective effect of pregnancy from breast cancer development [60–69]. It has been postulated [67, 70, 71] that the mechanism of pregnancy-induced protection is mediated by the induction of mammary gland differentiation driven by the hormonal milieu of pregnancy, which creates a specific genomic signature in the mammary gland that makes this organ permanently refractory to carcinogenesis [72, 73]. Alternative explanations attributed the protective effect of pregnancy to changes in the environmental milieu [74] and/or alterations in the immunological profile of the host [64]. A further refinement of the hypothesis of how pregnancy could be affecting cancer susceptibility through induction of differentiation of the mammary gland was first proposed by Russo and Russo [24], who postulated that the Lob 1 and the TEB found in the breast of nulliparous women or of young virgin rats, respectively, had not completed their differentiation into Lob 2, Lob 3 and Lob 4, retaining a high concentration of stem cells called Stem cells 1, which are susceptible to undergo neoplastic transformation when exposed to a carcinogenic agent [24]. After the postmenopausal involution of the mammary gland, the architecture of the parous breast is similar that of the nulliparous breast, containing predominantly Lob 1 composed of Stem cell 2, an epithelial cell population that is refractory to transformation. It was further postulated that the degree of differentiation acquired through early pregnancy permanently changes the "*genomic signature*" that differentiate the Lob 1 from early parous women from that of nulliparous women, shifting the Stem cell 1 to a Stem cell 2 that is refractory to carcinogenesis. These cells were called Stem cell 2 or HTN because after post-lactational involution, the mammary epithelium remains capable of responding with proliferation and differentiation to the stimulus of a new pregnancy; however, these cells are refractory to carcinogenesis, even though they are stimulated to proliferate and to regenerate the whole mammary gland. The Stem cell 2 or HTN is characterized by having a genomic signature that has been induced by the first cycle of differentiation. During the last twenty years, supporting evidence to this hypothesis has been generated. Studies by Smith and coworkers [75–77] using transgenic WAP-driven Cre and Rosa 26-fl-stop-fl-LacZ mice provided evidence of a new mammary epithelial cell population that originates from differentiated cells during pregnancy; 5–10 % of this parity-induced epithelium survives postlactational involution after the first pregnancy. With successive pregnancies, their percentage increases, reaching 60 % of the total epithelium in multiparous females. The parity-induced mammary epithelial cells (PI-MEC) is equivalent to the Stem cell 2 or HTN by Russo et al. [24] since these cells show capacity for self-renewal and contribute to mammary outgrowth in transplantation studies. PI-MEC can function as alveolar progenitors in subsequent pregnancies, and it is thought that they would be related to differences in response to hormonal stimulation and carcinogenic agents observed between nulliparous and parous females [75–77].

The crucial role of the number of mammary stem cells in breast cancer risk has also been postulated by Trichopoulos [78], number that would be reduced through the process of terminal differentiation after the first full-term pregnancy that in certain way is the same idea of shifting the number of Stem cell 1 (EUN) to another more differentiated cells or Stem cell 2 (HTN) postulated earlier by Russo et al. [24]. Several authors have focused in finding molecular changes as a mechanism of the pregnancy-induced protection [79–86]. Russo and coworkers have found that the post-pregnancy involuted rat mammary gland exhibits a genomic signature characterized by elevated expression of genes involved in the apoptotic pathways, such as testosterone repressed prostate message 2 (TRPM2), interleukin 1beta-converting enzyme (ICE), bcl-XL, bcl-XS, p53, p21, and c-myc, which can be from 3 to 5 fold up-regulated [79, 80,]. The activation of programmed cell death genes occurs through a p53-dependent process, modulated by c-myc and with partial dependence on the bcl2-family related genes. In addition, inhibin A and B, heterodimeric non-steroidal secreted glycoproteins with tumor suppressor activity are also upregulated [79, 80, 87, 88]. Genes whose level of expression progressively increases with time of pregnancy reaching their highest levels between 21 and 42 days post-partum are those coding for a fragment of glycogen phosphorylase, AMP activated kinase, bone morphogenetic protein 4 and vesicle-associated protein 1. G/T mismatch-specific thymine DNA glycosylase gene is also increased by five-fold in this model. These data indicate that the activation of genes involved in the DNA repair process is part of the signature induced in the mammary gland by pregnancy. These observations confirm previous findings that *in vivo* the ability of the cells to repair carcinogen-induced damage by unscheduled DNA synthesis and adduct removal is more efficient in the parous and animal mammary gland [70]. In concordance with the studies of Srivastava et al. 79], Siveraman et al. [83] observed that p53 can be implicated in the protective effect of parity, which can be mimicked by treatment of virgin rats with estrogen and progesterone. Studies by Medina et al. [81, 82] in the same hormonal model reported that the function of p53 is required for the hormone-mediated protection of DMBA-induced mammary tumorigenesis in mice. Genomic analysis of the mammary gland of virgin rats treated with estrogen and progesterone at doses that have been reported to mimic pregnancy, showed down-regulation of certain growth-promoting molecules, whereas markers involved in cell cycle control or the modulation of transforming growth factor beta (TGF-beta) signaling pathway were up-regulated in the post-treatment involuted mammary gland [84]. In this study, an unknown noncoding RNA (designated G.B7) and RbAp46, which has been implicated in a number of complexes involving chromatin remodeling, were found to be persistently up-regulated in the lobules of the regressed glands. Using gene profile analysis, D'Cruz and coworkers [86] also observed down regulation of growth factors potentially involved in epithelial proliferation as well as persistent upregulation of TGF-beta3 and several of its transcripts targets in the involuted gland of parous rats and mice. The proposed model of parity-induced specific changes [24] has been further confirmed by Ginger and Rosen [85], who reported that pregnancy induces multiple changes in the mammary epithelial cells, including nuclear accumulation of p53 and induction of whey acidic protein (WAP). During involution, a large component of the epithelium is eliminated through apoptosis, and a specific subpopulation of epithelial cells survives this process. The involuted mammary gland has persistent changes in gene expression, nuclear localization of p53, and an altered proliferative capacity in response to a carcinogen. Pregnancy would induce epigenetic changes, such as chromatin remodeling, DNA methylation/demethylation, and histone modifications, affecting cell fate in the parous mammary gland. As it has been previously published [31] all the genes that have been related to the Stem cell 2 or HTN seems to work in a different functional pathways than those described for the Stem cell 1 or EUN.

Collectively, the data described above present evidence that pregnancy, through the process of cell differentiation, shifts the Stem cell 1or EUN to Stem cell 2 or HTN cells that exhibit a specific genomic signature that could be responsible for the refractoriness of the mammary gland to carcinogenesis.

6.7 Isolation of the Stem Cells from the Rat Mammary Gland

The technique for isolating rat mammary stem cells was published elsewhere [89]. The methodology was crucial to understand the effect of hormones mainly human chorionic gonadotropin on the stem cell of the rat mammary gland [89].

Isolated the stem cells of the rat mammary gland of animals treated with hCG for 21 days and allowed to rest another additional 21 days and compared if there was a difference in mammospheres formation between the control and hCG treated cells have shown that the mammary epithelial cells from rats treated with hCG formed significantly less primary mammospheres supporting the concept in reduction of stem cells [89]. Immunofluorescence staining the mammospheres were centrifuged on glass slides and fixed with acetone-methanol (1:1) for 10 minutes at −20 °C. The staining of 2D differentiation assay was performed in 3 cm dishes. The following primary antibodies were used: ESA, CK14, CK18 and β-casein. Alexa fluor 555 donkey anti-rabbit, alexa fluor 488 donkey anti-mouse and alexa fluor 647 donkey anti-goat secondary antibodies were used to detect the staining. Primary mammospheres reacted with ESA (EpCAM) that is a marker of bipotent progenitor cells and differentiated luminal cells and also reacted with CK14 that is attributed to differentiated myoepithelial cells or to putative progenitor cells. Of interest is that the primary mammospheres are positive for CD24 that is significant reduced in the mammary gland of hCG treated animals. Immunostaining with CK14 showed CK14 is expressed in some of the cells in primary mammospheres from hCG treated rats, whereas is hard to observed in the spheres from control. Primary mammospheres-derived cells are capable of differentiation into different types of colonies under differentiating conditions. Using ESA, CK14 and beta casein as markers of cell differentiation it was found that there are more colonies with three lineages differentiation after hCG treatment . Of interest is the finding that hCG induces the formation of more colonies containing myoepithelial cells [89].

Comparison of the transcriptomic profile of the stems cells of the mammary gland from the control and hCG treated animals demonstrated by both Microarray and real time RT-PCR that CD24 is significantly reduced in the mammospheres from hCG treated rats [89]. MME (CD10) is significantly reduced in mammospheres from hCG treated rats based on the real time RT-PCR result, whereas is not significantly changed by microarray analysis (FDR 0.138). Krt (CK14) showed the trend of up-regulation in the mammospheres from hCG treated rats by microarray and real time RT-PCR analysis but did not reach the statistical significance ($P=0.086$, FDR 0.269). Another myoepithelial marker, ASMA, is also shown the trend of up-regulation in the mammospheres from hCG treated rats by microarray but did not reach the statistical significance ($P=0.032$, FDR 0.164). Analysis of genes showed in the literatures highly specific for myoepithelial cells and down-regulated in DCIS myoepithelial cells (by microarray) are up-regulated in the mammospheres from hCG treated rats. Among them are the adamalysin-thrombospondin (ADAMTS) proteinases that have been implicated in various cellular events, including cleavage of proteoglycans, extracellular matrix degradation, inhibition of angiogenesis, gonadal development, and organogenesis. The expression of ADAMTS1 in breast tissues is higher in the stromal fibroblast and myoepithelial cells. It is poorly expressed in luminal epithelial cells. Real time RT-PCR showed ADAMTS1 is significantly down regulated in invasive breast carcinoma compare to non-neoplastic mammary tissue [90]. ADAMTS1 is cancer-specific hypermethylated in colorectal tumors [91], and is expression is decreased in prostate cancer, and might be involved in the early steps of prostate cancer development [92].

IGFBP-3 had no direct inhibitory effect on breast cancer cells Hs578T but could accentuate apoptosis induced by the physiological trigger ceramide in an IGF-independent manner [93] and IGFBP-3 expression in MCF7 cells leads to the induction of apoptosis through the activation of caspases involved in a death receptor-mediated pathway and that IGFBP-3 functions as a negative regulator of breast cancer cell growth, independent of the IGF-IGF receptor axis [94]. Another gene that is upregulated in myoepithelial cells is Tropomyosin-1 that functions as a suppressor of transformation [95], and has been shown can resensitize breast cancer cells to anoikis [96]. Altogether mRNA expression profile showed that some genes specific for myoepithelial cells, such as Adamts1, Cyr61, Igfbp3 and Tpm1, are significantly up-regulated in the primary mammospheres from hCG treated rats. Adamts1, Igfbp3 and Tpm1 have been shown down-regulated in invasive breast cancer. Canonical pathways analyses of microarray data showed that pathways involved in immune system are down-regulated after hCG treatment, which is consistent to data from rat mammary gland tissues of in vivo experiment [97].

6.8 Role of the Mammary Gland Stem Cell in the Prevention of Breast Cancer

A comparative studies of human and rodent mammary development in relation to age and parity [71, 98–102] have revealed that the development of the mammary gland is a progressive process of growth initiated at childhood [103, 104]. Epidemiological studies in humans and experimental carcinogenesis models have provided wide evidence of the protective effect of pregnancy from breast cancer development [55–61, 105–107]. Russo and coworkers [70, 106, 108] have postulated that the mechanism of pregnancy-induced protection is mediated by the induction of mammary gland differentiation driven by the hormonal milieu of pregnancy, which creates a specific genomic signature in the mammary gland that makes this organ permanently refractory to carcinogenesis. A further refinement of the hypothesis of how pregnancy could be affecting cancer susceptibility through induction of differentiation of the mammary gland was first proposed by Russo and Russo [24], who postulated that the Lob 1 and the TEB found in the breast of nulliparous women or of young virgin rats, respectively, had not completed their differentiation into Lob 2, Lob 3 and Lob 4, retaining a high concentration of S/PC or Stem cell 1 or EUN, which is susceptible to undergo neoplastic transformation when exposed to a carcinogenic agent [24]. Although after the postmenopausal involution of the mammary gland the architecture of the breast in parous women is similar to that of the nulliparous women, both containing predominantly Lob 1, in parous women the Lob 1 is composed of Stem cells 2 or HTN, an epithelial cell population that through the completion of differentiation induced by pregnancy has permanently changed its "*genomic signature*" and becomes refractory to transformation [72, 73]. Nevertheless, the Stem cell 2 remains capable of responding with proliferation and differentiation to the stimulus of a new pregnancy. The Stem cell 2 of HTN is characterized by a genomic signature that has been induced by the first cycle of differentiation. The crucial role of the number of mammary stem cells in breast cancer risk has been postulated by Trichopoulos [105] but the concept that their number would be reduced through the process of terminal differentiation after the first full-term pregnancy has been indicated previously [24]. *However, an important concept in our study is that in the process of differentiation is not only the number of stem cells but the shift of stem cell 1 or EUN to stem cell 2 or HTN by changing the genomic and epigenomic profile of these cells, as it has been recently reported to be the case in the breast of postmenopausal parous women* [72, 73]. Several authors have focused in finding molecular changes as a mechanism of the pregnancy-induced protection. Russo and coworkers have found that the post-pregnancy involuted mammary gland exhibits a specific genomic signature as well as epigenetic changes, such as chromatin remodeling, DNA methylation/demethylation, and histone modifications that affect cell fate in the breast of parous women [72, 73]. *Collectively, the data described above present evidence that pregnancy, through the process of cell differentiation, shifts the Stem cell 1 or EUN to Stem cell 2 or HTN. These cells exhibit a specific genomic signature that could be responsible for the refractoriness of the mammary gland to carcinogenesis* [72, 73].

References

1. Deome KB, Faulkin LJ Jr, Bern HA, Blair PB (1959) Development of mammary tumors from hyperplastic alveolar nodules transplanted into gland-free mammary fat pads of female C3H mice. Cancer Res 19:515–520
2. Kordon EC, Smith GH (1998) An entire functional mammary gland may comprise the progeny from a single cell. Development 125:1921–1930
3. Russo J, Saby J, Isenberg WM, Russo IH (1977) Pathogenesis of mammary carcinomas induced in rats by 7,12-dimethylbenz(a)anthracene. J Natl Cancer Inst 59:436–445
4. Russo J, Isenberg W, Ireland M, Russo I (1976) Ultrastructural changes in the mammary epithelial cell population during neoplastic development induced by a chemical carcinogen. Electron Microscopy Society of America Proceedings, pp 250–251
5. Russo J, Tait L, Russo IH (1983) Susceptibility of the mammary gland to carcinogenesis. III. The cell of origin of rat mammary carcinoma. Am J Pathol 113:50–66
6. Russo J, Tay LK, Russo IH (1982) Differentiation of the mammary gland and susceptibility to carcinogenesis. Breast Cancer Res Treat 2:5–73
7. Bennett DC, Peachey LA, Durbin H, Rudland PS (1978) A possible mammary stem cell line. Cell 15:283–298
8. Rudland PS, Bennett DC, Warburton MJ (1980) Isolation and characterization of epithelial stem-cell cell lines from the rat mammary gland. Br J Cancer 41:666–668
9. Chepko G, Smith GH (1997) Three division-competent, structurally-distinct cellpopulations contribute to murine mammary epithelial renewal. Tissue Cell 29:239–253
10. Smith GH, Mehrel T, Roop DR (1990) Differential keratin gene expression in developing, differentiating, preneoplastic, and neoplastic mouse mammary epithelium. Cell Growth Differ 1:161–170
11. Stingl J, Eaves CJ, Kuusk U, Emerman JT (1998) Phenotypic and functional characterization in vitro of a multipotent epithelial cell present in the normal adult human breast. Differentiation 63:201–213
12. Stingl J, Eaves CJ, Zandieh I, Emerman JT (2001) Characterization of bipotent mammary epithelial progenitor cells in normal adult human breast tissue. Breast Cancer Res Treat 67:93–109

13. Hebbard L, Steffen A, Zawadzki V, Fieber C, Howells N, Moll J, Ponta H, Hofmann M, Sleeman J (2000) CD44 expression and regulation during mammary gland development and function. J Cell Sci 113(Pt 14):2619–2630
14. Kenney NJ, Smith GH, Lawrence E, Barrett JC, Salomon DS (2001) Identification of stem cell units in the terminal end bud and duct of the mouse mammary gland. J Biomed Biotechnol 1:133–143
15. Spangrude GJ, Aihara Y, Weissman IL, Klein J (1988) The stem cell antigens Sca-1 and Sca-2 subdivide thymic and peripheral T lymphocytes into unique subsets. J Immunol 141:3697–3707
16. Welm BE, Tepera SB, Venezia T, Graubert TA, Rosen JM, Goodell MA (2002) Sca-1(pos) cells in the mouse mammary gland represent an enriched progenitor cell population. Dev Biol 245:42–56
17. Zhou S, Schuetz JD, Bunting KD, Colapietro AM, Sampath J, Morris JJ, Lagutina I, Grosveld GC, Osawa M, Nakauchi H, Sorrentino BP (2001) The ABC transporter Bcrp1/ABCG2 is expressed in a wide variety of stem cells and is a molecular determinant of the side-population phenotype. Nat Med 7:1028–1034
18. Alvi AJ, Clayton H, Joshi C, Enver T, Ashworth A, Vivanco MM, Dale TC, Smalley MJ (2003) Functional and molecular characterisation of mammary side population cells. Breast Cancer Res 5:R1–R8
19. Gudjonsson T, Villadsen R, Nielsen HL, Ronnov-Jessen L, Bissell MJ, Petersen OW (2002) Isolation, immortalization, and characterization of a human breast epithelial cell line with stem cell properties. Genes Dev 16:693–706
20. Dontu G, Abdallah WM, Foley JM, Jackson KW, Clarke MF, Kawamura MJ, Wicha MS (2003) In vitro propagation and transcriptional profiling of human mammary stem/progenitor cells. Genes Dev 17:1253–1270
21. Jia Y, Qi C, Zhang Z, Zhu YT, Rao SM, Zhu YJ (2005) Peroxisome proliferator-activated receptor-binding protein null mutation results in defective mammary gland development. J Biol Chem 280:10766–10773
22. Couse JF, Korach KS (1999) Estrogen receptor null mice: what have we learned and where will they lead us? Endocr Rev 20:358–417
23. Russo J, Ao X, Grill C, Russo IH (1999) Pattern of distribution of cells positive for estrogen receptor alpha and progesterone receptor in relation to proliferating cells in the mammary gland. Breast Cancer Res Treat 53:217–227
24. Russo J, Russo IH (1997) Role of differentiation in the pathogenesis and prevention of breast cancer. Endocr Relat Cancer 4:7–21
25. Petersen OW, Gudjonsson T, Villadsen R, Bissell MJ, Ronnov-Jessen L (2003) Epithelial progenitor cell lines as models of normal breast morphogenesis and neoplasia. Cell Prolif 36(Suppl 1):33–44
26. Clarke RB, Anderson E, Howell A, Potten CS (2003) Regulation of human breast epithelial stem cells. Cell Prolif 36(Suppl 1):45–58
27. Clarke RB, Spence K, Anderson E, Howell A, Okano H, Potten CS (2005) A putative human breast stem cell population is enriched for steroid receptor-positive cells. Dev Biol 277:443–456
28. Clayton H, Titley I, Vivanco M (2004) Growth and differentiation of progenitor/stem cells derived from the human mammary gland. Exp Cell Res 297:444–460
29. Russo J, Fernandez SV, Russo PA, Fernbaugh R, Sheriff FS, Lareef HM, Garber J, Russo IH (2006) 17 beta estradiol induces transformation and tumorigenesis in human breast epithelial cells. FASEB J. 20:1622–1634
30. Huang Y, Fernandez S, Goodwin S, Russo PA, Russo IH, Sutter T, Russo J (2007) Epithelial to mesenchymal transition in human breast epithelial cells transformed by 17- beta- estradiol. Cancer Res 67:11147–11157
31. Russo J, Balogh GA, Chen J, Fernandez SV, Fernbaugh R, Heulings R, Mailo DA, Moral R, Russo PA, Sheriff F, Vanegas JE, Wang R, Russo IH (2006) The concept of stem cell in the

mammary gland and its implication in morphogenesis, cancer and prevention. Front Biosci 11:151–172
32. Shim GJ, Wang L, Andersson S, Nagy N, Kis LL, Zhang Q, Makela S, Warner M, Gustafsson JA (2003) Disruption of the estrogen receptor beta gene in mice causes myeloproliferative disease resembling chronic myeloid leukemia with lymphoid blast crisis. Proc Natl Acad Sci U S A 100:6694–6699
33. Imamov O, Morani A, Shim GJ, Omoto Y, Thulin-Andersson C, Warner M, Gustafsson JA (2004) Estrogen receptor beta regulates epithelial cellular differentiation in the mouse ventral prostate. Proc Natl Acad Sci U S A 101:9375–9380
34. Pais V, Leav I, Lau KM, Jiang Z, Ho SM (2003) Estrogen receptor-beta expression in human testicular germ cell tumors. Clin Cancer Res 9:4475–4482
35. Palmieri C, Saji S, Sakaguchi H, Cheng G, Sunters A, O'Hare MJ, Warner M, Gustafsson JA, Coombes RC, Lam EW (2004) The expression of oestrogen receptor (ER)-beta and its variants, but not ERalpha, in adult human mammary fibroblasts. J Mol Endocrinol 33:35–50
36. Russo J, Hasan Lareef M, Balogh G, Guo S, Russo IH (2003) Estrogen and its metabolites are carcinogenic agents in human breast epithelial cells. J Steroid Biochem Mol Biol 87:1–25
37. Hu YF, Lau KM, Ho SM, Russo J (1998) Increased expression of estrogen receptor beta in chemically transformed human breast epithelial cells. Int J Oncol 12:1225–1228
38. Russo J, Lareef MH, Tahin Q, Hu YF, Slater C, Ao X, Russo IH (2002) 17Beta-estradiol is carcinogenic in human breast epithelial cells. J Steroid Biochem Mol Biol 80:149–162
39. Lareef MH, Garber J, Russo PA, Russo IH, Heulings R, Russo J (2005) The estrogen antagonist ICI-182-780 does not inhibit the transformation phenotypes induced by 17-beta-estradiol and 4-OH estradiol in human breast epithelial cells. Int J Oncol 26:423–429
40. Fernandez SV, Russo IH, Lareef M, Balsara B, Russo J (2005) Comparative genomic hybridization of human breast epithelial cells transformed by estrogen and its metabolites. Int J Oncol 26:691–695
41. Kehrl JH (1995) Hematopoietic lineage commitment: role of transcription factors. Stem Cells 13:223–241
42. Shivdasani RA, Orkin SH (1995) Erythropoiesis and globin gene expression in mice lacking the transcription factor NF-E2. Proc Natl Acad Sci U S A 92:8690–8694
43. Tsai S, Bartelmez S, Sitnicka E, Collins S (1994) Lymphohematopoietic progenitors immortalized by a retroviral vector harboring a dominant-negative retinoic acid receptor can recapitulate lymphoid, myeloid, and erythroid development. Genes Dev 8:2831–2841
44. Briggs JA, Burrus GR, Stickney BD, Briggs RC (1992) Cloning and expression of the human myeloid cell nuclear differentiation antigen: regulation by interferon alpha. J Cell Biochem 49:82–92
45. Briggs R, Dworkin L, Briggs J, Dessypris E, Stein J, Stein G, Lian J (1994) Interferon alpha selectively affects expression of the human myeloid cell nuclear differentiation antigen in late stage cells in the monocytic but not the granulocytic lineage. J Cell Biochem 54:198–206
46. Briggs RC, Briggs JA, Ozer J, Sealy L, Dworkin LL, Kingsmore SF, Seldin MF, Kaur GP, Athwal RS, Dessypris EN (1994) The human myeloid cell nuclear differentiation antigen gene is one of at least two related interferon-inducible genes located on chromosome 1q that are expressed specifically in hematopoietic cells. Blood 83:2153–2162
47. Briggs RC, Kao WY, Dworkin LL, Briggs JA, Dessypris EN, Clark J (1994) Regulation and specificity of MNDA expression in monocytes, macrophages, and leukemia/B lymphoma cell lines. J Cell Biochem 56:559–567
48. Hamada T, Tashiro K, Tada H, Inazawa J, Shirozu M, Shibahara K, Nakamura T, Martina N, Nakano T, Honjo T (1996) Isolation and characterization of a novel secretory protein, stromal cell-derived factor-2 (SDF-2) using the signal sequence trap method. Gene 176:211–214
49. Lindahl T, Satoh MS, Poirier GG, Klungland A (1995) Post-translational modification of poly(ADP-ribose) polymerase induced by DNA strand breaks. Trends Biochem Sci 20:405–411

50. Menissier-de Murcia J, Molinete M, Gradwohl G, Simonin F, de Murcia G (1989) Zinc-binding domain of poly(ADP-ribose)polymerase participates in the recognition of single strand breaks on DNA. J Mol Biol 210:229–233
51. Boulikas T (1991) Relation between carcinogenesis, chromatin structure and poly(ADP-ribosylation) (review). Anticancer Res 11:489–527
52. Althaus FR, Hofferer L, Kleczkowska HE, Malanga M, Naegeli H, Panzeter PL, Realini CA (1994) Histone shuttling by poly ADP-ribosylation. Mol Cell Biochem 138:53–59
53. Malanga M, Pleschke JM, Kleczkowska HE, Althaus FR (1998) Poly(ADP-ribose) binds to specific domains of p53 and alters its DNA binding functions. J Biol Chem 273:11839–11843
54. Pleschke JM, Kleczkowska HE, Strohm M, Althaus FR (2000) Poly(ADP-ribose) binds to specific domains in DNA damage checkpoint proteins. J Biol Chem 275:40974–40980
55. Zahradka P, Ebisuzaki K (1982) A shuttle mechanism for DNA-protein interactions. The regulation of poly(ADP-ribose) polymerase. Eur J Biochem 127:579–585
56. Le Rhun Y, Kirkland JB, Shah GM (1998) Cellular responses to DNA damage in the absence of Poly(ADP-ribose) polymerase. Biochem Biophys Res Commun 245:1–10
57. Cosi C, Marien M (1999) Implication of poly (ADP-ribose) polymerase (PARP) in neurodegeneration and brain energy metabolism. Decreases in mouse brain NAD+ and ATP caused by MPTP are prevented by the PARP inhibitor benzamide. Ann N Y Acad Sci 890:227–239
58. Ha HC, Snyder SH (2000) Poly(ADP-ribose) polymerase-1 in the nervous system. Neurobiol Dis 7:225–239
59. Pieper AA, Verma A, Zhang J, Snyder SH (1999) Poly (ADP-ribose) polymerase, nitric oxide and cell death. Trends Pharmacol Sci 20:171–181
60. MacMahon B, Cole P, Lin TM, Lowe CR, Mirra AP, Ravnihar B, Salber EJ, Valaoras VG, Yuasa S (1970) Age at first birth and breast cancer risk. Bull World Health Organ 43:209–221
61. Lambe M, Hsieh CC, Chan HW, Ekbom A, Trichopoulos D, Adami HO (1996) Parity, age at first and last birth, and risk of breast cancer: a population-based study in Sweden. Breast Cancer Res Treat 38:305–311
62. Kelsey JL, Gammon MD, John EM (1993) Reproductive factors and breast cancer. Epidemiol Rev 15:36–47
63. Moon RC (1981) Influence of pregnancy and lactation on experimental mammary carcinogenesis. In: Pike MC, Siiteri PK, Welsch CW (eds) Banbury Report 8 hormones and breast cancer. Cold Spring Harbor Laboratory Press, Cold Spring Harbor, pp 353–394
64. Sinha DK, Pazik JE, Dao TL (1988) Prevention of mammary carcinogenesis in rats by pregnancy: effect of full-term and interrupted pregnancy. Br J Cancer 57:390–394
65. Yang J, Yoshizawa K, Nandi S, Tsubura A (1999) Protective effects of pregnancy and lactation against N-methyl-N-nitrosourea-induced mammary carcinomas in female Lewis rats. Carcinogenesis 20:623–628
66. Welsch CW (1985) Host factors affecting the growth of carcinogen-induced rat mammary carcinomas: a review and tribute to Charles Brenton Huggins. Cancer Res 45:3415–3443
67. Russo J, Russo IH (1980) Influence of differentiation and cell kinetics on the susceptibility of the rat mammary gland to carcinogenesis. Cancer Res 40:2677–2687
68. Swanson SM, Whitaker LM, Stockard CR, Myers RB, Oelschlager D, Grizzle WE, Juliana MM, Grubbs CJ (1997) Hormone levels and mammary epithelial cell proliferation in rats treated with a regimen of estradiol and progesterone that mimics the preventive effect of pregnancy against mammary cancer. Anticancer Res 17:4639–4645
69. Rajkumar L, Guzman RC, Yang J, Thordarson G, Talamantes F, Nandi S (2001) Short-term exposure to pregnancy levels of estrogen prevents mammary carcinogenesis. Proc Natl Acad Sci U S A 98:11755–11759
70. Tay LK, Russo J (1981) Formation and removal of 7,12-dimethylbenz[a]anthracene--nucleic acid adducts in rat mammary epithelial cells with different susceptibility to carcinogenesis. Carcinogenesis 2:1327–1333
71. Russo IH, Koszalka M, Russo J (1991) Comparative study of the influence of pregnancy and hormonal treatment on mammary carcinogenesis. Br J Cancer 64:481–484

72. Russo J, Santucci-Pereira J, de Cicco RL, Sheriff F, Russo PA, Peri S, Slifker M, Ross E, Mello ML, Vidal BC, Belitskaya-Levy I, Arslan A, Zeleniuch-Jacquotte A, Bordas P, Lenner P, Ahman J, Afanasyeva Y, Hallmans G, Toniolo P, Russo IH (2011) Pregnancy-induced chromatin remodeling in the breast of postmenopausal women. Int J Cancer. doi:10.1002/ijc.27323
73. Belitskaya-Levy I, Zeleniuch-Jacquotte A, Russo J, Russo IH, Bordas P, Ahman J, Afanasyeva Y, Johansson R, Lenner P, Li X, de Cicco-Lopez RL, Peri S, Ross E, Russo PA, Santucci-Pereira J, Sheriff FS, Slifker M, Hallmans G, Toniolo P, Arslan AA (2011) Characterization of a genomic signature of pregnancy identified in the breast. Cancer Prev Res 4:1457–1464
74. Thordarson G, Jin E, Guzman RC, Swanson SM, Nandi S, Talamantes F (1995) Refractoriness to mammary tumorigenesis in parous rats: is it caused by persistent changes in the hormonal environment or permanent biochemical alterations in the mammary epithelia? Carcinogenesis 16:2847–2853
75. Wagner KU, Boulanger CA, Henry MD, Sgagias M, Hennighausen L, Smith GH (2002) An adjunct mammary epithelial cell population in parous females: its role in functional adaptation and tissue renewal. Development 129:1377–1386
76. Henry MD, Triplett AA, Oh KB, Smith GH, Wagner KU (2004) Parity-induced mammary epithelial cells facilitate tumorigenesis in MMTV-neu transgenic mice. Oncogene 23:6980–6985
77. Boulanger CA, Wagner KU, Smith GH (2005) Parity-induced mouse mammary epithelial cells are pluripotent, self-renewing and sensitive to TGF-beta1 expression. Oncogene 24:552–560
78. Trichopoulos D, Lagiou P, Adami HO (2005) Towards an integrated model for breast cancer etiology: the crucial role of the number of mammary tissue-specific stem cells. Breast Cancer Res 7:13–17
79. Srivastava P, Russo J, Russo IH (1997) Chorionic gonadotropin inhibits rat mammary carcinogenesis through activation of programmed cell death. Carcinogenesis 18:1799–1808
80. Srivastava P, Russo J, Russo IH (1999) Inhibition of rat mammary tumorigenesis by human chorionic gonadotropin associated with increased expression of inhibin. Mol Carcinog 26:10–19
81. Medina D, Kittrell FS (2003) p53 function is required for hormone-mediated protection of mouse mammary tumorigenesis. Cancer Res 63:6140–6143
82. Medina D (2004) Breast cancer: the protective effect of pregnancy. Clin Cancer Res 10:380S–384S
83. Sivaraman L, Conneely OM, Medina D, O'Malley BW (2001) p53 is a potential mediator of pregnancy and hormone-induced resistance to mammary carcinogenesis. Proc Natl Acad Sci U S A 98:12379–12384
84. Ginger MR, Gonzalez-Rimbau MF, Gay JP, Rosen JM (2001) Persistent changes in gene expression induced by estrogen and progesterone in the rat mammary gland. Mol Endocrinol 15:1993–2009
85. Ginger MR, Rosen JM (2003) Pregnancy-induced changes in cell-fate in the mammary gland. Breast Cancer Res 5:192–197
86. D'Cruz CM, Moody SE, Master SR, Hartman JL, Keiper EA, Imielinski MB, Cox JD, Wang JY, Ha SI, Keister BA, Chodosh LA (2002) Persistent parity-induced changes in growth factors, TGF-beta3, and differentiation in the rodent mammary gland. Mol Endocrinol 16:2034–2051
87. Russo J, Russo IH (2004) Endocrine control of breast development. In: Russo J, Russo IH (eds) Molecular basis of breast cancer: prevention and treatment, 1st edn. Springer, Berlin, pp 64–67
88. Russo J, Russo IH (1997) Role of hCG and inhibin in breast cancer. Int J Cancer 4:297–306
89. Santucci-Pereira J, Russo J (2016) Biological basis of breast cancer. In: Russo J (ed) The pathobiology of breast cancer, chapter 10
90. Porter S, Scott SD, Sassoon EM, Williams MR, Jones JL, Girling AC, Ball RY, Edwards DR (2004) Dysregulated expression of adamalysin-thrombospondin genes in human breast carcinoma. Clin Cancer Res 10:2429–2440

91. Lind GE, Kleivi K, Meling GI, Teixeira MR, Thiis-Evensen E, Rognum TO, Lothe RA (2006) ADAMTS1, CRABP1, and NR3C1 identified as epigenetically deregulated genes in colorectal tumorigenesis. Cell Oncol 28:259–272
92. Gustavsson H, Wang W, Jennbacken K, Welen K, Damber JE (2009) ADAMTS1, a putative anti-angiogenic factor, is decreased in human prostate cancer. BJU Int 104:1786–1790
93. Gill ZP, Perks CM, Newcomb PV, Holly JM (1997) Insulin-like growth factor-binding protein (IGFBP-3) predisposes breast cancer cells to programmed cell death in a non-IGF-dependent manner. J Biol Chem 272:25602–25607
94. Kim HS, Ingermann AR, Tsubaki J, Twigg SM, Walker GE, Oh Y (2004) Insulin-like growth factor-binding protein 3 induces caspase-dependent apoptosis through a death receptor-mediated pathway in MCF-7 human breast cancer cells. Cancer Res 64:2229–2237
95. Mahadev K, Raval G, Bharadwaj S, Willingham MC, Lange EM, Vonderhaar B, Salomon D, Prasad GL (2002) Suppression of the transformed phenotype of breast cancer by tropomyosin-1. Exp Cell Res 279:40–51
96. Bharadwaj S, Thanawala R, Bon G, Falcioni R, Prasad GL (2005) Resensitization of breast cancer cells to anoikis by tropomyosin-1: role of Rho kinase-dependent cytoskeleton and adhesion. Oncogene 24:8291–8303
97. Russo J, Ruso IH (eds) (2013) Methodological Approach to studying the human breast. In: Role of the trasnscriptome in breast cancer prevention. Springer, New York, pp 243–268
98. Russo J, Russo IH (1978) DNA labeling index and structure of the rat mammary gland as determinants of its susceptibility to carcinogenesis. J Natl Cancer Inst 61:1451–1459
99. Russo IH, Russo J (1978) Developmental stage of the rat mammary gland as determinant of its susceptibility to 7,12-dimethylbenz[a]anthracene. J Natl Cancer Inst 61:1439–1449
100. Russo J, Gusterson BA, Rogers AE, Russo IH, Wellings SR, van Zwieten MJ (1990) Comparative study of human and rat mammary tumorigenesis. Lab Invest 62:244–278
101. Russo J (1983) Basis of cellular autonomy in the susceptibility to carcinogenesis. Toxicol Pathol 11:149–166
102. Russo J, Russo IH (1998) Role of pregnancy and chorionic gonadotropin in breast cancer prevention. In: Birkhauser MH, Rozenbaum H (eds) Proc IV European Congress on Menopause. ESKA, Paris, pp 133–142
103. Russo J, Russo IH (2004) Biological and molecular basis of breast cancer. Springer Verlag, Heidelberg
104. Russo J, Rivera R, Russo IH (1992) Influence of age and parity on the development of the human breast. Breast Cancer Res Treat 23:211–218
105. Vessey MP, McPherson K, Roberts MM, Neil A, Jones L (1985) Fertility in relation to the risk of breast cancer. Br J Cancer 52:625–628
106. Kelsey JL, Horn-Ross PL (1993) Breast cancer: magnitude of the problem and descriptive epidemiology. Epidemiol Rev 15:7–16
107. Russo I, Russo J (1994) Role of hCG and inhibin in breast-cancer (review). Int J Oncol 4:297–306
108. Russo J, Balogh GA, Russo IH (2008) Full-term pregnancy induces a specific genomic signature in the human breast. Cancer Epidemiol Biomarkers Prev 17:51–66

Chapter 7
The Mechanisms of Breast Cancer Metastasis

7.1 Introduction

Primary tumors are responsible for only 10 % of deaths from cancer and most all primary breast carcinomas do not compromise survival while they are confined to the breast. Breast carcinoma spreads by direct invasion, by the lymphatic and by the blood vessel route [1]. Breast cancers often disseminate metastatic colonies in many tissues throughout the body, including the brain, liver, bones, and lungs. Some of these metastases are already present at the time of diagnosis, and others become manifest clinically months or years, or decades after the initial therapy [2]. The whole process from primary to metastasis has been called invasion-metastasis cascade [3].

7.2 Route of Metastasis

The primary breast tumor can invade locally in the parenchyma, nipple, skin fascia and adjacent muscle of the chest wall (Fig. 7.1). According to Rosai [4] the degree of local invasion is generally greater in invasive lobular carcinoma and its variants, presumably aided by the lack of E-cadherin in the tumor cells. In a study performed by Rosen [5] in 18 mastectomies for carcinoma measuring less than 1 cm, residual invasive carcinoma was found in 11 % and residual in situ carcinoma in an additional 22 %. In the reported literature [6–8] nipple invasion has been found in 23 % to 31 % of all clinically detectable invasive carcinomas. As depicted in Fig. 7.1 the lymph node stations typically involved with metastatic breast carcinoma are the axilla and the internal mammary region, with the supraclavicular. Axillary node metastases [4] are present in 40 % to 50 % of clinically detectable cases and are divided into levels according to their topographic relation with the insertion of the pectoralis minor muscle: low or proximal,

J. Russo, *The Pathobiology of Breast Cancer*,
DOI 10.1007/978-3-319-40815-6_7

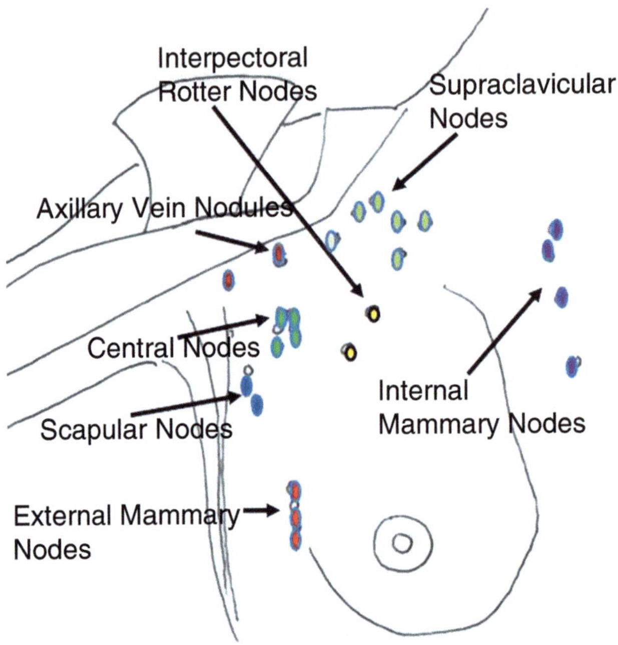

Fig. 7.1 Routes of breast cancer metastasis

medium, and high or distal. As reported by Veronesi and collaborators [9]. Supraclavicular lymph node involvement is present approximately in 20 % of patients with axillary lymph node involvement but is almost zero in cases with negative axillae. Twenty two percent [10] of breast cancer metastasize in the internal mammary chain.

7.3 The Mechanism of Metastasis

Carcinoma in situ are growing in the confined of the ductules and when loss of the basement membrane (BM) takes place the dissemination of cells from the primary tumor define the invasive process (see Chaps. 2 and 4). It is well known that even before carcinoma cells breach the basement membrane, they often succeed in stimulating angiogenesis on the stromal side of the membrane and when they are in the stromal compartment they gain direct access to the blood and lymphatic vessels.

This invasion into vessels is often termed intravasation. The invasiveness-and associated intravasation-by the cancer cells is stimulated by epidermal growth factor (EGF) released by their macrophage partners. Importantly, counts of the density of these triads in histopathological sections of human breast cancers provide a strong prognostic factor of eventual metastatic relapse of the cancer patients [3]. The blood, in particular, represents an actively hostile environment for metastasizing cancer cells. These circulating tumor cells (CTCs) have been the objects of intensive investigation in recent years [11]. A variety of techniques are being developed to measure the concentrations of CTCs in the circulation of cancer patients [12]. There are clear indications that the levels of CTCs in patients with metastatic breast cancer provide some indication of clinical progression [11, 12] (see Sect. 7.6). CTCs may persist for only a short time in the circulation, in large part because, unlike red and white blood cells, they are ill suited for passaging through micro vessels in various tissues [3]. Once lodged in the blood vessels of various tissues, cancer cells must extravasate. Close associations of disseminated cancer cells and macrophages have been documented at sites of extravasation, suggesting that, as is the case with intravasation, cancer cells recruit macrophages to help them escape from the circulation into the tissue parenchyma [3].

Metastasizing cancer cells may form micro metastases that are small clumps of disseminated cancer cells also termed colonization. More than 30 % of breast cancer patients harbor hundreds, likely thousands of micro metastases in their bone marrow at the time of initial clinical presentation, but only half of the women in this group will ever develop metastatic disease [13]. Using antibodies reactive with cytokeratins is possible to detect micro metastasis in lymph node (LN) (Fig. 7.2). Using cytokeratin-specific antibodies make it possible to detect a single-cell micro metastasis among 105 or even 106 surrounding mesenchymal cells in the blood, bone marrow, or lymph node. Larger micro metastases can often be detected in the lymph nodes that are connected with a primary tumor via draining lymphatic ducts (Fig. 7.3) [13].

The number of disseminated tumor cells in the marrow, often termed DTCs, seems to represent a far more useful prognostic marker than the concentration of circulating tumor cells (CTCs) in the blood. This may reflect the fact that the number of DTCs represents the accumulation of disseminated cancer cells over an extended period of time, whereas the concentration of CTCs may be dictated by complex kinetic processes governing their lifetimes-and thus their *steady-state* concentration in the circulation [11, 12].

7.4 The Lymphatic Vessels as a Path for Metastatic Dissemination

Distant metastases from the primary breast cancer are seen most commonly in the skeletal system, lung and pleura, liver, ovary, adrenal gland, and central nervous system [14–16]. Invasive lobular carcinoma has a particular tendency to metastasize to the abdominal cavity, particularly to the gastrointestinal tract, ovaries, and serosal surfaces [17–19].

The pattern of metastatic spread of breast carcinoma as evaluated by Fisher et al. [20] in a large randomized series of patients treated with various modalities brought them to the following conclusions: "*there is no orderly pattern of tumor dissemination; regional nodes are ineffective as barriers to tumor spread and, when positive, are more an indicator of a particular host-tumor relationship than the instigator of dis-*

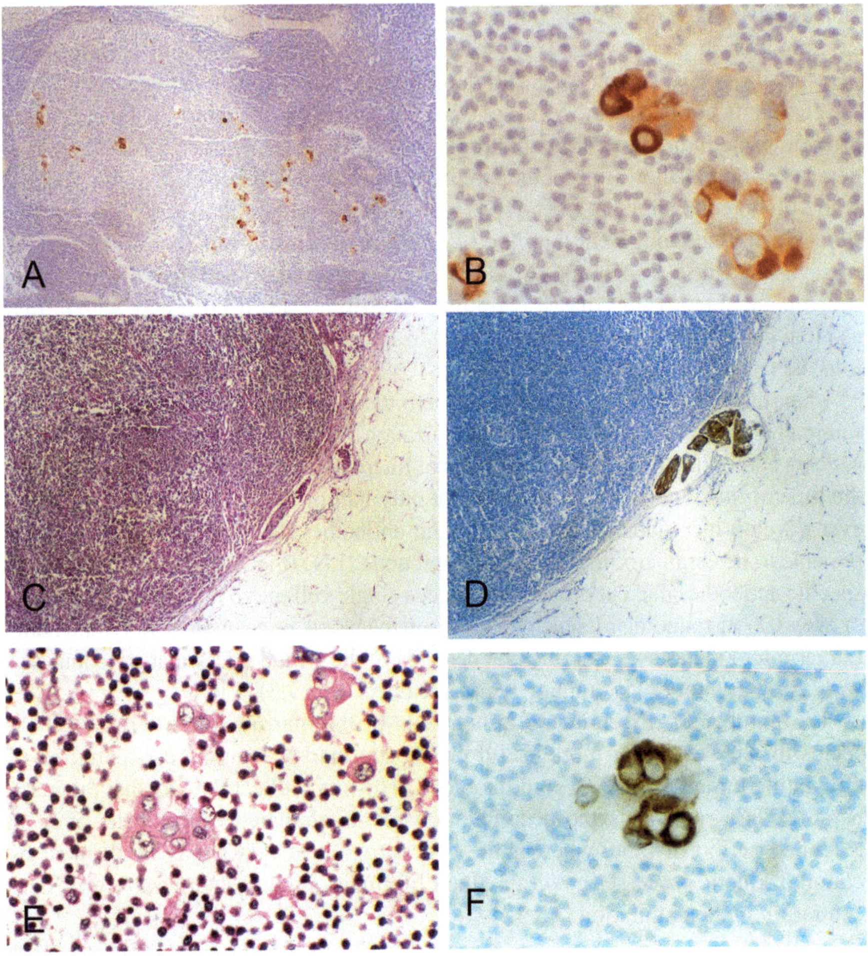

Fig. 7.2 (**a**) Lymph node presenting isolated and small cluster of breast cancer cells stained with cytokeratin, 4×. (**b**) Higher magnification of breast cancer cells stained with keratin, 40×. (**c**) Cluster of neoplastic cells trapped in the capsule of the lymph node, H&E, 4×; (**d**) Similar image than the one depicted in (**c**) but stained with keratin. (**e**) Higher magnification of breast cancer cells in the lymph node, H&E staining, 40×. (**f**) Same as (**e**) but immune-stained with an antibody against keratin, 40×

tant metastases; the bloodstream is of considerable importance in tumor dissemination; complex host-tumor interrelationships affect every facet of the disease; operable breast carcinoma is a systemic disease; and variations in local-regional therapy are unlikely to substantially affect survival".

The technique of sentinel lymph node biopsy for the evaluation and management of breast carcinoma [21–26] is based on the concept that if the sentinel node is negative, the other nodes of that group will also be negative in nearly all instances, whereas if it is positive, the chance that there will be additional metastases in that

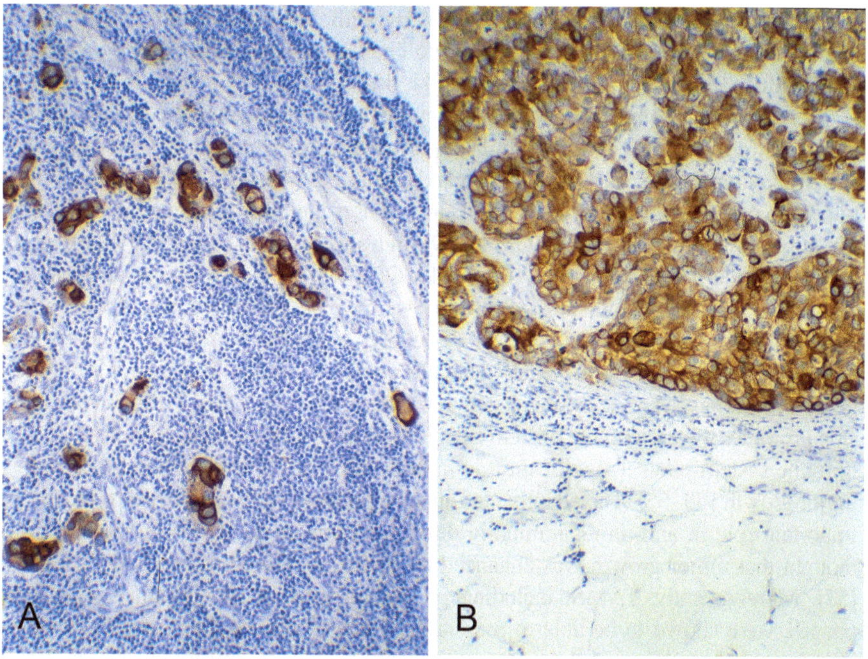

Fig. 7.3 (**a**) Micro-metastatic foci of breast cancer cells in the Lymph node immuno-stained with an antibody against keratin, 20×. (**b**) Most of the lymph node is occupied by metastatic breast cancer cells immuno-stained with an antibody against keratin, 40×

nodal group is about one third. The study of the sentinel node that proves negative on frozen section should include at least three step sections stained with H&E plus at least one section immunostained for keratin [27–30]. The immunostain of choice is a keratin cocktail, such as AEl/AE3 [31] (Figs. 7.2 and 7.3]. Cluster of metastatic cells in sentinel lymph node highlighted with keratin stain. Diagnosis of metastatic carcinoma on the basis of keratin-positive cells that is not evident in the H&E preparations needs to be very carefully considered.

7.5 The Concept of Stem Cells and Metastasis

The concept of the cancer stem cells (CSC) (see Chap. 6) was first hypothesized in human acute myeloid leukemia (AML) [32, 33] and this concept has been extended to many solid tumors, and in particular breast cancer [34–37]. CSCs play a key role in not only the original tumorigenicity but also in their ability for local invasion and migration [38–40] exhibiting the ability to metastasize to specific parts of the body [41–43]. In 2003 it was reported that breast cancer can originate from BCSCs [44]. The authors identified and isolated small subset of cells within primary breast cancer cells of which a few cancer cells were able to form palpable tumors in the mammary fat pad of non-obese diabetic/severe combined immunodeficient (NOD/SCID) mice. Such cells express CD44 or CD44 with epithelial specific antigen (ESA), but not CD24, consistent with the phenotypic characteristics of mammary stem cells with multi-potent differentiation ability [44, 45].

The tumor microenvironment plays a significant role in regulating chemo resistance and radiation resistance via inflammatory cytokines to stimulate CSC self-renewal, which may promote tumor proliferation and metastasis [42, 46]. Among them, Wnt is known to cause self-renewal of CSCs at early phases and plays a significant role in the initiation and maintenance of CSCs [47–49]. In a small patient cohort exhibiting tumor metastasis, the population of EPAM+CD44+CD47+MET+ correlated with increased metastasis and low overall survival [50]. As we have described in Chap. 3, the epithelial–mesenchymal transition (EMT) is a phenotypic process converting polarized and adjacent epithelial cells to mesenchymal cells conferring motile and migratory properties [51] and may serve as a critical step to the metastatic processes. HER2 overexpression promotes the enrichment of stem cells in both normal and malignant cells [52]. The immune system plays a significant role in preventing tumor initiation and controlling tumor growth [53–55]. The adaptive immune response is well established to exhibit an important role in anti-tumor immunity through macrophages [56]. Previously, it was thought that tumor growth was enhanced by tumor-associated macrophages (TAMS) [57]. More recently, TAMS, including pro-tumorigenic or anti-tumorigenic macrophages, were shown to be able to present antigens, produce inflammatory cytokines, initiate angiogenesis, and tolerate cytotoxic activity [58]. In physiological conditions, however, the macrophage phagocytosis is significant for clearing damaged or foreign cells. Pro-phagocytic signals on the target cells enable macrophage-mediated phagocytic engulfment and clearance. Tumors including breast cancer may evade TAMS through the expression of anti-phagocytic signals, including CD47 and CD200 [59]. CD47 is a widely distributed membrane protein that interacts with the signal-regulatory protein α (SIRPα), an inhibitory receptor on myeloid cells that gives a so-called do-not-eat-me signal in tumor cells. Thus, multiple strategies may be developed to modulate CD47-SIRPα signaling in reducing the aggressiveness of BCSCs.

7.6 Role of Circulating Tumor Cells in Breast Cancer Metastasis

Detection of CTCs has shown an association of CTC counts with clinical outcome in breast cancer [60–62]. The FDA-cleared Cell Search® system (Veridex LLC) has become the gold standard [63–65]. Due to the heterogeneity between both the primary tumor and its metastases as well as between different metastases of an individual patient [66, 67], molecular CTC analysis could serve as an easily accessible liquid biopsy for metastatic disease. More than 260 studies are listed in clinicaltrials.gov lists, which explore the utility of CTC detection, enumeration and targeted molecular analysis. The most reliable data have been provided by the CellSearch® system [60] and analysis of diagnostic leukapheresis samples [68]. The CellSearch® system is the only FDA-cleared CTC detection and enumeration device currently available and has therefore been used in the majority of studies. Its usefulness in breast cancer has been established [62], although its dependence on positive selection of EpCAM-positive CTCs may lead to false-negative or false-low CTC results in some patients [66]. By enrichment-independent analysis of leukapheresis samples, however, it was shown that CTC numbers are generally low ranging from 1 to 15 cells per ml of blood [68, 69] at least for breast cancer. There is considerable efforts in the scientific community to adjust standard molecular methods to few or even single cells.

7.7 Metastasis of Breast Cancer to Specific Sites

The propensity to develop metastases depends on tumor size, histopathological grade, and lymph node status. Development of distant metastases usually occurs in 10–15 % of all breast cancer patients within 3 years of the detection of the primary tumor [70]. However, some breast cancers show a trend for late metastatic recurrence even 10 years or more after initial diagnosis [70]. The skeleton, liver, lung, and brain account for the most common sites affected by breast cancer cell colonization. Metastatic lesions are found at the highest frequency in bone (83 %) whereas liver and lung are usually affected to a lesser extent (27 %) [71]. The median survival after diagnosis of bone metastases is 24–40 months compared to 3 months in patients with liver metastasis [72]. Overt bone metastases are classified according to their radiographic appearance as either osteolytic or osteoblastic. Both types of lesions result from an imbalance of the normal bone remodeling process. Osteolytic lesions show an increase in bone resorption caused by activated osteoclasts, while compensatory bone formation is impaired [73]. In contrast, osteoblastic lesions are characterized by disorganized new bone formation and insufficient bone resorption [73]. This classification represents two extremes, as patients can have mixed lesions containing both osteolytic and osteoblastic elements. The majority of breast cancer-related bone metastases are characterized as osteolytic, whereas approximately 15 % are of osteoblastic or mixed entity [74, 75].

Current treatment options for cancer-related bone disease are rarely curative, and advanced pain management is often the major treatment avenue. Palliative treatment with antiresorptive drugs, such as bisphosphonates and the antireceptor activator of nuclear factor kappa-B ligand (RANKL) antibody Denosumab have been found to reduce the frequency of skeletal-related events [76, 77].

7.8 Role of P53 and Metastasis

One of the master regulators of metastasis is p53, which directly controls the transcription of genes that are involved in canonical metastasis pathways by activation of downstream target genes, including *CDKN1A*, *PCNA*, *GADD45*, *BAX*, *NOXA*, *MDM2*, and *miR-34a* which are responsible for inducing cell adhesion, motility, invasion, EMT, stemness, ECM interactions, and anoikis [78]. *TP53* mutation is associated with poor prognosis in breast cancer [79]. P53 is thought to inhibit metastasis by transcriptionally regulating targets that are implicated in key metastasis pathways [80–86]. *TP53* is mutated in about 40 % of all breast cancers [87]. This rate varies among subtypes, with the highest frequency in basal-like (80 %) and HER2-enriched (72 %) subtypes and the lowest in the Luminal A (12 %) and Luminal B (29 %) subtypes [87]. Restoring p53 function in established tumors leads to tumor regression [88–91]. However, incomplete tumor regression was observed when p53 was reactivated in late-stage tumors [88]. One therapeutic strategy has been to develop small-molecule inhibitors of MDM2 to restore the function of the p53 pathway [92] and the adenoviral delivery of WT p53 cDNA (Advexin) [59].

7.9 Problems with the Treatment of Metastasis

A major obstacle facing systemic treatment of advanced or metastasis breast cancer is drug resistance, which leads to treatment failure and finally mortality [93]. This problem has driven research to look for other alternative mechanisms in due that more than half of the protein coding genes are predicted to be modulated by miRNAs, which makes it contribute to nearly all the physiological and pathological processes [94]. The association of miRNAs with tumor biology was first determined assuming observed deletions and down-regulation of miR-15 and miR-16 in B-cell chronic lymphocytic leukemia [95]. It has been identified eight differentially expressed miRNAs in breast cancer tissues as compared to normal breast tissue, including miR-200b, miR-200c, miR-21, miR-378, let-7a, miR-320, miR-23a, and miR-22 [96], therefore miRNAs is a new promising molecule to be targeted against breast cancer and the developed drug resistance in metastatic disease. Five members of miRNA-200 family (miR-200a, miR200b, miR200c, miR-141 and miR-429) and miR-205 were markedly downregulated in cells that had undergone epithelial mesenchymal transition or EMT (see Chap. 3). EMT was prevented in response to TGF-β by enforced expression of miR-200 [97]. miRNA deregulation is implicated in drug resistance to chemotherapeutic agents, for example miR-221 and miR-222 overexpression contributes to tamoxifen resistance through negative regulation of ER-α, whereas knockdown of miR-221 and/or miR-222 restores ER-α expression and tamoxifen sensitivity [98]. miR-451 and miR-27 were involved in resistance of MCF-7 breast cancer cells to chemotherapeutic drug doxorubicin mediated by MDR-1 [99, 100]. Both miR-125a and miR-125b function as tumor suppressors in SKRB3 cells, a HER2-overexpressing human breast cancer cell line, by suppressing HER2 mRNA and protein levels, which results in reduced cell growth, motility and invasiveness [101, 102]. miRNA 17/20 downregulates cell migration and invasion through inhibiting the secretion of a subset of cytokines and secreted plasminogen activators [103]. miR-17-5p has extensive complementarities to the mRNA of AIB1 (named for "amplified in breast cancer 1"). Overexpression of miR-17-5P in MCF-7 cells induced the invasive and migratory abilities by targeting HBP1/b–catenin pathway [104]. miR-31 prevents metastasis at multiple steps by inhibiting the expression of pro-metastatic genes including RhoA [105]. miR-31 alters both the capacity of cancer cell to exit from a primary tumor and the ability of already disseminated neoplastic cells to survive in the foreign microenvironment in the site of metastasis. Metastasis requires the formation of new blood vessels (angiogenesis) which is driven by the overexpression of angiogenic factors such as the vascular endothelial growth factor (VEGF). miR-126 expression is lost in primary breast carcinoma of patients who relapse, however, when it is restored tumor growth and proliferation are reduced. Moreover miR-126 plays a role in angiogenesis through targeting VEGF [106]. miR-151-5p up-regulation may suppress metastasis in primary breast tumors while miR-9 levels were elevated in invasive lobular carcinomas as compared with invasive ductal carcinomas [107], whereas miR-23b regulates focal adhesion, cell spreading, cell–cell junctions and the formation of lamellipodia in breast cancer, suggesting its central role for it in cytoskeletal dynamics [108]. All these examples emphasize the therapeutic potential of miRNAs in breast cancer treatment.

References

1. Chaubert P, Hurlimann J (1992) Mammary origin of metastases. Immunohistochemical determination. Arch Pathol Lab Med 116:1181–1188
2. Brinkley O, Haybittle JL (1975) The curability of breast cancer. Lancet 2:95–97
3. Weinberg RA (2014) The biology of cancer, 2nd edn. Garland Science, Taylor and Francis Group, New York
4. Rosai J (2004) Breast. Chapter 20. In: Rosai and Ackerman's Surgical Pathology. Mosby, New York, pp 1802–1818
5. Rosen PP, Fracchia AA, Urban JA, Schottenfeld D, Robbins GF (1975) "Residual" mammary carcinoma following simulated partial mastectomy. Cancer 35:739–747
6. Lagios MD, Gates EA, Westdahl PR, Richards V, Alpert BS (1979) A guide to the frequency of nipple involvement in breast cancer. A study of 149 consecutive mastectomies using a serial subgross and correlated radiographic technique. Am J Surg 138:135–142
7. Morimoto T, Komaki K, Inui I, Umemoto A, Yamamoto H, Harada K, Inoue K (1985) Involvement of nipple and areola in early breast cancer. Cancer 55:2459–2463
8. Wertheim U, Ozzello L (1980) Neoplastic involvement of nipple and skin flap in carcinoma of the breast. Am J Surg Pathol 4:543–549
9. Veronesi U, Cascinelli N, Bufalino R, Morabito A, Greco M, Galluzzo D, Donne VD, DeLellis R, Piotti P, Sacchini V, Conti R, Clemente C (1983) Risk of internal mammary lymph node metastases and its relevance on prognosis of breast cancer patients. Ann Surg 198:681–684
10. Donegan WL (1977) The influence of untreated internal mammary metastases upon the course of mammary cancer. Cancer 39:533–538
11. Giuliano M, Giordano A, Jackson S, De Giorgi U, Mego M, Cohen EN, Gao H, Anfossi S, Handy BC, Ueno NT, Alvarez RH, De Placido S, Valero V, Hortobagyi GN, Reuben JM, Cristofanilli M (2014) Circulating tumor cells as early predictors of metastatic spread in breast cancer patients with limited metastatic dissemination. Breast Cancer Res 16(5):440
12. Toss A, Mu Z, Fernandez S, Cristofanilli M (2014) CTC enumeration and characterization: moving toward personalized medicine. Ann Transl Med 2(11):108
13. Weigelt B, Peterse JL, van't Veer LJ (2005) Breast cancer Metastasis markers and models. Nat Rev Cancer 5:591–602

14. Cifuentes N, Pickren JW (1979) Metastases from carcinoma of mammary gland. An autopsy study. J Surg Oncol 11:193–205
15. Lamovec J, Zidar A (1991) Association of leptomeningeal carcinomatosis in carcinoma of the breast with infiltrating lobular carcinoma. An autopsy study. Arch Pathol Lab Med 115:507–510
16. Merrill CF, Kaufman DI, Dimitrov NV (1991) Breast cancer metastatic to the eye is a common entity. Cancer 68:623–627
17. Cohn M, Middleton L, Valero V, Sahin A (2003) Gastrointestinal metastases of carcinoma of the breast (Abstract). Mod Pathol 16:26a
18. Gagnon Y, Tetu B (1989) Ovarian metastases of breast carcinoma. A clinicopathologic study of 59 cases. Cancer 64:892–898
19. Merino MJ, Li Volsi VA (1981) Signet ring carcinoma of the female breast. A clinicopathologic analysis of 24 cases. Cancer 48:1830–1837
20. Fisher B, Montague E, Redmond C, Barton B, Borland D, Fisher ER, Deutsch M, Schwarz G, Margolese R, Donegan W, Volk H, Honvolinka C, Gardner B, Cohen I Jr, Lesnick G, Cruz AB, Lawrence W, Nealon T, Butcher H, Lawton R (1977) Comparison of radical mastectomy with alternative treatment s for primary breast cancer. A first report of results from a prospective randomized clinical trial. Cancer 39:2827–2839
21. Jani AB, Basu A, Heimann R, Hellman S (2003) Sentinel lymph node versus axillary lymph node dissection for early-stage breast carcinoma: a comparison using a utility-adjusted number needed to treat analysis. Cancer 97:359–366
22. Krag D, Weaver D, Ashikaga T, Moffat F, Klimberg VS, Shriver C, Feldman S, Kusminsky R, Gadd M, Kuhn J, Harlow S, Beitsch P (1998) The sentinel node in breast: a multicenter validation study. N Engl J Med 339:941–946
23. McMasters KM, Giuliano AE, Ross MI, Reintgen DS, Hunt KK, Byrd DR, Klimberg VS, Whitworth PW, Tafra LC, Edwards MJ (1998) Sentinel-lymph-node biopsy for breast cancer -not yet the standard of care. N Engl J Med 339:990–995
24. Reintgen D, Giuliano R, Cox CE (2000) Sentinel node biopsy in breast cancer: an overview. Breast J 6:299–305
25. Schwartz GF, Giuliano AE, Veronesi U, Consensus Conference Committee (2001) Proceedings of the consensus conference on the role of sentinel lymph node biopsy in carcinoma of the breast, April 19 to 22, Philadelphia, Pennsylvania. Hum Pathol 2002(33): 579–589
26. Veronesi U, Paganelli G, Viale G, Galimberti V, Luini A, Zurrida S, Robertson C, Sacchini V, Veronesi P, Orvieto E, de Cicco C, Intra M, Tosi G, Scarpa D (1999) Seiltinellymph node biopsy and axillary dissection in breast cancer: results in a large series. J Natl Cancer Inst 91:368–373
27. Creager AJ, Geisinger KR (2002) Intraoperative evaluation of sentinel lymph nodes for breast carcinoma: current methodologies. Adv Anat Pathol 9:233–243
28. Freneaux P, Nos C, Vincent-Salomon A, Genin P, Sigal-Zafrani B, Al Ghuzian A, Birolini MJ, Clough K, Sastre-Garau X (2002) Histological detection of minimal metastatic involvement in axillary sentinel nodes: a rational basis for a sensitive methodology. Mod Pathol 15:641–646
29. Silverberg SG (2002) Sentinel node processing: recommendations for pathologists. Am J Surg Pathol 26:383–385
30. Tumer RR, Ollila DW, Stern S, Giuliano AE (1999) Optimal histopathologic examination of the sentinel lymph node for breast carcinoma staging. Am J Surg Pathol 23:263–267
31. Cohen C, Alazraki N, Styblo T, Waldrop SM, Grant SF, Larsen T (2002) Immunohistochemical evaluation of sentinel lymph nodes in breast carcinoma patients. Applmmunohistochem Mol Morphol 10:296–303
32. Bonnet D, Dick JE (1997) Human acute myeloid leukemia is organized as a hierarchy that originates from a primitive hematopoietic cell. Nat Med 3:730–737
33. Qian Z, Fernald AA, Godley LA, Larson RA, Le Beau MM (2002) Expression profiling of CD34+ hematopoietic stem/progenitor cells reveals distinct subtypes of therapy-related acute myeloid leukemia. Proc Natl Acad Sci U S A 99:14925–14930

34. Wicha MS (2007) Breast cancer stem cells: the other side of the story. Stem Cell Rev 3:110–112
35. Vlashi E, Kim K, Lagadec C, Donna LD, McDonald JT, Eghbali M, Sayre JW, Stefani E, McBride W, Pajonk F (2009) In vivo imaging, tracking, and targeting of cancer stem cells. J Natl Cancer Inst 101:350–359
36. Cho RW, Wang X, Diehn M, Shedden K, Chen GY, Sherlock G, Gurney A, Lewicki J, Clarke MF (2008) Isolation and molecular characterization of cancer stem cells in MMTV-Wnt-1 murine breast tumors. Stem Cells 26:364–371
37. Geng SQ, Alexandrou AT, Li JJ (2014) Breast cancer stem cells: multiple capacities in tumor metastasis. Cancer Lett 349(1):1–7. doi:10.1016/j.canlet.2014.03.036, Epub 2014 Apr 12
38. Clarke MF (2005) A self-renewal assay for cancer stem cells. Cancer Chemother Pharmacol 56(Suppl. 1):64–68
39. Moncharmont C, Levy A, Gilormini M, Bertrand G, Chargari C, Alphonse G, Ardail D, Rodriguez-Lafrasse C, Magne N (2012) Targeting a cornerstone of radiation resistance: cancer stem cell. Cancer Lett 322:139–147
40. Chen D, Bhat-Nakshatri P, Goswami C, Badve S, Nakshatri H (2013) ANTXR1, a stem cell-enriched functional biomarker, connects collagen signaling to cancer stem-like cells and metastasis in breast cancer. Cancer Res 73:5821–5833
41. Liu H, Patel MR, Prescher JA, Patsialou A, Qian D, Lin J, Wen S, Chang YF, Bachmann MH, Shimono Y, Dalerba P, Adorno M, Lobo N, Bueno J, Dirbas FM, Goswami S, Somlo G, Condeelis J, Contag CH, Gambhir SS, Clarke MF (2010) Cancer stem cells from human breast tumors are involved in spontaneous metastases in orthotopic mouse models. Proc Natl Acad Sci U S A 107:18115–18120
42. Sansone P, Storci G, Tavolari S, Guarnieri T, Giovannini C, Taffurelli M, Ceccarelli C, Santini D, Paterini P, Marcu KB, Chieco P, Bonafe M (2007) IL-6 triggers malignant features in mammospheres from human ductal breast carcinoma and normal mammary gland. J Clin Invest 117:3988–4002
43. Wicha MS, Liu S, Dontu G (2006) Cancer stem cells: an old idea–a paradigm shift. Cancer Res 66:1883–1890, Discussion 1895–1886
44. Al-Hajj M, Wicha MS, Benito-Hernandez A, Morrison SJ, Clarke MF (2003) Prospective identification of tumorigenic breast cancer cells. Proc Natl Acad Sci U S A 100:3983–3988
45. Van Pham P, Vu NB, Duong TT, Nguyen TT, Truong NH, Phan NL, Vuong TG, Pham VQ, Nguyen HM, Nguyen KT, Nguyen NT, Nguyen KG, Khat LT, Van Le D, Truong KD, Phan NK (2012) Suppression of human breast tumors in NOD/SCID mice by CD44 shRNA gene therapy combined with doxorubicin treatment. Onco Targets Ther 5:77–84
46. Ginestier C, Liu S, Diebel ME, Korkaya H, Luo M, Brown M, Wicinski J, Cabaud O, Charafe-Jauffret E, Birnbaum D, Guan JL, Dontu G, Wicha MS (2010) CXCR1 blockade selectively targets human breast cancer stem cells in vitro and in xenografts. J Clin Invest 120:485–497
47. Jamieson CH, Ailles LE, Dylla SJ, Muijtjens M, Jones C, Zehnder JL, Gotlib J, Li K, Manz MG, Keating A, Sawyers CL, Weissman IL (2004) Granulocyte–macrophage progenitors as candidate leukemic stem cells in blast-crisis CMLN. Engl J Med 351:657–667
48. Ginestier C, Hur MH, Charafe-Jauffret E, Monville F, Dutcher J, Brown M, Jacquemier J, Viens P, Kleer CG, Liu S, Schott A, Hayes D, Birnbaum D, Wicha MS, Dontu G (2007) ALDH1 is a marker of normal and malignant human mammary stem cells and a predictor of poor clinical outcome. Cell Stem Cell 1:555–567
49. Abrahamsson AE, Geron I, Gotlib J, Dao KH, Barroga CF, Newton IG, Giles FJ, Durocher J, Creusot RS, Karimi M, Jones C, Zehnder JL, Keating A, Negrin RS, Weissman IL, Jamieson CH (2009) Glycogen synthase kinase 3beta missplicing contributes to leukemia stem cell generation. Proc Natl Acad Sci U S A 106:3925–3929
50. Baccelli I, Schneeweiss A, Riethdorf S, Stenzinger A, Schillert A, Vogel V, Klein C, Saini M, Bauerle T, Wallwiener M, Holland-Letz T, Hofner T, Sprick M, Scharpff M, Marme F, Sinn HP, Pantel K, Weichert W, Trumpp A (2013) Identification of a population of blood circulating tumor cells from breast cancer patients that initiates metastasis in a xenograft assay. Nat Biotechnol 31:539–544

51. Thiery JP, Acloque H, Huang RY, Nieto MA (2009) Epithelial–mesenchymal transitions in development and disease. Cell 139:871–890
52. Duru N, Fan M, Candas D, Menaa C, Liu HC, Nantajit D, Wen Y, Xiao K, Eldridge A, Chromy BA, Li S, Spitz DR, Lam KS, Wicha MS, Li JJ (2012) HER2-associated radioresistance of breast cancer stem cells isolated from HER2-negative breast cancer cells. Clin Cancer Res 18:6634–6647
53. Chao MP, Majeti R, Weissman IL (2012) Programmed cell removal: a new obstacle in the road to developing cancer. Nat Rev Cancer 12:58–67
54. Lamagna C, Aurrand-Lions M, Imhof BA (2006) Dual role of macrophages in tumor growth and angiogenesis. J Leukoc Biol 80:705–713
55. Fang LY, Izumi K, Lai KP, Liang L, Li L, Miyamoto H, Lin WJ, Chang C (2013) Infiltrating macrophages promote prostate tumorigenesis via modulating androgen receptor-mediated CCL4–STAT3 signaling. Cancer Res 73:5633–5646
56. Jaiswal S, Chao MP, Majeti R, Weissman IL (2010) Macrophages as mediators of tumor immunosurveillance. Trends Immunol 31:212–219
57. Mantovani A, Bottazzi B, Colotta F, Sozzani S, Ruco L (1992) The origin and function of tumor-associated macrophages. Immunol Today 13(7):265–270
58. Allavena P, Sica A, Solinas G, Porta C, Mantovani A (2008) The inflammatory microenvironment in tumor progression: the role of tumor-associated macrophages. Crit Rev Oncol Hematol 66:1–9
59. Parker LP, Wolf JK, Price JE (2000) Adenoviral-mediated gene therapy with Ad5CMVp53 and Ad5CMVp21 in combination with standard therapies in human breast cancer cell lines. Ann Clin Lab Sci 30:395–405
60. Cristofanilli M, Budd GT, Ellis MJ, Stopeck A, Matera J, Miller MC, Reuben JM, Doyle GV, Allard WJ, Terstappen LW et al (2004) Circulating tumor cells, disease progression, and survival in metastatic breast cancer. N Engl J Med 351:781–791
61. Liu MC, Shields PG, Warren RD, Cohen P, Wilkinson M, Ottaviano YL, Rao SB, Eng-Wong J, Seillier-Moiseiwitsch F, Noone AM et al (2009) Circulating tumor cells: a useful predictor of treatment efficacy in metastatic breast cancer. J Clin Oncol 27:5153–5159
62. Bidard FC, Peeters DJ, Fehm T, Nole F, Gisbert-Criado R, Mavroudis D, Grisanti S, Generali D, Garcia-Saenz JA, Stebbing J et al (2014) Clinical validity of circulating tumour cells in patients with metastatic breast cancer: a pooled analysis of individual patient data. Lancet Oncol 15:406–414
63. Cristofanilli M, Hayes DF, Budd GT, Ellis MJ, Stopeck A, Reuben JM, Doyle GV, Matera J, Allard WJ, Miller MC et al (2005) Circulating tumor cells: a novel prognostic factor for newly diagnosed metastatic breast cancer. J Clin Oncol 23:1420–1430
64. de Bono JS, Scher HI, Montgomery RB, Parker C, Miller MC, Tissing H, Doyle GV, Terstappen LW, Pienta KJ, Raghavan D (2008) Circulating tumor cells predict survival benefit from treatment in metastatic castration-resistant prostate cancer. Clin Cancer Res 14:6302–6309
65. Cohen SJ, Punt CJ, Iannotti N, Saidman BH, Sabbath KD, Gabrail NY, Picus J, Morse M, Mitchell E, Miller MC et al (2008) Relationship of circulating tumor cells to tumor response, progression-free survival, and overall survival in patients with metastatic colorectal cancer. J Clin Oncol 26:3213–3221
66. Sieuwerts AM, Kraan J, Bolt J, van der Spoe LP, Elstrodt F, Schutte M, Martens JW, Gratama JW, Sleijfer S, Foekens JA (2009) Anti-epithelial cell adhesion molecule antibodies and the detection of circulating normal-like breast tumor cells. J Natl Cancer Inst 101:61–66
67. Almendro V, Kim HJ, Cheng YK, Gonen M, Itzkovitz S, Argani P, van Oudenaarden A, Sukumar S, Michor F, Polyak K (2014) Genetic and phenotypic diversity in breast tumor metastases. Cancer Res 74:1338–1348
68. Fischer JC, Niederacher D, Topp SA, Honisch E, Schumacher S, Schmitz N, Zacarias Fohrding L, Vay C, Hoffmann I, Kasprowicz NS et al (2013) Diagnostic leukapheresis

enables reliable detection of circulating tumor cells of nonmetastatic cancer patients. Proc Natl Acad Sci U S A 110:16580–16585
69. Stoecklein NH, Klein CA (2010) Genetic disparity between primary tumours, disseminated tumour cells, and manifest metastasis. Int J Cancer 126:589–598
70. Psaila B, Kaplan RN, Port ER, Lyden D (2006) Priming the 'soil' for breast cancer metastasis: the pre-metastatic niche. [Research Support, N.I.H., Extramural. Research Support, Non--U.S. Gov't. Review]. Breast Dis 26:65–74 (PubMed)
71. Yoneda T, Hiraga T (2005) Crosstalk between cancer cells and bone microenvironment in bone metastasis. [Review]. Biochem Biophys Res Commun 328(3):679–687. doi:10.1016/j.bbrc.2004.11.070.PubMedCrossRef
72. Coleman RE (1997) Skeletal complications of malignancy [Review]. Cancer 80(8 Suppl):1588–1594 (PubMedCrossRef)
73. Kozlow W, Guise TA (2005) Breast cancer metastasis to bone: mechanisms of osteolysis and implications for therapy. J Mammary Gland Biol Neoplasia 10(2):169–180 (PubMedCrossRef)
74. Buijs JT, van der Pluijm G (2009) Osteotropic cancers: from primary tumor to bone. Cancer Lett 273(2):177–193. doi:10.1016/j.canlet.2008.05.044.PubMedCrossRef
75. Thibaudeau L, Quent VM, Holzapfel BM, Taubenberger AV, Straub M, Hutmacher DW (2014) Mimicking breast cancer-induced bone metastasis in vivo: current transplantation models and advanced humanized strategies. Cancer Metastasis Rev 33(2-3):721–735. doi:10.1007/s10555-014-9499
76. Brown JE, Neville-Webbe H, Coleman RE (2004) The role of bisphosphonates in breast and prostate cancers [Review]. Endocr Relat Cancer 11(2):207–224 (PubMedCrossRef)
77. Kostenuik PJ, Nguyen HQ, McCabe J, Warmington KS, Kurahara C, Sun N et al (2009) Denosumab, a fully human monoclonal antibody to RANKL, inhibits bone resorption and increases BMD in knock-in mice that express chimeric (murine/human) RANKL. J Bone Miner Res 24(2):182–195. doi:10.1359/jbmr.081112.PubMedCrossRef
78. Powell E, Piwnica-Worms D, Piwnica-Worms H (2014) Contribution of p53 to metastasis. Cancer Discov 4(4):405–414. doi:10.1158/2159-8290.CD-13-0136, Epub 2014 Mar 21
79. Brosh R, Rotter V (2009) When mutants gain new powers: news from the mutant p53 field. Nat Rev Cancer 9:701–713
80. Muller PA, Vousden KH, Norman JC (2011) p53 and its mutants in tumor cell migration and invasion. J Cell Biol 192:209–218
81. Xia M, Land H (2007) Tumor suppressor p53 restricts Ras stimulation of RhoA and cancer cell motility. Nat Struct Mol Biol 14:215–223
82. Mashimo T, Watabe M, Hirota S, Hosobe S, Miura K, Tegtmeyer PJ et al (1998) The expression of the KAI1 gene, a tumor metastasis suppressor, is directly activated by p53. Proc Natl Acad Sci U S A 95:11307–11311
83. Liu WM, Zhang XA (2006) KAI1/CD82, a tumor metastasis suppressor. Cancer Lett 240:183–194
84. Scheel C, Weinberg RA (2012) Cancer stem cells and epithelial–mesenchymal transition: concepts and molecular links. Semin Cancer Biol 22:396–403
85. Chang CJ, Chao CH, Xia W, Yang JY, Xiong Y, Li CW et al (2011) p53 regulates epithelial-mesenchymal transition and stem cell properties through modulating miRNAs. Nat Cell Biol 13:1467
86. Frisch SM, Francis H (1994) Disruption of epithelial cell–matrix interactions induces apoptosis. J Cell Biol 124:619–626
87. Cancer Genome Atlas Network (2012) Comprehensive molecular portraits of human breast tumours. Nature 490:61–70
88. Feldser DM, Kostova KK, Winslow MM, Taylor SE, Cashman C, Whittaker CA et al (2010) Stage-specific sensitivity to p53 restoration during lung cancer progression. Nature 468:572–575
89. Martins CP, Brown-Swigart L, Evan GI (2006) Modeling the therapeutic efficacy of p53 restoration in tumors. Cell 127:1323–1334

90. Ventura A, Kirsch DG, McLaughlin ME, Tuveson DA, Grimm J, Lintault L et al (2007) Restoration of p53 function leads to tumour regression in vivo. Nature 445:661–665
91. Xue W, Zender L, Miething C, Dickins RA, Hernando E, Krizhanovsky V et al (2007) Senescence and tumour clearance is triggered by p53 restoration in murine liver carcinomas. Nature 445:656–660
92. Secchiero P, Bosco R, Celeghini C, Zauli G (2011) Recent advances in the therapeutic perspectives of Nutlin-3. Curr Pharm Des 17:569–577
93. Tian W, Chen J, He H, Deng Y (2013) MicroRNAs and drug resistance of breast cancer: basic evidence and clinical applications. Clin Transl Oncol 15:335–342
94. Bartel DP (2009) MicroRNAs: target recognition and regulatory functions. Cell 136:215–233
95. Calin GA, Dumitru CD, Shimizu M, Bichi R, Zupo S, Noch E et al (2002) Frequent deletions and downregulation of micro-RNA genes miR15 and miR16 at 13q14 in chronic lymphocytic leukemia. Proc Natl Acad Sci U S A 99:15524–15529
96. Nygaard S, Jacobsen A, Lindow M, Eriksen J, Balslev E, Flyger H et al (2009) Identification and analysis of miRNAs in human breast cancer and teratoma samples using deep sequencing. BMC Med Genomics 2:35
97. Christoffersen NR, Shalgi R, Frankel LB, Leucci E, Lees M, Klausen M et al (2010) p53-independent upregulation of miR-34a during oncogene-induced senescence represses MYC. Cell Death Differ 17:236–245
98. Zhao JJ, Lin J, Yang H, Kong W, He L, Ma X et al (2008) MicroRNA-221/222 negatively regulates estrogen receptor alpha and is associated with tamoxifen resistance in breast cancer. J Biol Chem 283:31079–31086, View Record in Scopus
99. Zhu H, Wu H, Liu X, Evans BR, Medina DJ, Liu CG et al (2008) Role of MicroRNA miR-27a and miR-451 in the regulation of MDR1/P-glycoprotein expression in human cancer cells. Biochem Pharmacol 76:582–588
100. Kovalchuk O, Filkowski J, Meservy J, Ilnytskyy Y, Tryndyak VP, Chekhun VF et al (2008) Involvement of microRNA-451 in resistance of the MCF-7 breast cancer cells to chemotherapeutic drug doxorubicin. Mol Cancer Ther 7:2152–2159
101. Scott GK, Goga A, Bhaumik D, Berger CE, Sullivan CS, Benz CC (2007) Coordinate suppression of ERBB2 and ERBB3 by enforced expression of micro-RNA miR-125a or miR-125b. J Biol Chem 282:1479–1486
102. Tang F, Zhang R, He Y, Zou M, Guo L, Xi T (2012) MicroRNA-125b induces metastasis by targeting STARD13 in MCF-7 and MDA-MB-231 breast cancer cells. PLoS One 7, e35435
103. Yu Z, Willmarth NE, Zhou J, Katiyar S, Wang M, Liu Y et al (2010) microRNA 17/20 inhibits cellular invasion and tumor metastasis in breast cancer by heterotypic signaling. Proc Natl Acad Sci U S A 107:8231–8236
104. Li H, Bian C, Liao L, Li J, Zhao RC (2011) miR-17-5p promotes human breast cancer cell migration and invasion through suppression of HBP1. Breast Cancer Res Treat 126:565–575
105. Valastyan S, Reinhardt F, Benaich N, Calogrias D, Szász AM, Wang ZC et al (2009) A pleiotropically acting microRNA, miR-31, inhibits breast cancer metastasis. Cell 137:1032–1046
106. Zhu N, Zhang D, Xie H, Zhou Z, Chen H, Hu T et al (2011) Endothelial-specific intron-derived miR-126 is down-regulated in human breast cancer and targets both VEGFA and PIK3R2. Mol Cell Biochem 351:157–164
107. Krell J, Frampton AE, Jacob J, Pellegrino L, Roca-Alonso L, Zeloof D et al (2012) The clinico-pathologic role of microRNAs miR-9 and miR-151-5p in breast cancer metastasis. Mol Diagn Ther 16(3):167–172
108. Pellegrino L, Stebbing J, Braga VM, Frampton AE, Jacob J, Buluwela L et al (2013) miR-23b regulates cytoskeletal remodeling, motility and metastasis by directly targeting multiple transcripts. Nucleic Acids Res 41(10):5400–5412

Chapter 8
How to Build Up Adequate Prognostic Markers in the Molecular Biology Context of Breast Cancer

8.1 Introduction

Using the tumor histologic pattern, degree of nuclear pleomorphism, number of mitoses, presence of inflammatory response, blood vessel invasion, lymph node involvement, and the presence of hormone receptors [1–27] attempts have been made to identify patients at high risk of early disease recurrence and thus to more effectively target aggressive adjuvant chemotherapy and intensive follow-up protocols. However, many breast carcinomas having similar histologic, nuclear, or mitotic grades or estrogen receptor (ER) levels differ markedly in their recurrence behavior, thus affecting a patient's survival. It is therefore obvious that single morphologic or biologic characteristics are insufficient to predict the biologic behavior of a tumor; therefore, a combination of various criteria may be necessary to accurately separate subpopulations of patients with breast cancer at increased risk of recurrence or shortened survival [28].

A major focus of this chapter is how to use single characteristics of primary breast tumors, such as ER status, histologic grade (HG), nuclear grade (NG), mitotic grade (MG), and the combinations of HG, MG, and NG to give a final grade (FG) similar to that of Bloom and Richardson [7] and lymph node status (LN) of the patient at the time of surgery to identify subpopulations of patients at increased risk of early disease recurrence or shorter survival.

J. Russo, *The Pathobiology of Breast Cancer*,
DOI 10.1007/978-3-319-40815-6_8

8.2 Tumor Grading

Tumors in general are graded according to the criteria established by Bloom and Richardson (7, 28) and selecting for grading the least differentiated area of the tumors, instead of the average, in order to determine the weight that the most undifferentiated areas have in tumor prognosis (Table 8.1).

The nuclear grade 1 is characterized by nuclei that are uniform in size and shape and the chromatin is uniformly granular (Fig. 8.1). Instead the nuclear grade 2 is characterized by a moderate variation in size and shape, with mild chromatin condensation and rarely the nucleolus is prominent (Fig. 8.2). The nuclear grade 3 is characterized by marked degree of pleomorphism by variation in size and shape (Figs. 8.3 and 8.4). The nuclear grade 3 is also characterized by having a more irregular chromatin and a prominent nucleolus (Figs. 8.3 and 8.4).

The histological grade is determined by the conservation of the ductular or glandular pattern. In the histologic grade 1 most of the tumor is formed by well-preserved ductular pattern (Fig. 8.5a, b), whereas in histological grade 2 there is a moderate formation of tubules and solid masses of tumor cells that can be intermingled with other containing lumen formation (Fig. 8.5c, d, and e). When the tubular or glandular formation has been completely lost the histological grade 3 is given (Fig. 8.5).

The mitotic grade is determined as indicated in Table 8.1 by the number of mitosis per 10 high power fields (HPF) with the adequate correction for the specific microscope, since there are small variations depending on the brand. It is accepted to count mitosis with a 40× objective and a 10× ocular, with a field area of 0,152 mm^2. The most important is to recognize the mitotic figures in the different phases of the cycle to have a consistent counting of mitosis. In Fig. 8.6 are examples of prophases, metaphases, anaphases and telophases. Abnormal mitoses should also be counted.

The Final Grade or FG is determined by adding the values obtained for HG, MG, and NG. A total score of 3–5 represented an FG of I; 6–7, an FG of II; and score of 8–9, an FG of III (Table 8.2) [28]. The higher mitotic grade is usually seemed in tumors with nuclear grade 3 (Fig. 8.7) however this is not a rule.

Table 8.1 Criteria for tumor grading

Tubule formation	Nuclear pleomorphism	Mitotic count for a 40× objective with a field area of 0.152 mm^2
Majority of tumors >75 % (score 1)	Small regular nuclei (score 1)	0–5 mitoses per 10 HPF (score 1)
Moderate 10 % to 75 % (score 2)	Moderate increase in size (score 2)	6–10 mitoses per 10 HPF (score 2)
Minimal <10 % (score 3)	Marked variation in size, nucleoli, chromatin clumping (score 3)	>10 mitoses per 10 HPF (score 3)

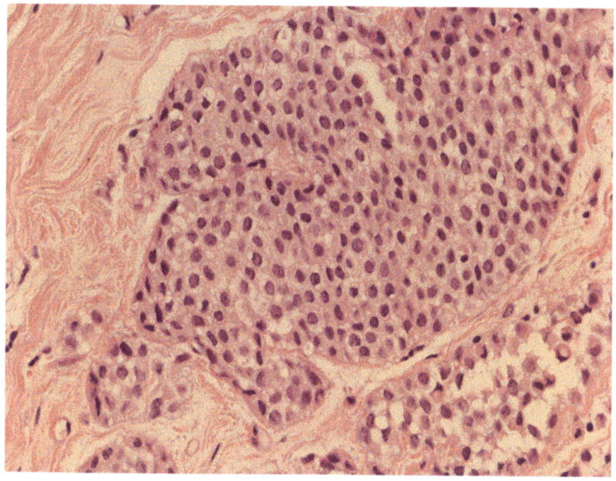

Fig. 8.1 The nuclear grade 1 is characterized by nuclei that are uniform in size and shape and the chromatin is uniformly granular. Hematoxylin and eosin, 40×

8.3 Tumor Grading and Prognosis of Breast Cancer

In our series of six hundred forty-six patients [28] we have found that grading the tumors by histologic pattern, nuclear pleomorphism, and number of mitoses according to the criteria listed in Table 8.1 yielded subsets of patients with significant differences in recurrence and survival patterns. Life table analyses of HG, MG, and NG individually showed these differences (Table 8.3). The proportions free of recurrent disease at 60 months after surgery were as follows: 100 %, 79 %, and 60 % for HG I, 2, and 3, respectively *(P*=0.0029); 65 %, 54 %, and 50 %, respectively for MG 1, 2, and 3 *(P*<0.000 I); 83 %, 68 %, and 52 % for NG I, 2, and 3, respectively *(P*=0.0002). The grade variables had similar effects on time to death. Significant differences in recurrence and death patterns were found between HG2 and HG3 *(P*<0.005), NG2 and NG3 *(P*<0.0005), and MGI and MG2 *(P*<0.0003). No significant differences were found in the recurrence or death patterns between HG I and HG2 *(P > 0040),* NG I and NG2 *(P > 0040),* and MG2 and MG3 *(P* > 0.50) [28].

To test the importance of tumor HG, MG, NG, and FG when these were used in conjunction with LN, ER, and other commonly used predictors, Cox proportional-hazards regressions were run for recurrence and breast cancer death (Tables 8.5 and 8.6). Because MG was always a significant prognostic factor in survival curves, Race it was combined with LN, E2R status, and tumor size (Figs. 8.8, 8.9 and 8.10). HG and NG were found to be significant in the recurrence and death regressions, respectively (Tables 8.4 and 8.5). FG was also a significant prognostic factor. In other regressions not reported, the coefficient on FG was significant for recurrence *(P*<0.00 I) and for death *(P*<0.000 I). Tumor size was correlated with LN status *(P*=0.0001), but size and LN status each contained useful information not contained in the other. Both were highly significant in all regressions. E2R status was significant in predicting death (Table 8.5) [28].

Age was cut at 55 years to serve as a proxy for the patient's menopausal status. It was felt that this was better than using reported menopausal status, because many

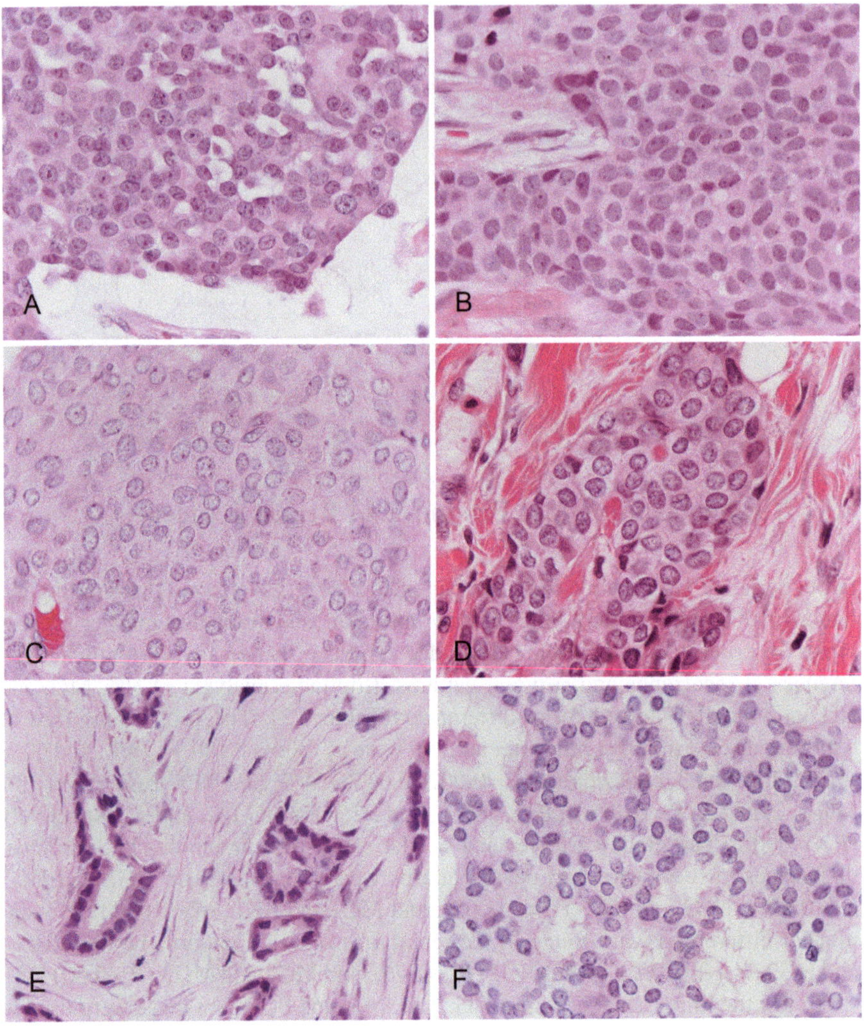

Fig. 8.2 (**a**, **b**, **c**, **d**, **e** and **f**) are depicting nuclear grade 2 characterized by a moderate variation in size and shape, with mild chromatin condensation and rarely the nucleolus is prominent. Hematoxylin and eosin, 40×

women had had hysterectomies and did not know whether oophorectomies had been performed at the time of that surgery. However, the coefficients on age and tumor type variables did not differ significantly from zero in any of these regressions.

For the life table analyses the patients were placed in six subgroups, based on two tumor grade groups (MG 1 vs MG2,3) and the three LN status groups (OLN+,1-3LN+, and 4+LN+). Separate analyses were per formed for patients who did not receive adjuvant therapy (Fig. 8.11a) and those who did (Fig. 8.11b*)*. As expected, most of the LN patients received no adjuvant therapy. The untreated MG2,3/1-

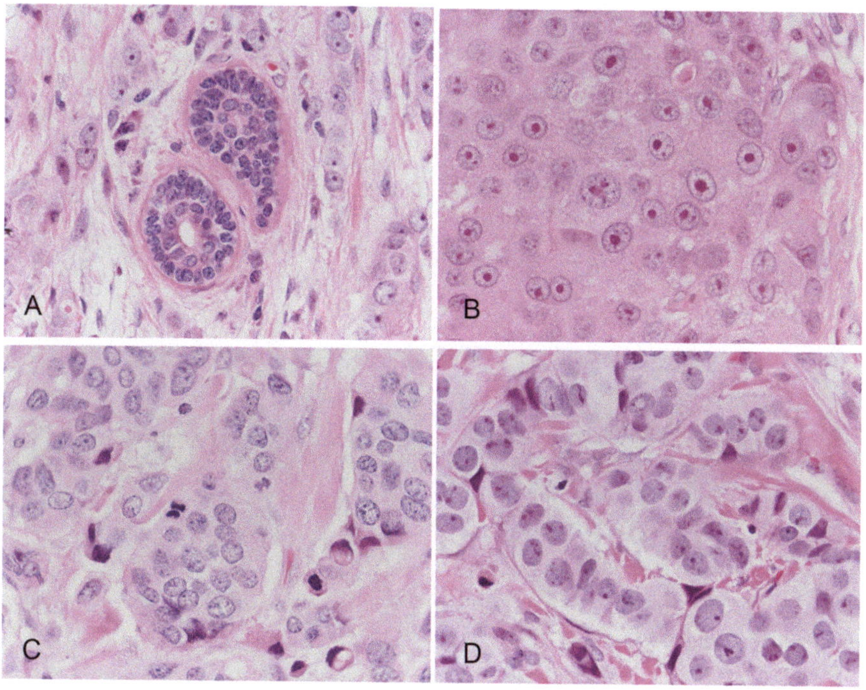

Fig. 8.3 (**a**) Normal nucleus of a mammary duct. (**b**, **c** and **d**) Nuclear grade 3 characterized by marked degree of pleomorphism by variation in size and shape, irregular chromatin and a prominent nucleolus. Hematoxylin and eosin, 40×

3LN+ with 12 patients and the untreated MG2,3/4+LN+ with 2 patients had an insufficient number of patients to draw meaningful conclusions, therefore, they are not shown in Fig. 8.11a [28].

There were significant differences in prognoses of the corresponded to their death experience (without therapy $P<0.0001$, with therapy $P<0.0001$). The recurrence experience of these patients corresponded to their death experience (without therapy $p<0.0001$), with therapy $p<0.0001$). The estimated proportions free of recurrent disease after 48 months are given in Table 8.6 [28].

The differences found among the six subgroups could not be attributed to either LN status or to tumor MG alone. Within each LN group, higher MG resulted in worse recurrence and survival experience. Among the 270 patients who were OLN+ and who were not selected for adjuvant therapy, life table analysis showed 86.3 % of the MG 1 were disease free at 48 months after surgery, compared with 63.6 % of the MG2,3 (life table $P=0.0002$). The 122 patients who had 4+ LN+ and did receive adjuvant therapy had recurrences and died more quickly; at 12 months after surgery, 88.5 % of the MGl were disease free, compared with 71.9 % of MG2,3 (life-table $P=0.013$). Similar results were found for the proportions dead as a result of breast cancer. Among the 270 OLN+ who received no adjuvant therapy, the estimated 60-month breast cancer survival rate was 88.8 % for MG I, whereas the rate was only

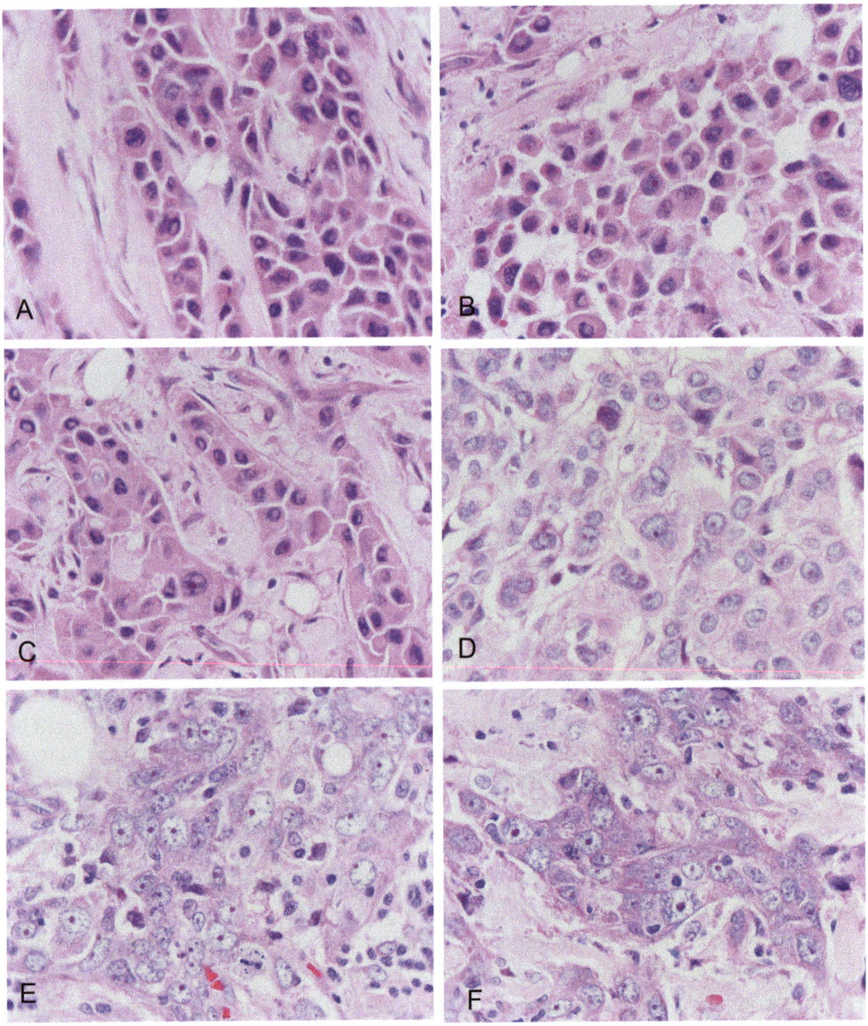

Fig. 8.4 (**a**, **b**, **c**, **d**, **e** and **f**) Invasive ductal carcinomas containing nuclear grade 3 with marked degree of pleomorphism, irregular chromatin and prominent nucleolus. Hematoxylin and eosin, 40×

75.6 % for the MG2,3 (life table $P = 0.0008$). The 24-month breast cancer survival rates for the 122 4+LN+ who did receive adjuvant therapy were 86.1 % for the MG I and 53.5 % for the MG2,3 (life table $P < 0.0001$) (Fig. 8.11b) [28].

Furthermore, within each MG group, more severe LN involvement resulted in earlier recurrences and deaths. For example, of the 205 MG 1 who received adjuvant therapy, 75.5 % of the OLN+, 75.0 % of 1-3LN+, and 53.9 % of 4+LN+ were disease free at 48 months (life table $P = 0.0001$). The corresponding rates for the 291patients with MG I tumors who did not receive adjuvant therapy were 86.3 %, 65.4 %, and 40.4 % for those with OLN+, 1-3LN+, or 4+LN+, respectively (life table $P = 0.0001$) [28].

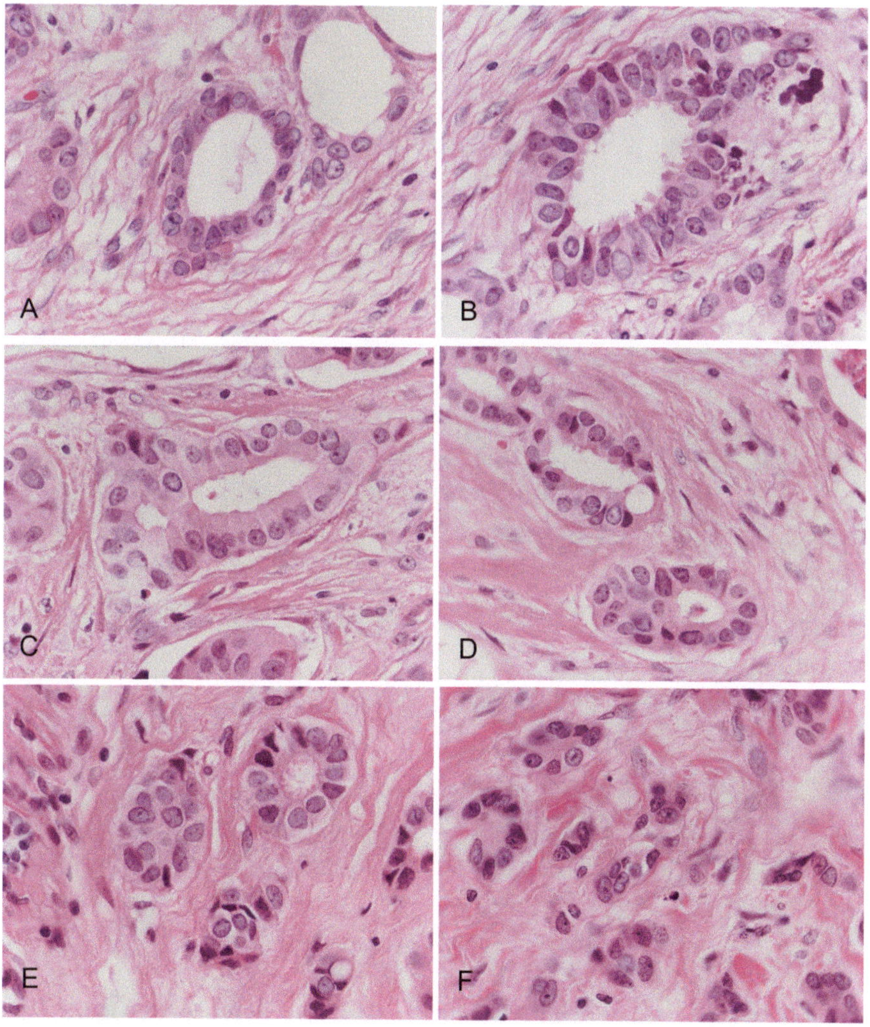

Fig. 8.5 (**a** and **b**) Histological grade 1; (**c**, **d** and **e**) Histological grade 2 and E: histological grade 3. Hematoxylin and eosin, 40×

The life table analyses suggested that an increase in MG had about the same impact on prognosis of recurrence and death as a one-step increase in LN, compare, for example, the treated MG1/4+LN+ with the treated MG2,3/1-3LN+. This is consistent with the Cox regressions in Tables 8.4 and 8.5 in that the regression coefficients on LN and MG had similar magnitudes [28].

Life table analysis of ER and MG also confirmed the Cox regressions in Tables 8.4 and 8.5. The two ER groups and the two MG groups divided the cases into four subgroups that had significantly different patterns of time to recurrence and time to breast cancer-related death. Figure 8.12 show that both ER and MG had

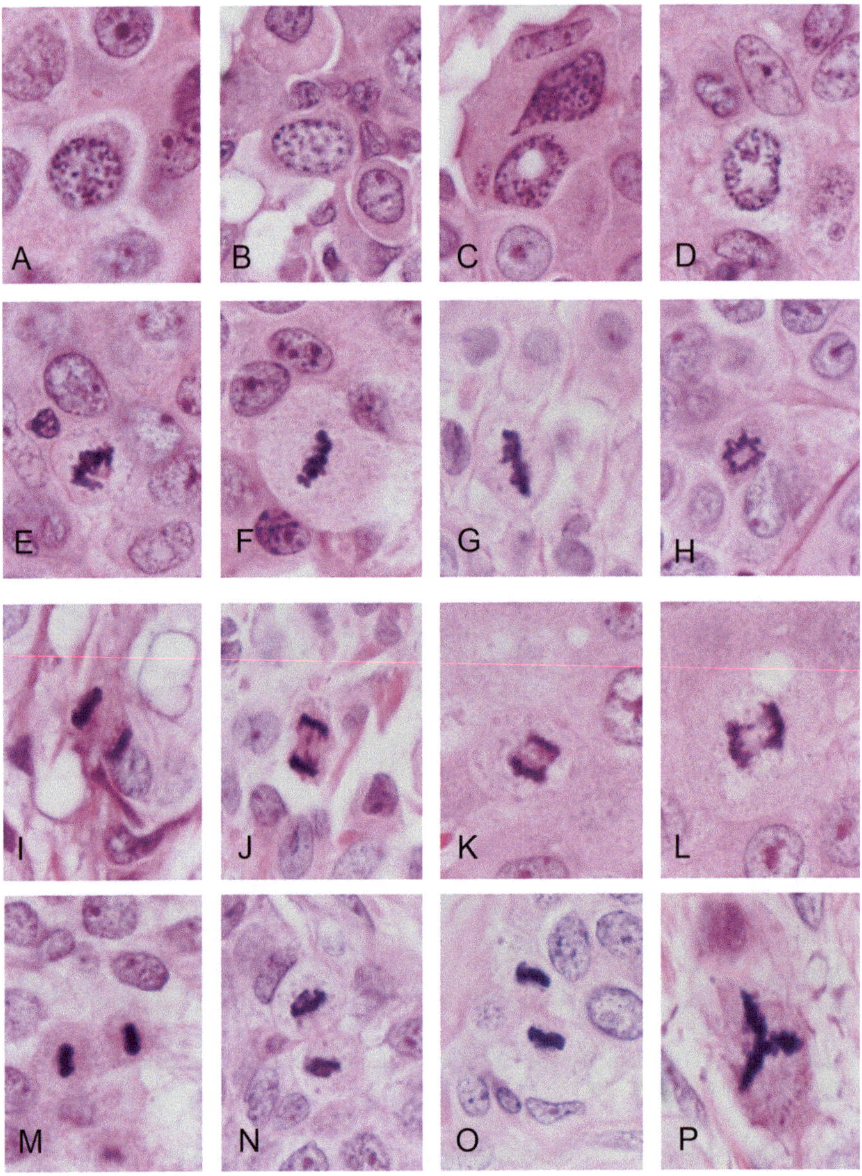

Fig. 8.6 (**a**, **b**, **c** and **d**) Prophase. (**e**, **f**, **g** and **h**) Metaphase. (**i**, **j**, **k** and **l**) Anaphase. (**m**, **n**, and **o**) Telophase. (**p**) Tripolar mitosis. Hematoxylin and eosin, 40×

significant effects on the time of death. Among the 233 patients with ER+ tumors who did not receive adjuvant therapy. The MG 1 had an estimated 36-month breast cancer survival rate of 94.7 % *versus* 83.1 % for the MG2,3 (life table $P=0.0035$). Among the 161 ER+ patients who did receive adjuvant therapy, the 36-month rates

Table 8.2 Tumor grading

Total score	Points
Grade 1	3–5 points
Grade 2	6–7 points
Grade 3	8–9 points

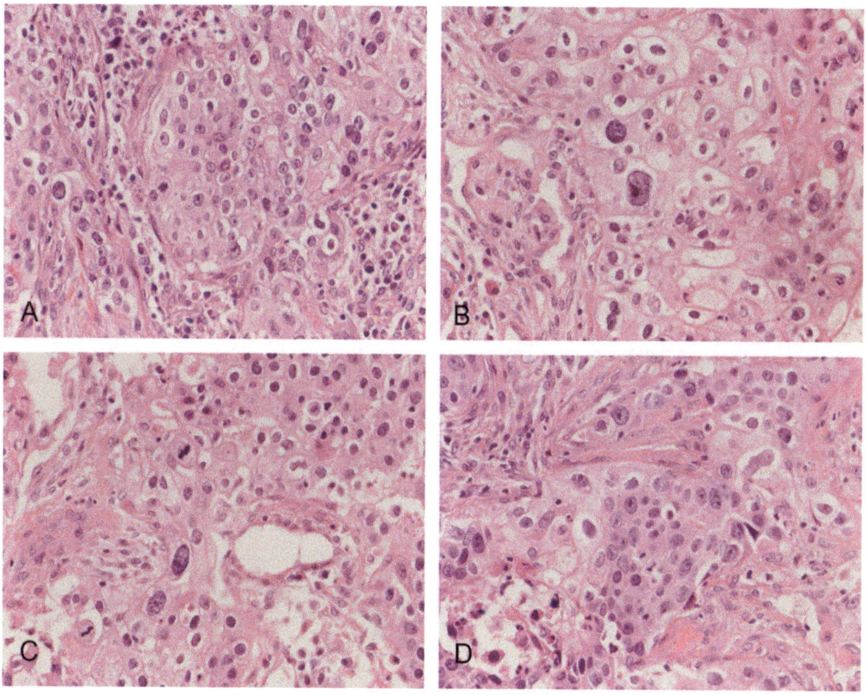

Fig. 8.7 (**a**, **b**, **c** and **d**) Invasive ductal carcinoma with high nuclear grade 3 and numerous mitosis (NG3). Hematoxylin and eosin, 40×

were 89.7 % and 71.0 % for MG I and MG2,3, respectively (lifetable $P=0.0001$). The differences resulting from ER were only significant among the MG 1 (life table $P=0.0028$ without treatment, $P=0.0205$ with treatment). In Fig. 8.12b, the MG2,3/ER+appear to fare worse than the MG2,3/ER -, but the difference between the two curves is not significant ($P=0.6699$) [28].

The Cox regression of Table 8.4 found no significant effect of ER on the time to recurrence. Using the four MG/ER subgroups (and analyzing treated and untreated patients separately), life table analysis of the same data set confirmed this finding. There were no significant differences between the recurrence patterns of subgroups that differed only in their E,R status. Differences were observed among the ER+ between the MG I and the MG2,3. (life table $P=0.0006$ without therapy, $P=0.065$ with therapy). The estimated proportions free of recurrent disease at 36 months after

Table 8.3 Percentage of patients nonrecurrent for single variables (lifetable estimates)

	%	SE	%	SE	%	SE
Histologic grade ($P=0.0029$)	1		2		3	
n	7		62		577	
12 mo	100	–	96.8	2.3	91.6	1.2
24 mo			94.9	2.9	77.6	1.8
36 mo			87.9	4.7	70.1	2.1
48 mo			87.9	4.7	66.7	2.2
60 mo			78.8	6.5	60.1	2.6
72 mo			78.8	6.5	57.7	2.9
Mitotic grade ($P<0.0001$)	1		2		3	
n	496		79		71	
12 mo	94.7	1.0	80.5	4.5	87.2	4.0
24 mo	83.2	1.8	65.2	5.5	69.4	5.6
36 mo	75.5	1.8	58.5	5.9	63.9	6.0
48 mo	73.8	2.1	54.1	6.2	53.7	6.9
60 mo	65.5	2.8	54.1	6.2	49.7	7.4
72 mo	62.9	3.1				
Nuclear grade ($P=0.0002$)	1		2		3	
n	21		389		236	
12 mo	95.2	4.6	93.7	1.2	89.2	2.0
24 mo	89.8	6.9	83.6	2.0	71.8	3.0
36 mo	82.6	9.4	77.2	2.4	63.1	3.4
48 mo			74.5	2.5	59.6	3.5
60 mo			67.6	3.1	52.3	4.0
72 mo			65.1	3.5	50.9	4.2
Lymph node status ($P<0.0001$)	0		1-3		4+	
n	314		185		147	
12 mo	95.2	1.2	92.3	2.0	85.4	3.0
24 mo	88.6	1.8	79.5	3.1	58.5	4.3
36 mo	82.2	2.3	71.7	3.6	49.3	4.6
48 mo	80.2	2.5	65.8	4.1	47.8	4.7
60 mo	72.2	3.2	60.4	4.8	42.6	5.1
72 mo	72.2	3.2	60.4	4.8	36.0	5.6
Estrogen receptor status ($P=0.0440$)	(–)		(+)			
n	252		394			
12 mo	90.3	1.9	93.3	1.3		
24 mo	75.6	2.8	81.9	2.0		
36 mo	67.7	3.2	75.0	2.4		
48 mo	63.8	3.4	72.7	2.5		
60 mo	59.4	3.7	64.0	3.3		
72 mo	59.4	3.7	60.0	3.8		

Table 8.4 Cox proportional-hazard regression of recurrence

	Coefficient (std. error)	Relative risk[a]
Mitotic grade	0.464[b] (0.173)	1.59
Nuclear grade	0.218 (0.159)	1.24
Histologic grade	0.643[b] (0.319)	1.90
LN status	0.588[c] (0.111)	1.80
E_2R	0.050 (0.150)	1.05
Adjuvant therapy	–0.146 (0.183)	0.86
Age	–0.044 (0.147)	0.96
Tumor size		
2–5 cm	0.434[b] (0.205)	1.54
>5 cm	0.896[c] (0.241)	2.45
Tumor type infiltrating ductal Ca	0.155 (0.164)	1.17
Race	–0.059 (0.282)	0.94
Likelihood-ratio tests to reject H_0		
All coeff = 0	$^2=99.33$ df = 11	($P<0.0001$)
Grade coeff = 0	$^2=17.8$ df = 3	($P<0.0005$)

[a]The figures in this column give the relative risk for each unit increase in the descriptive variable
[b]Coefficient 0 ($P<0.025$). Two-sided for adjuvant therapy, age and race
[c]Coefficient 0 ($P<0.001$). One sided for all others

surgery among the MGI/ER+ and the MG2,3/ER+ were 85.3 % and 57.3 % for the untreated and 67.6 % and 61.5 % for the treated groups [28].

8.4 Characteristics of the Primary Tumor, Such as ER Status, Tumor Size, and Histologic Grade, and Lymph Node Status at the Time of Surgery Served Significantly to Predict the Outcome of the Disease with Regard to Both Recurrence and Patient Survival

The study [28] of 646 breast cancer patients followed for a mean of 44 months after modified radical mastectomy revealed that the characteristics of the primary tumor, such as ER status (using the charcoal method), tumor size, and histologic grade, and

Table 8.5 Cox proportional-hazard regression of death

	Coefficient (std. error)	Relative risk[a]
Mitotic grade	0.752[b] (0.203)	2.12
Nuclear grade	0.411[c] (0.194)	1.51
Histologic grade	0.700 (0.469)	2.01
LN status	0.447[b] (0.139)	1.56
E_2R	0.456[c] (0.184)	1.58
Adjuvant therapy	–0.062 (0.226)	1.06
Age	–0.058 (0.181)	0.94
Tumor size		
2–5 cm	0.452 (0.266)	1.57
>5 cm	1.160[b] (0.302)	3.19
Tumor type infiltrating ductal Ca	0.235 (0.193)	1.26
Race	–0.04 (0.337)	0.99
Likelihood–ratio tests to reject H_0		
All coeff=0	$^2=103.4$ df=11	($P<0.0001$)
Grade coeff=0	$^2=29.9$ df=3	($P<0.0001$)

[a]The figures in this column give the relative risk for each unit increase in the descriptive variable
[b]Coefficient 0 ($P<0.001$). One-sided for all others
[c]Coefficient 0 ($P<0.025$). Two sided for adjuvant therapy, age and race

Table 8.6 Estimated proportions of patients free of recurrent disease at 48 months after surgery

	Without therapy		With therapy	
	MG1	MG2,3	MG1	MG2,3
By MG and LN				
0	0.863	0.636	0.755	(0.806)
LN 1-3	0.654	(0.470)	0,750	(0.466)
4+	(0.404)	(0.500)	0.539	(0.381)
By MG and E_2R				
E_2R+	0.835	(0.573)	0.676	(0.439)
E_2R-	0.711	(0.616)	0.630	(0.520)

Estimates in parentheses refer to subgroups that had fewer than 15 patients remaining at risk at 48

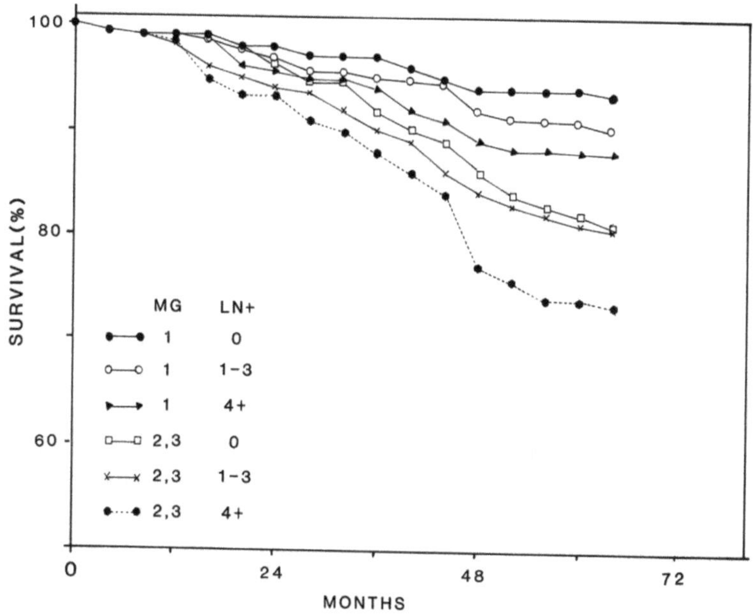

Fig. 8.8 Comparison of MG/LN groups. Simulated survival curves generated from the first estimated Cox regression equation in Table 8.5 (n=646)

lymph node status at the time of surgery served significantly to predict the outcome of the disease with regard to both recurrence and patient survival. Each unit increase in LN status, ER status, MG, or tumor size increased the risk of death by factors of 1.6, 1.6, 2.1, and approximately 2, respectively. The importance of these variables was not due to any correlation with age at time of surgery, tumor type, or adjuvant therapy. Because the latter variables were also included in the regressions, whatever effects they had were captured separately by the coefficients on these variables [28].

Predictions of time to recurrence or death were considerably more accurate when MG, LN, ER, and tumor size were used together, rather than individually. Although many of the high-risk patients identified by MG were also ER -, and many of the 4+LN+ had tumors larger than 5cm, each of the four measures identified a significant number of high-risk patients not identified by the other three individually. Of course, when two or three of these measures agreed, the patients' risk was especially severe. Each of the four variables contained significant and useful information not provided by the others. It has been demonstrated that the detection of axillary lymph node metastases on examination of the mastectomy specimen has a predictive value. Indeed, this has proven to be the single most useful prognostic factor in patients with invasive cancer and has been the basis for selection of patients for adjuvant chemotherapy. The detection of four or more nodes with metastatic involvement was associated with a greater incidence of short term treatment failure and less five-year survival than when one to three nodes contained metastases. Similar observations have been reported by other authors [29, 30]. In addition, when LN status was

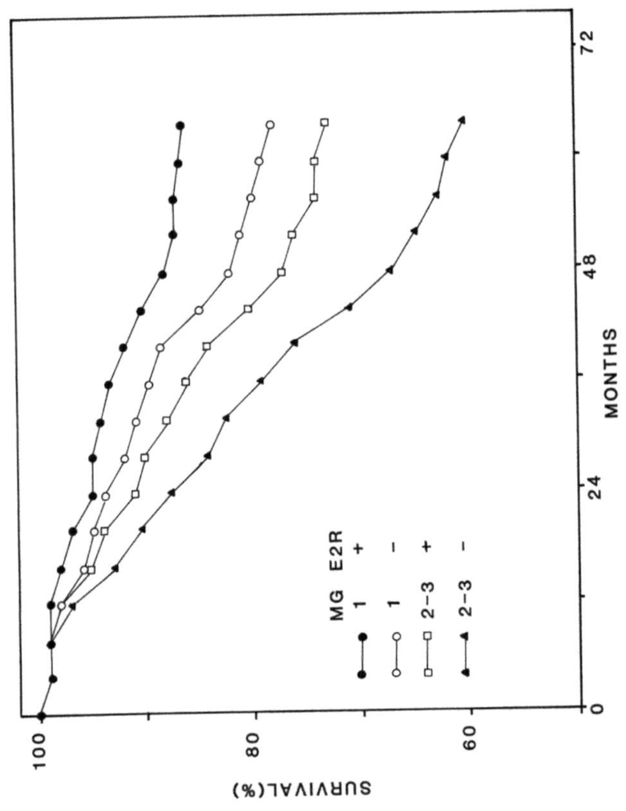

Fig. 8.9 Comparison of MG/E2R groups. Simulated survival curves generated from the first estimated Cox regression equation in Table 8.5 (n=646)

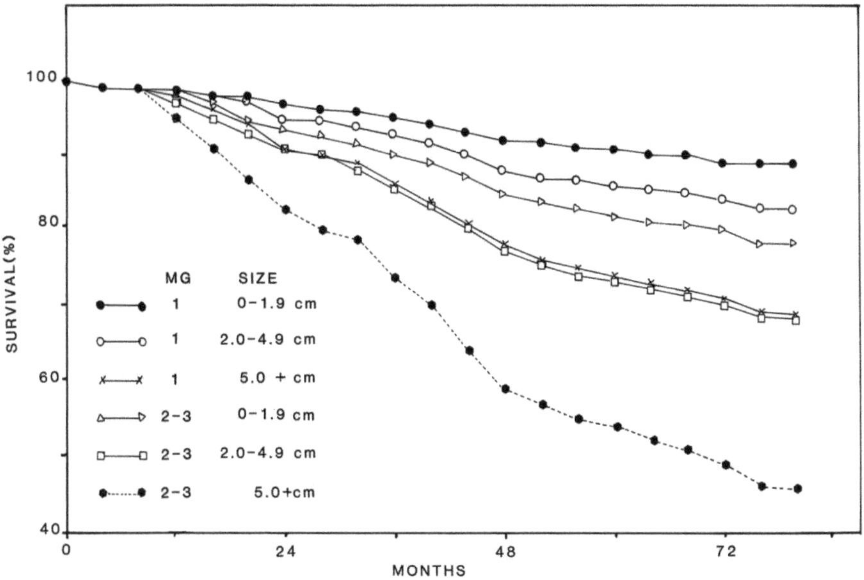

Fig. 8.10 Comparison of MG/tumor size groups. Simulated survival curves generated from the first estimated Cox regression equation in Table 8.5 (n = 646)

considered in combination with MG, it allowed the identification of groups of patients having different biologic characteristics and prognosis. For example, patients who have OLN+ seldom receive any therapy beyond surgery. However, we found that patients who had MG2, 3/0LN+ had recurrence and death patterns similar (after allowing for treatment differences) to those of the MGIJI-3LN+, who currently were more likely to receive treatment. The MG I/0LN+ fared much better than these groups. Similarly, while the 4+LN+ as a group had bad prognosis, the MG I tended to behave more like the *MG2,3/l*-3LN+ and might be less likely candidates for adjuvant treatment [28].

HG and NG have been variously reported to be related to or independent of breast cancer prognosis [7–10, 31–33]. Our results [28] show that, while HG and NG were not as important as MG, they had a significant influence on the risk of recurrence and breast cancer death, respectively. Although it is generally accepted that ER-negative carcinomas have a poorer prognosis than do ER-positive ones, patients with ER-negative tumors are more likely to develop early recurrence, and both their average disease-free interval and average survival are shorter [3, 18, 21, 34, 35] our observations do not show an influence of ER status on recurrence. Previous studies have suggested that the curves for the proportion non recurrent among patients with E,R+ and E,R− tumors start to converge at 54 months after removal of the primary tumor [6]. A similar tendency starting at about 48 months after surgery was observed in the present study.

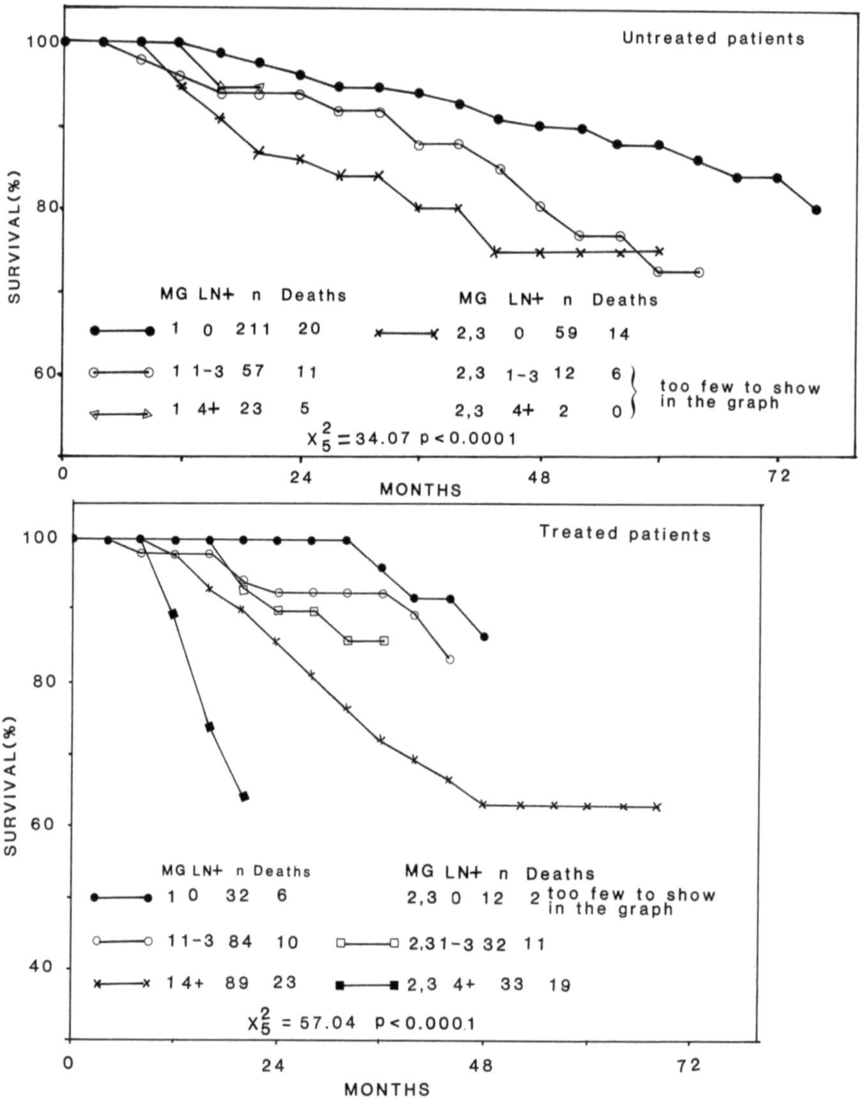

Fig. 8.11 Estimated rates of death (Survival %) due to breast cancer by MG and LN status. Life-table analysis for untreated (**a**, *upper*) and treated patients (**b**, *lower*). There were no patients with four or more positive LN and MG2, 3 who did not receive adjuvant treatment

It has been reported that the E,R content of human breast cancer specimens is related to the degree of differentiation and final grade of the tumor [25, 32, 36] as well as the DNA labeling index.[37–41]. Our results indicate that in the comparison of ER status with MG, this latter parameter was a better discriminant for predicting survival than E,R status.

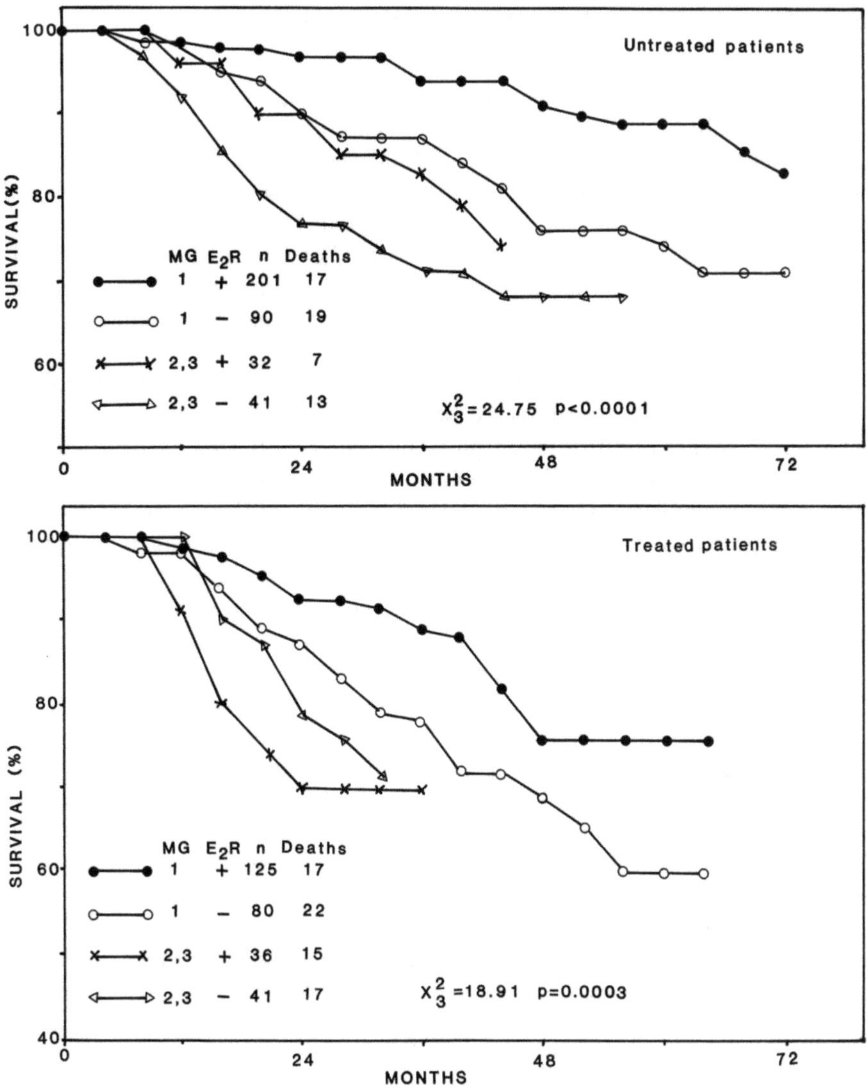

Fig. 8.12 Estimated rates of breast cancer death (Survival %) by MG and E2R status. Life-table analysis for untreated (**a**, *upper*) and treated patients (**b**, *lower*)

These results are important in light of the recent consensus for adjuvant chemotherapy for breast cancer'" in which it was concluded that the adequate evaluation of a given adjuvant treatment requires new criteria for grouping patients. We believe that MG is a variable that has a significant effect on the course of the disease and should be considered when deciding whether to give adjuvant therapy.

8.5 ER and PR as Biomarkers of Prognosis

The standard evaluation of breast cancer for clinical purposes involves IHC characterization of ER, PR and HER2 status. Hormone receptor-positive breast cancers account for around 75–80 % of all cases and standardized IHC assays for the routine testing of ER and PR are used to guide the selection of patients for hormonal-based therapies. HER2 represents the only additional predictive marker currently in routine use and 10–15 % of breast cancers have HER2 overexpression and/or amplification with around half of these co-expressing hormone receptors. The remaining 10–15 % of breast cancers are negative for the three hormone receptors and has been called triple negative breast cancer.

Estrogen receptor has, for a long time, been known to play a crucial role in the carcinogenesis of most breast cancers. More than 50 % of breast cancers have an abundance of estrogen receptors, particularly ER-alpha, the presence of which correlates with a positive prognosis for the patient. The absence of an overexpression of ER-alpha has the inverse prognosis [42]. The effects of estrogens are primarily carried out by their interaction with estrogen receptors, which are part of a large super family of nuclear membrane bound, ligand-activated transcription factors [43]. Estrogen receptor expression is present in 10–20 % of normal human mammary tissue and plays an important role in physiological events like puberty and pregnancy. Estrogen receptors come in two isoforms, ER-alpha and ER-beta, both of which appear in homology in breast cancer, yet play a different role [44]. It has also been found that the tissue distribution and function of both ER-alpha and ER-beta are similar in rats, the model to be used in this study, as they are found in human. Both receptors have similar binding regions, but have different gene expression characteristics [45]. ER-beta is mostly involved the inhibition of tumor growth, though not much is understood about this receptor isoform [44].

Estrogen receptors have two paths of function, both of which regulate gene expression, direct and indirect [45]. The direct, or classical pathway, is when a ligand (estrogen or anti-estrogens) activated ER binds cis-regulatory elements on the DNA called estrogen response elements (EREs) [45]. The indirect pathway is where the ER binds transcriptional factors to the N-terminal activating region, and then the transcriptional factors bind the DNA; this is called transcriptional crosstalk [46]. In both cases the estrogen receptors dimerize and interact with a co-regulator complex that help make the DNA accessible from the chromatin [45]. Estrogen receptors may also have nongenomic actions by interacting with scaffold proteins or various signaling molecules [46]. In breast cancer cells some nongenomic actions of an estradiol activated ER-alpha include activation of the MAP-K pathway, interaction with ErbB2 (HER-s/neu), and activation of epidermal growth factor receptors, all of which are involved in cell proliferation [46].

The molecular composition of mammary carcinomas plays a deciding role in the form of treatment to be used. Today a more precise way of classifying human breast cancer is to use molecular diagnostics [47]. Simply diagnosing based on morphological features can be inaccurate and lead to misdiagnoses because many tumors might deviate from the typical morphology of that type of carcinoma [47]. Breast cancer is typically divided into two main groups: estrogen receptor positive, which includes luminal A and B sub types, and estrogen receptor negative, which includes HER2, normal-like and basal-like cancers [48]. Depending on the expression of these markers, as well as other genes and proteins, the patient will receive a different therapy regimen. Once the estrogen/progesterone status is determined, a course of treatment can be mapped out. If the carcinoma is positive for these receptors, then endocrine therapy is an option, and typically has the best survival outcome. Selective estrogen receptor modulators, or SERMs, act directly on the estrogen receptor but in a way that allows it to be selective with the types of tissues it targets, therefore not completely shutting down ER activity [45]. Tamoxifen is a SERM and acts as an estrogen antagonist by binding to the same ligand binding region as estrogen, therefore blocking further estrogen binding [43], however, transcription is not activated because there is an incomplete conformational change that occurs upon binding. Raloxifen is another SERM similar to Tamoxifen and can be further beneficial in preventative treatment because of a lower risk than Tamoxifen of causing endometrial carcinomas and complications [45]. Fulvestrant, a pure anti-estrogen, binds ER in a similar way to Tamoxifen, however it does not do so in a tissue selective manner, effectively shutting down ER activity completely [45]. Because of the full body nature for which Fulvestrant functions, it is typically used only once a resistance has been developed against Tamoxifen [45].

Resistance to Tamoxifen is common, 40 % of ER alpha positive tumors, with patients receiving endocrine therapy and is the main cause when this therapy becomes ineffective [44, 45]. Strangely enough, 80 % of hormone therapy resistant mammary tumors will still have the same expression levels of estrogen receptors as they did prior to hormone treatment [44]. At this point, aromatase inhibitors can be started to try and treat the resistant tumors [45]. By blocking aromatization of androgens you are blocking the supply of estrogen and indirectly blocking the function of estrogen receptors [45]. However, when a mammary carcinoma lacks an overexpression of estrogen receptors, progesterone receptors and HER2, endocrine therapies will be ineffective because the proteins the drugs interact with are not present. The so called triple-negative breast cancer (very similar in characteristics to basal-like carcinoma) has a very poor prognosis due to the inability to receive effective hormone therapies. In 2010, 12–17 % of women diagnosed with breast cancer were triple-negative [42]. For this group of carcinoma type, treatment options include chemotherapy in addition to platinum salts, PARP inhibitors or antiangiogenic agents [42].

The presence of estrogen and progesterone receptors in the tumor tissue correlates well with response to hormone therapy and chemotherapy [49, 50]. It is presently accepted the immune-histochemical method that has the versatility to be performed in minute amount of fixed tissue. The two parameters evaluated in immune-histochemical preparations of hormone receptors are the number of tumor cell nuclei stained and the intensity of the reaction. The first is expressed as a percentage of the entire tumor cell nuclei population, and the second is graded as negative, weak, moderate, and strong. The two parameters are sometimes combined into a scoring system, of which three major versions exist [51, 52]. The receptors are expressed in both ductal and lobular type of carcinomas; however medullary carcinomas and intraductal carcinomas of the comedocarcinoma type are negative, whereas mucinous carcinomas have the highest rates of positivity.

The Allred scoring system is used for evaluation of the presence of hormone receptor in the human breast cancer. Allred scoring has become the standard in the immune-histochemical analysis of mammary carcinomas, and reduces the borderline or weak positive groups significantly [53] (Figs. 8.13 and 8.14). The grading

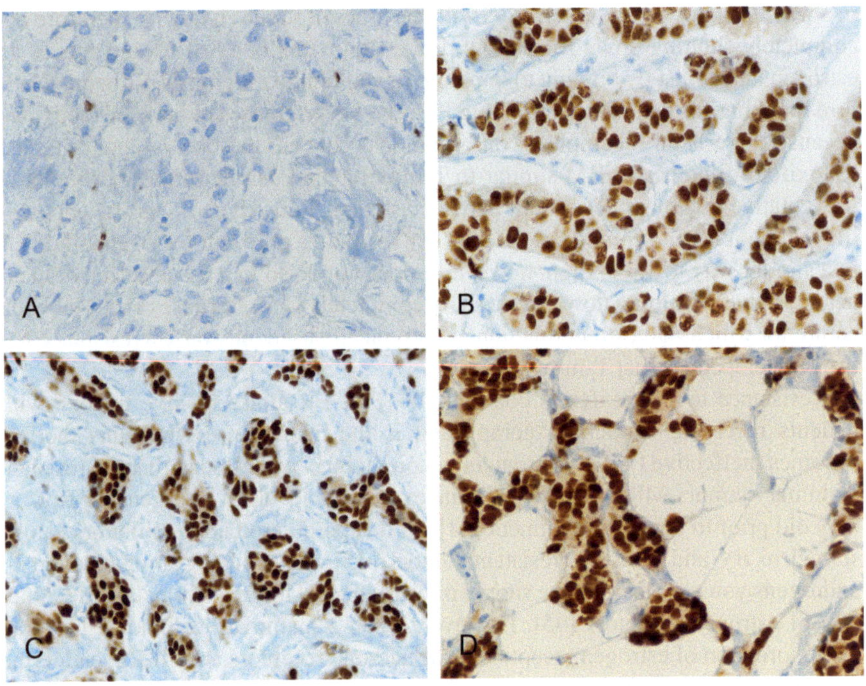

Fig. 8.13 Invasive ductal carcinoma reacting with an antibody against ER. (**a**) Negative; (**b, c, d**) Positive estrogen receptor cells, 40×

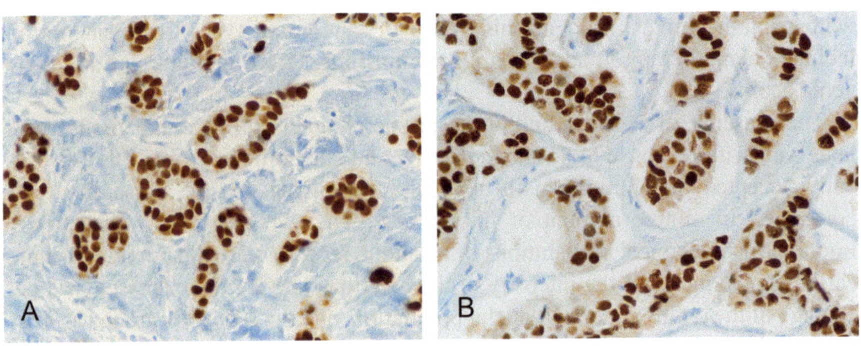

Fig. 8.14 (**a** and **b**) Invasive ductal carcinoma reacting with an antibody against PR, 40×

system looks at the intensity of the positive stained cells and the proportion of the cells that are positive. Each of these characteristics receives a score, which then make up the total score. The total score will be between 0 and 8, 8 being the highest positive scoring. A total score, greater than 2, results in a positive overall scoring for the tumor sample.

8.6 Role of HER2 in Breast Cancer Diagnosis and Treatment

Research into the nature of breast cancer has led to understanding the importance of certain genes and proteins, specifically how they affect the development and attribute unique features to the cancers. One of the most influential genes is the ERBB2 gene, also known as Human Epidermal Growth Factor Receptor 2 (HER2), which codes for a protein called HER2 [53]. The ERBB2 gene itself was first identified in rats and humans the early 1980s. The rat gene (identified first) was referred to as the *neu* oncogene after the transforming, cancer-producing gene found in a cell-line of rat neural tumors [54]. It was found to be homologous to the avian erythroblastosis virus, or v-ERBB2 [55]. Thus, it was given the full name of v-erb-b2 avian erythroblastic leukemia viral oncogene homolog 2, symbolized by ERBB2 [56]. The human homologue gene was called HER2 as it related to the Epidermal Growth Factor Receptor (EGFR, aka ERBB1), forming a family of receptors [57]. Subsequent studies identified the two other members of the ERBB family: ERBB3 and ERBB4, also called HER3 and HER4, respectively. [58, 59].

HER2 is a cell-surface receptor and a member of the EGFR family (Figs. 8.15, 8.16, 8.17 and 8.18), consisting of structurally similar Receptor-Tyrosine Kinases (RTKs, which are specifically activated by phosphorylating their tyrosine amino acid) [60, 61]. It contains three structural components characteristic to its EGFR family: an extracellular ligand-binding domain, a transmembrane domain, and an intracellular domain residing within the cytoplasm of the cell

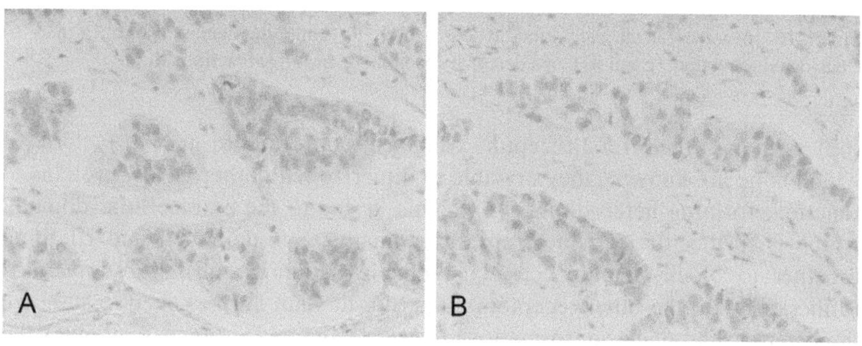

Fig. 8.15 Invasive ductal carcinoma reacting with an antibody against Her2. (**a** and **b**) Negative reaction, 40×

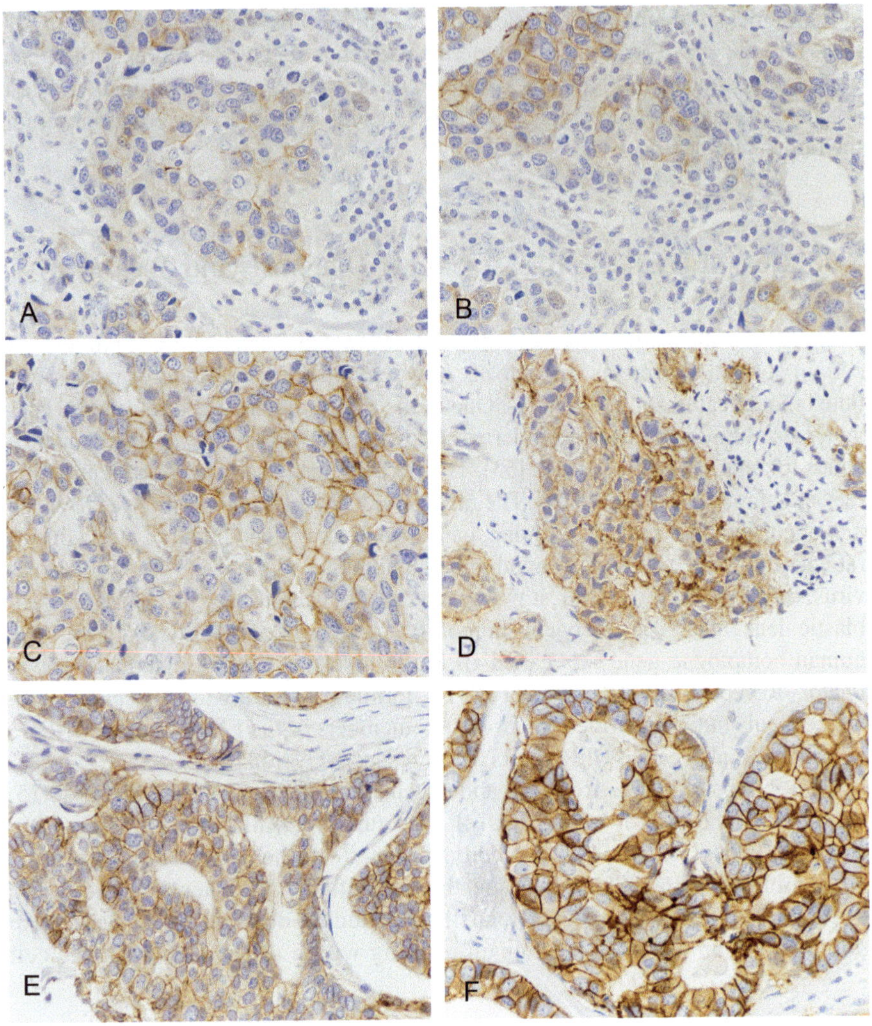

Fig. 8.16 Invasive ductal carcinoma reacting with an antibody against Her2. (**a** and **b**) Considered+reaction, (**c** and **d**) Considered 2+ reaction, (**e** and **f**) 3+reaction, 40×

[62]. Although the HER2 receptors cannot directly bind to growth factors (of which none are known), they are able to dimerize with other EGFR proteins to function, forming heterodimers [63]. This is due to the extracellular domain, which is further divided into four sub-domains: Sub-domain I and III (bind together to block ligand access and enables dimerization), Sub-domain II (dimerizes with the other receptors), and Sub-domain IV (believed to stabilize the receptor and open-conformation) [64, 65]. The entire process involves the other receptors becoming activated via a growth factor ligand, resulting in their

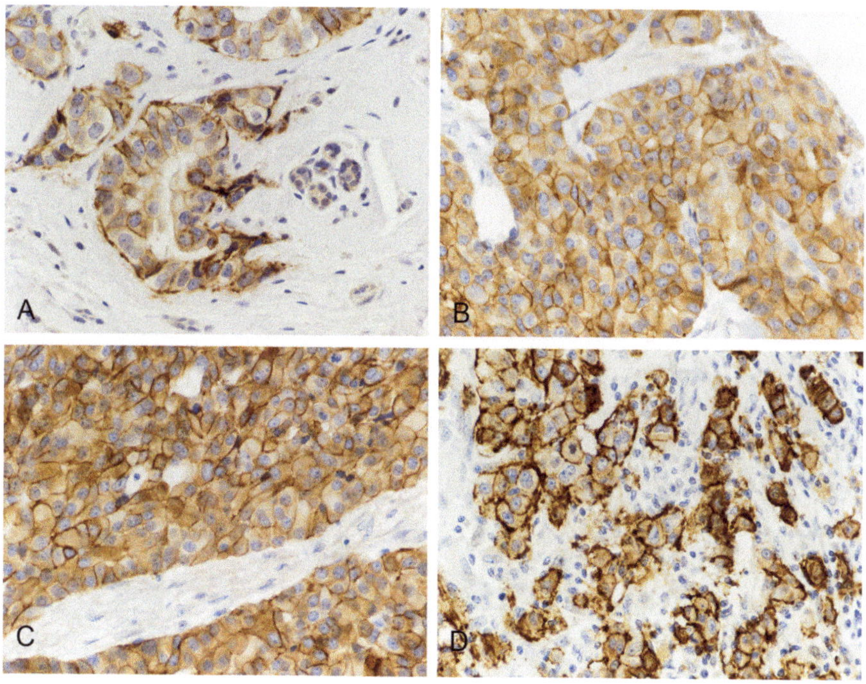

Fig. 8.17 Invasive ductal carcinoma reacting with an antibody against Her2. (**a, b, c** and **d**) Considered 3+ reaction, 40×

conformational change needed to form dimers on the extracellular domains [66]. Upon forming the dimer, the intracellular domains reposition, allowing for exposure of several specific amino acids to be phosphorylated and adaptor proteins to be attached to the to a carboxyl-terminal tail of the domain [67]. In addition to this method, HER2 receptors are also able to autophosphorylate for ligand-independent signaling [68].

The general function of HER2 receptor, once dimerized is to activate pathways leading to cellular proliferation and growth [69]. Of these heterodimers, the most active and potent is the HER2/HER3 dimer [70]. Among the pathways activated by HER2 are those of the Mitogen-Activated Protein Kinase (MAPK). A MAPK group known as Extracellular signal-regulated Kinases (ERKs), upon activation, is able to increase their activity levels, relocate to the cellular nucleus to alter transcription and gene expression, and ultimately promote mitosis and differentiation while inhibiting apoptosis [71, 72]. The Phosphoinositide 3-Kinase/Protein Kinase B (PI3K/Akt) pathway displays similar functions, where activated PI3K catalyzes signaling molecules leading to Akt activation, which in turn inhibits apoptosis and promoting cell growth [73]. As HER2 is a proto-oncogene, over-expression or amplification of HER2 thus leads to uncontrolled cellular proliferation, known to

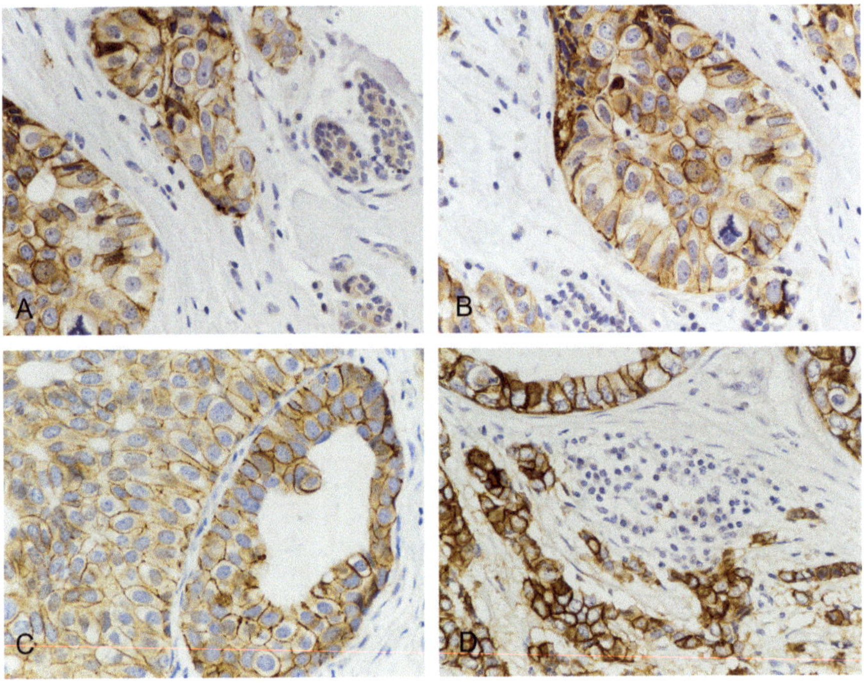

Fig. 8.18 Invasive ductal carcinoma reacting with an antibody against Her2. (**a, b, c** and **d**) Considered 3+ reaction, 40×

lead to mammary tumorigenesis, aggressive tumor growth (due to angiogenesis), and even metastasis [74–76].

Clinically, HER2 expression breast cancers occur in 20–30 % of all breast cancer cases [77]. The most common carcinomas occur within the milk ducts of the breast, known as ductal carcinoma, while others occur within the milk-secreting glands, or lobules, of the breast, known as lobular carcinoma [78] (see Chap. 4). These carcinomas can be identified by their cellular proliferation invading into other organs and tissues as invasive carcinomas, but can also remain situated within the original tissue as in situ carcinomas [79]. HER2 amplification has been found mostly in about 40 % of ductal carcinomas [80]. Lobular carcinomas rarely express HER2, and often only in the pleomorphic variant [81] It is also found.

Several different methods utilize HER2 overexpression and amplification to obtain diagnostic information. Some of the most commonly used are Immunohistochemistry (IHC) tests, which utilize specific antibodies to detect antigens in the cells of a tissue section, allowing for diagnostics of abnormal cells in cancerous tumors [82]. Thus, expression levels of HER2 can be detected on the cell surfaces (Figs. 8.15, 8.16, 8.17 and 8.18).

HER2 is expressed in both HER2/neu+ and Luminal B (HER2+) breast cancers. These subtypes are generally regarded to be more aggressive (actively spread into other organs and systems), have poorer tumor grades (poorly-differentiated cells), affect the lymph nodes, have higher recurrence rates, and have an overall poorer prognosis than the other subtypes [83, 84]. Despite the poor prognosis associated with these breast cancers, several treatments are available, including several drugs specific to the HER2/neu receptor. These drugs include Trastuzumab (commercially known as Herceptin) and Pertuzumab. Both are monoclonal antibodies, or antibodies produced by cloned immune cells from a single parent cell, specifically cancer cells over expressing the HER2/neu receptor [85]. Trastuzumab antibodies are able to readily bind to the HER2/neu receptor on the extracellular component, inhibiting the intracellular domain to down-regulate the downstream pathways, and ultimately reduce proliferation of the breast cancer cells [86]. Pertuzumab functions in the same manner, but instead targets the HER2 receptor directly to prevent dimerization with other receptors [87]. However, these treatments are not without risk: Trastuzumab itself is known, in rare cases, to cause cardiac disorders such as cardiomyopathy [88]. Other treatments less specific to HER2 have also shown to be effective. This includes chemotherapeutic regiments utilizing an anthracycline agent, such as doxorubicin, and a combination of other agents [89]. Unfortunately, hormone therapies do not seem to be effective against HER2+

In summary,HER2 / neu (c-erbB-2) is an oncogene that encodes a transmembrane glycoprotein with tyrosine kinase activity known as p185, which belongs to the family of epidermal growth factor receptors [90, 91]. Its overexpression can be measured by immune-histochemistry or FISH [92] and a good correlation exists between these methods [93–95]. When the results are either 3+ or 0 by immunohystochemisty no further validation is needed, but if the immunotest gives instead a result of 1+ or 2+, the performance of FISH is recommended. Overexpression of HER2/ neu by either technique is a very good predictor of response to Herceptin, but not a very good predictor of response to chemotherapy or overall survival.

8.7 Evaluation of KI67

Ki-67 is a protein found in the nucleus of the cell. The protein was defined by an antibody that was generated by the nuclei of the Hodgkin Lymphoma cell line in immunized mice. The antigen was later discovered to be a protein, but the name Ki-67, after the location of discovery (Kiel University, Germany) and the clone number in the 96-well plate remained the same [96, 97].

Ki-67 is expressed in the G1, S, and G2 phases of cell division and mitosis, but not during the G0 phase, also known as the resting phase [97]. The expression levels change throughout the different phases, with low expression in the G1 and S phases, the highest expression during mitosis [96]. This is followed by a decrease in expression during the anaphase and telophase at the end of mitosis [96]. Due to the presence of Ki-67 in all human cell types during the cell cycle and its quick disappearance from the cell after the end of the proliferative state, it is used as a marker for cell proliferation [96]. While much is known about Ki-67's location in the cell, its expression in relation to cell proliferation, and its primary structure, little is known about its actual function in the cell [97].

The usual methods of finding the most effective treatment for breast cancer are determining the grade, stage, and endocrine status of the tumor. While not required, it is recommended that a proliferation maker such as Ki-67 is used. It also has predictive and prognostic applications. Ki-67 levels are low in healthy breast tissue. In the case of ductal carcinoma in situ, high expression of Ki-67 is associated with a negative prognosis and higher chance for recurrence [96]. Similarly, in triple negative breast cancer, a high ki-67 index is an indicator for a good response to chemotherapy, but a poor overall prognosis [98].

Ki-67 levels are used to indicate the chance of tumor recurrence in relation to time. Tumor types with higher Ki-67 expression, such as HER-2 enriched tumors and triple negative tumors, have a significantly higher initial risk of recurrence. In the Luminal A and Luminal B subtypes, the expression of Ki-67 is low, and therefore the risk of recurrence is much lower [99].

8.8 Molecular Profiling of Breast Cancer

After the publication of the new genomic classification of breast Cancer [100] several adaptation compiling few representative genes have been introduced in the practice of oncology. Among them are the Oncotype Dx that contain 21 gene expression signature and the first major trial was published in 2004 [101]. This study comprises women that are ER+ and LN negative breast cancer. This test was recommended by NCCN and ASCO. The Oncotype was developed from a FFPE compliant assay to predict distant recurrence of ER+ breast cancer and originally selected 250 candidate genes to test on NSABP B-14 and B-20 trials. At the end they refined a 16+5 gene panel that could predict recurrence [101].

The Mammaprint contains 70 gene expression signature and the major trial was from 2002 comprising women <61 years, T1-T2, N0 disease [102].

The Prosigna from Nanostring contains 50 gene expression signature + 5 control genes and represents the old PAM50 assay. The PAM or prediction analysis of microarrays recapitulated the microarray classifier using RT-PCR-based PAM50 assay in comparison to standard clinical molecular markers. The major trial was in 2013 using stage I–IIIi cancer population and cleared by the FDA in 2013. Thirteen years after Perou's paper [100], the PAM50 is entering the clinical arena. The PAM50 gene signature has been transferred to a novel and robust method for mRNA quantification [103]. The methods works well in FFPE, does not rely on amplification of nucleic acids and is intended for kit use in local labs with the proper instruments. The PAM50 expressions results are used to calculate a risk of recurrence score (ROR) and provide Low, intermediate and high risk groups. The score is based on the intrinsic

Table 8.7 Major features of three genomic tests in breast cancer

	Mammaprint	Oncotype Dx	Prosigna
Input material	Fresh frozen FFPE	FFPE	FFPE
Platform	Microarray	qPCR	nCounter
# genes analyzed	70	21	50+5
Target patient population	Stage I-II	ER+ Stage I-II	Stage I-II (Stage 1-III)
Regulatory	FDA cleared (frozen)	NCCN/ASCO guidelines	FDA cleared
Performance site	Central	Central	Decentralized kit format
Features	Binary stratification; molecular subtypes; lots of data	Gold standard; ER+ cancer; now in DCIS	ROR compares well; innovative technology; *intrinsic subtypes*

subtype and pathologic characteristics (T,N), with special weighting given to a set of proliferation associated genes. PAM50 correlates well the Oncotype and the use of the 4 IIC parameters (ER, PR, Her2 and ki67). In Table 8.7 is summarized the major features of these three genomic tests in breast cancer

8.9 General Considerations

For the prognosis of breast cancer, oncologists find it helpful to know the tumor grade as well as the size of the tumor and involvement of the lymph nodes and there is no doubt that the practical oncologist want something more than ER, PR, and HER2 results such Ki-67, a nuclear protein associated with cellular proliferation; PI3K (phosphatidylinositol 3-kinase) mutations; and tumor-infiltrating lymphocytes and how to use next-generation sequencing panels.

ER, PR, and HER2 have established clinical value and are used widely and with confidence. However PR is considered minor when compared with ER that determines most of the time the receptivity for hormone therapy. It is well known that tumors that are ER and PR positive do better than those that are ER positive/PR negative, but PR alone may be sufficient to confer some sensitivity to anti-estrogen therapy and those cases that are PR positive but ER negative remain candidates for endocrine anti estrogen therapy. Therefore the evaluation of PR cannot be underestimated. The biological basis of this is that PR gene is regulated by ER and usually is detected in tumor cells with an activated ER signaling pathway. An important problem that needs to be acknowledged is that there as not good antibodies for PR like the ones that are offered for ER. The values of ER status or hormonal status of the tumors are important in view that some patients with ER-positive tumors benefit from extended hormonal therapy for 10 years instead of the five years typically recommended now.

Whereas HER2 positive tumors are managed by the oncologist the resistance of those tumors to the treatment with HER2-targeted drugs has push to develop the T-DM1, also known as trastuzumab emtansine, a conjugate of Herceptin and a chemotherapeutic agent that appears to be remarkably effective in patients whose tumors are HER2 positive and who have become resistant to Herceptin.

A problem that oncologists and pathologist are facing is the evaluation of Ki67. High Ki-67 is linked with high risk of recurrence in ER-positive cancers and is associated with greater responsiveness to chemotherapy. However the evaluation of Ki67 is not well standardized with large discordance in how labs assign Ki-67 results.

The new generation of prognostic markers that are commercially available are Prosigna (NanoString Technologies), Breast Cancer Index (bioTheranostics), Oncotype DX (Genomic Health), and Mamma-Print (Agendia) already provide prognostic information however they are not widely used and these next-generation tests are not helping the oncologist to make the right decisions about which drug or specific therapy to choose. Even the problem is more severe with the sequencing-based tests will likely be most useful in cases involving patients with metastatic disease and here the problem is how to treat a mutation and is the right drug for the right target available or ready to be used?

In this chapter we wanted that the reader be aware that still we do not have the ultimate prognostic and predictive test and in some cases the right interpretation of the available test is confusing for the oncologist, the pathologist and the patient.

References

1. Aamdal S, Bormer O, Jorgensen O et al (1984) Estrogen receptor and long-term prognosis in breast cancer. Cancer 53:2525–2529
2. Antomiades K, Spector H (1979) Correlation of estrogcn receptor levels with histology and cytomorphology in human mammary cancer. Am J Clin Pathol 71:497–503
3. Bishop HM, Blarney RW, Elston CW, Hoybittle JL, Nicholson RI, Griffiths K (1979) Relationship of estrogen-receptor status to survival in breast cancer. Lancet 1:283–284
4. Black MM, Opler SR, Speer FD (1955) Survival in breast cancer cases in relation to the structure of the primary tumor and regional lymph nodes. Surg Gynecol Obstet 100:543–551
5. Black MM, Speer FD, Opler SR (1956) Structural representations of tumor-host relationships in mammary carcinoma. Biologic prognostic significance. Am J Clin Pathol 26:250–265
6. Black MM, Speer FD (1957) Nuclear structure in cancer tissues. Surg Gynecol Obstet 105:97–102
7. Bloom HJG, Richardson WW (1957) Histologic grading and prognosis in breast cancer: a study of 1409 cases of which 359 have been followed for 15 years. Br J Cancer 11:359–377
8. Bloom HJG, Richardson WW, Field JR (1970) Host resistance and survival in carcinoma of breast. Study of 104 cases of medullary carcinoma in a series of 1411 cases of breast cancer followed for 20 years. Br Med J 3:181–188
9. Cutler SJ, Black MM, Goldenberg IS (1963) Prognostic factor in cancer of the female breast. Cancer 16:1589–1597
10. Cutler SJ, Black MM, Friedell GH, Vidone RA, Goldenberg IS (1966) Prognostic factors in cancer of the female breast. II, Reproducibility of histopathologic classification. Cancer 19:75–82

11. Fisher B, Slack NH (1970) Number oflymph nodcs examined and the prognosis of breast cancer. Surg Gynecol Obstet 131:79–88
12. Fisher B, Glass A, Redmond C et al (1977) L-Phenylalanine mustard (L-PAM) in the management of primary breast cancer: an update of earlier findings and a comparison of the utilizing L.PAM plus 5 fluorouracil (5FU). Cancer 39:2883–2903
13. Fisher ER, Gregorio RM, Fishcr B (1975) The pathology of invasive breast cancer-a syllabus derived from findings of the National Surgical Adjuvant Breast Project (Protocol N. 4). Cancer 36:1–84
14. Fisher ER, Palekar AS, Redmond C, Barton N, Fisher B (1980) Pathologic findings from the National Surgical Adjuvant Breast Project. (Protocol N. 4). VI. Invasive papillary cancer. Am J Clin Pathol 73:313–322
15. Fisher ER, Redmond CK, Lieu H, Rockette H, Fisher B (1980) Collaborating NSABP Investigators: correlation of estrogen receptor and pathologic characteristics of invasive breast cancer. Cancer 45:349–353
16. Foote FW Jr (1959) Surgical pathology of cancer of the breast. In: Parsons WH (ed) Cancer of the breast. Charles C Thomas, Springfield, pp 37–38
17. Hahnel R, Wooding T, Vivian AB (1979) Prognostic value of estrogen receptor in primary breast canccr. Cancer 44:671–675
18. Knight WA, Livingston RB, Bregory EJ, McGuire WL (1977) Estrogen receptor as an independent prognostic factor for early recurrence In breast cancer. Cancer Res 37:4669–4671
19. Lane N, Gokiel H, Salerno RA, Haagensen CD (1961) Clinocopatholo·VoI.88·No.2 RECURRENCE AND SURVIVAL IN BREAST CANCER 131 gic analysis of the surgical curability of breast cancers: a minimum ten year study of a personal series. Am Surg 153:483–498
20. Masters JRW, Sangster K, Hawkins RA, Shivas AA (1976) Elastosis and estrogen receptors in human breast cancer. Br J Cancer 33:342–343
21. Maynard PV, Blaney RN, Elston CW, Haybittle JL, Griffiths K (1978) Oestrogen receptor assay in primary breast cancer and early recurrence of the disease. Cancer Res 38:4292–4295
22. McDivitt RW, Stewart FW, Berg JW (1968) Tumors of the breast, Atlas of tumor pathology, second series. Armed Forces Institute of Pathology, Washington, DC
23. McGuire WL (1975) Current status of estrogen receptors in human breast cancer. Cancer 36:638–644
24. Patchefsky A, Shaber GS, Schwartz GF, Feig SA, Norlinger RE (1977) The pathology of breast cancer detected by mass population screening. Cancer 40:1659–1670
25. Rich MA, Furmanski P, Brooks SC (1978) The Breast Cancer Prognostic Study Surgery and Pathology Associates: prognostic value of estrogen receptor. Cancer Res 38:4296–4298
26. Rosen PP (1979) The pathological classification of human mammary carcinoma, Past, present and future. Ann Clin Lab Sci 144:156
27. Rosen PP, Lieberman PH, Braun DW Jr, Koslolf C, Adair F (1978) Lobular carcinoma *in situ* of the breast. Detailed analysis of 99 patients with average follow up of24 years. Am J Surg Pathol 2:225–251
28. Russo J, Frederick J, Ownby HE, Fine G, Husain M, Krickstein HI, Robbins TO, Rosenberg BF (1987) Predictors of recurrence and survival of breast cancer patients. Am J Clin Pathol 88:132–138
29. Fisher ER, Gregorio R, Redmond C (1976) Pathologic findings from the National Surgical Adjuvant Breast Project (Protocol N.4). II. The significance of regional node histology other than sinus histiocytosis in invasive mammary cancer. Am J Clin Pathol 65:21–30
30. Primary Therapy of Breast Cancer Study Group (1978) Identification of breast cancer patients with high risk of early recurrence after radical mastectomy. II. Clinical and pathological correlation. Cancer 42:2809–2826
31. Kister SJ, Sommers SC, Haagensen CD, Friedell GH, Cooley E, Varma A (1969) Nuclear grade and sinus histiocytosis in cancer of the breast. Cancer 23:570–575

32. McCarthy KS, Barton TK, Fetter BF et al (1980) Correlation of estrogen and progesterone receptors with histologic differentiation in mammary carcinoma. Cancer 46:2851–2858
33. Parl FF, Dupont WD (1982) A retrospective cohort study of histologic risk factors in breast cancer patients. Journal Cancer 50:2410–2416
34. Rich MA, Brennan MG, The Scientific, Pathology and Surgical Associates of the Michigan Cancer Foundation, Breast Cancer Prognostic Study (1980) The breast cancer prognostic program: a study of the metastatic process. In: Brennan MJ, McGrath CL, Rich MA (eds). Academic, New York, pp 29–54
35. Walt AJ, Singhakowinta A, Brooks SC, Cortez A (1976) The surgical implications of estrophile protein estimations in carcinoma of the breast. Surgery 80:506–512
36. Chua DYF, Pang MWY, Rauff A, Aw S, Chan S (1985) Correlation of steroid receptors with histologic differentiation in mammary carcinoma. A Singapore experience. Cancer 56:2228–2234
37. McGuire WL, Carbone PP, Sears MH, Escher GC (1975) Overview on estrogen receptors in human breast cancers. In: McGuire WL, Carbone PP, Vollmer EP (eds) Estrogen receptors on human breast cancer. Raven Press, New York, pp 37–49
38. McGuire WL, Dressler LG (1985) Emerging impact of flow cytometry in predicting recurrence and survival in breast cancer patients. J Natl Cancer Inst 75:405–510
39. Meyer JS, Bauer WC (1977) Thymidine labelling index of human breast carcinomas, First annual meeting of cell kinetics society, Birmingham, Alabama, p 45
40. Meyer JS, Nixon B (1979) Advance stage and early relapse of breast carcinoma associated with high thymidine labeling index. Cancer Res 39:4042–4047
41. Sanfilippo O, Daidone MG, DiFronzo G, Silvestrini R (1979) Short term tissue culture of human breast cancer. Cancer 43:2365–2369
42. Foulkes WD, Smith IE, Reis-Filho JS (2010) Triple-negative breast cancer. N Engl J Med 363(20):1938–1948
43. Nelson LR, Bulun SE (2001) Estrogen production and action. J Am Acad Dermatol 45(3 Suppl):S116–S124
44. Scherbakov AM, Krasil'nikov MA, Kushlinskii NE (2013) Molecular mechanisms of hormone resistance of breast cancer. Bull Exp Biol Med 155(3):384–395
45. Williams C, Lin CY (2013) Oestrogen receptors in breast cancer: basic mechanism and clinical implications. Ecancermedicalscience 7:370
46. Björnström L, Sjöberg M (2005) Mechanisms of estrogen receptor signaling: convergence of genomic and nongenomic actions on target genes. Mol Endocrinol 19(4):833–842
47. Ramaswamy S et al (2001) Multiclass cancer diagnosis using tumor gene expression signatures. Proc Natl Acad Sci 98(26):15149–15154
48. Marchia C, Reis-Filho JS (2008) Molecular diagnosis in breast cancer. Diagn Histopathol 14(5):202–213
49. Barnes DM, Hanby AM (2001) Estogens and progesterone receptors in breast cancer, past, present and future. Histopathology 38:271–274
50. Hawkins RA, Roberts MM, Forrest APM (1980) Estrogen receptors and breast cancer. Current status. Br J Surg 67:162–165
51. Harvey JM, Clark GM, Osborne CK, Allred DC (1999) Estrogen receptor status by immunohistochemistry is superior to the ligand-binding assay for predicting response to adjuvant endocrine therapy in breast cancer. J Clin Oncol 17:1474–1481
52. Regitnig P, Reiner A, Dinges HP, Hofler G, Muller-Holzner E, Lax S, Obrist P, Rudas M, Quehenberger F (2002) Quality assurance for detection of estrogen and progesterone receptor by immunohistochemistry in Austrian pathology laboratories. Virchows Arch 441:328–334
53. Akiyama T, Sudo C et al (1986) The product of the human C-Erbb-2 gene - a 185-kilodalton glycoprotein with tyrosine kinase-activity. Science 232(4758):1644–1646
54. Shih C, Padhy LC et al (1981) Transforming genes of carcinomas and neuroblastomas introduced into mouse fibroblasts. Nature 290(5803):261–264

55. Schechter AL, Stern DF et al (1984) The neu oncogene: an erb-B-related gene encoding a 185,000-Mr tumour antigen. Nature 312(5994):513–516
56. Muleris M, Almeida A et al (1997) Assignment of v-erb-b2 avian erythroblastic leukemia viral oncogene homolog 2 (ERBB2) to human chromosome band 17q21.1 by in situ hybridization. Cytogenet Cell Genet 76(1-2):34–35
57. King CR, Kraus MH et al (1985) Amplification of a novel v-erbB-related gene in a human mammary carcinoma. Science 229(4717):974–976
58. Kraus MH, Issing W et al (1989) Isolation and characterization of ERBB3, a third member of the ERBB/epidermal growth factor receptor family: evidence for overexpression in a subset of human mammary tumors. Proc Natl Acad Sci U S A 86(23):9193–9197
59. Plowman GD, Culouscou JM et al (1993) Ligand-specific activation of HER4/p180erbB4, a fourth member of the epidermal growth factor receptor family. Proc Natl Acad Sci U S A 90(5):1746–1750
60. Robinson DR, Wu YM et al (2000) The protein tyrosine kinase family of the human genome. Oncogene 19(49):5548–5557
61. Herbst RS (2004) Review of epidermal growth factor receptor biology. Int J Radiat Oncol Biol Phys 59(2 Suppl):21–26
62. Hubbard SR, Till JH (2000) Protein tyrosine kinase structure and function. Annu Rev Biochem 69:373–398
63. Sundaresan S, Roberts PE et al (1998) Biological response to ErbB ligands in nontransformed cell lines correlates with a specific pattern of receptor expression. Endocrinology 139(12):4756–4764
64. Sliwkowski MX (2003) Ready to partner. Nat Struct Biol 10(3):158–159
65. Landgraf R (2007) HER2 therapy. HER2 (ERBB2): functional diversity from structurally conserved building blocks. Breast Cancer Res 9(1):202
66. Garrett TPJ, McKern NM et al (2002) Crystal structure of a truncated epidermal growth factor receptor extracellular domain bound to transforming growth factor α. Cell 110(6):763–773
67. Jones RB, Gordus A et al (2006) A quantitative protein interaction network for the ErbB receptors using protein microarrays. Nature 439(7073):168–174
68. Roy V, Perez EA (2009) Beyond trastuzumab: small molecule tyrosine kinase inhibitors in HER-2-positive breast cancer. Oncologist 14(11):1061–1069
69. Olayioye MA, Neve RM et al (2000) The ErbB signaling network: receptor heterodimerization in development and cancer. EMBO J 19(13):3159–3167
70. Pinkas-Kramarski R, Soussan L, Waterman H, Levkowitz G, Alroy I, Klapper L, Lavi S, Seger R, Ratzkin BJ, Sela M et al (1996) Diversification of Neu differentiation factor and epidermal growth factor signaling by combinatorial receptor interactions. EMBO J 15:2452–2467
71. Zhang W, Liu HT (2002) MAPK signal pathways in the regulation of cell proliferation in mammalian cells. Cell Res 12(1):9–18
72. Pearson G, Robinson F et al (2001) Mitogen-activated protein (MAP) kinase pathways: regulation and physiological functions. Endocr Rev 22(2):153–183
73. West KA, Castillo SS et al (2002) Activation of the PI3K/Akt pathway and chemotherapeutic resistance. Drug Resist Updat 5(6):234–248
74. Yano T, Ohtsu A, Boku N, Hashizume K, Nakanishi M, Ochiai A (2006) Comparison of HER2 gene amplification assessed by fluorescence in situ hybridization and HER2 protein expression assessed by immunohistochemistry in gastric cancer. Oncol Rep 15(1):65–71
75. Neve RM, Lane HA et al (2001) The role of overexpressed HER2 in transformation. Ann Oncol 12:9–13
76. Howe LR, Chang SH et al (2005) HER2/neu-induced mammary tumorigenesis and angiogenesis are reduced in cyclooxygenase-2 knockout mice. Cancer Res 65(21):10113–10119
77. Mitri Z, Constantine T et al (2012) The HER2 receptor in breast cancer: pathophysiology, clinical use, and new advances in therapy. Chemother Res Pract 2012:743193

78. Maxwell WL (1994) Wheater's functional histology–a text and colour atlas by H.G. Burkitt, B. Young, and J.W. Heath, 3rd edition, 416 pp; 1993. Churchill Livingston, Edinburgh
79. Allred DC et al (1992) Overexpression of HER-2/< i>neu and its relationship with other prognostic factors change during the progression of in situ to invasive breast cancer. Hum Pathol 23(9):974–979
80. Collins LC, Schnitt SJ (2005) HER2 protein overexpression in estrogen receptor-positive ductal carcinoma in situ of the breast: frequency and implications for tamoxifen therapy. Mod Pathol 18(5):615–620
81. Hoff ER, Tubbs RR et al (2002) HER2/neu amplification in breast cancer: stratification by tumor type and grade. Am J Clin Pathol 117(6):916–921
82. Ramos-Vara JA (2005) Technical aspects of immunohistochemistry. Vet Pathol 42(4):405–426
83. Tan M, Yu D (2007) Molecular mechanisms of ErbB2-mediated breast cancer chemoresistance. Adv Exp Med Biol 608:119–129
84. Creighton CJ (2012) The molecular profile of luminal B breast cancer. Biologics 6:289–297
85. Fendly BM, Winget M et al (1990) Characterization of murine monoclonal-antibodies reactive to either the human epidermal growth-factor receptor or Her2 Neu gene-product. Cancer Res 50(5):1550–1558
86. Goldenberg MM (1999) Trastuzumab, a recombinant DNA-derived humanized monoclonal antibody, a novel agent for the treatment of metastatic breast cancer. Clin Ther 21(2):309–318
87. Scheuer W, Friess T et al (2009) Strongly enhanced antitumor activity of trastuzumab and pertuzumab combination treatment on HER2-positive human xenograft tumor models. Cancer Res 69(24):9330–9336
88. Zeglinski M, Ludke A et al (2011) Trastuzumab-induced cardiac dysfunction: a 'dual-hit'. Exp Clin Cardiol 16(3):70–74
89. Pritchard KI, Shepherd LE et al (2006) HER2 and responsiveness of breast cancer to adjuvant chemotherapy. N Engl J Med 354(20):2103–2111
90. Hung MC, Lau YK (2001) Basic science of HER-2/neu and breast cancer. Diagn Mol Pathol 10:139–152
91. Suo Z, Risberg B, Karlsson MG, Villman K, Skovlund E, Nesland JM (2002) The expression of EGFR family ligands in breast carcinomas. Int J Surg Pathol 10:91–99
92. Zhao J, Wu R, Au A, Marquez A, Yu Y, Shi Z (2002) Determination of HER2 gene amplification by chromogenic in situ hybridization (CISH) in archival breast carcinoma. Mod Pathol 15:657–665
93. Rhodes A, Jasani B, Anderson E, Dodson AR, Balaton AJ (2002) Evaluation of HER2/neu immuohistochemical assay sensitivity and scoring in formalin-fixed and paraffin-processed cell lines and breast tumors. A comparative study involving results from laboratories in 21 countries. Am J Clin Pathol 118:408–417
94. Smith KL, Robbins PD, Dawkins HJ, Papadimitriou JM, Redmond SL, Carrello S, Harvey JM, Sterrett GF (1994) C-erbB-2 amplification in breast cancer. Detection in formalin-fixed paraffin-embedded tissue by in situ hybridization. Hum Pathol 25:413–418
95. Yerushalmi R et al (2010) Ki67 in breast cancer: prognostic and predictive potential. Lancet Oncol 11(2):174–183
96. Scholzen T, Gerdes J (2000) The Ki-67 protein: from the known and the unknown. J Cell Physiol 182(3):311–322
97. Rao C, Shetty J, Prasad KH (2013) Immunohistochemical profile and morphology in triple - negative breast cancers. J Clin Diagn Res 7(7):1361–1365
98. Ribelles N et al (2013) Pattern of recurrence of early breast cancer is different according to intrinsic subtype and proliferation index. Breast Cancer Res 15(5):R98
99. Sørlie T, Perou CM, Fan C, Geisler S, Aas T, Nobel A, Anker G, Akslen LA, Botstein D, Børresen-Dale AL, Lønning PE (2006) Gene expression profiles do not consistently predict the clinical treatment response in locally advanced breast cancer. Mol Cancer Ther 5(11):2914–2918

100. Tang G, Shak S, Paik S, Anderson SJ, Costantino JP, Geyer CE Jr, Mamounas EP, Wickerham DL, Wolmark N (2011) Comparison of the prognostic and predictive utilities of the 21-gene Recurrence Score assay and Adjuvant! for women with node-negative, ER-positive breast cancer: results from NSABP B-14 and NSABP B-20. Breast Cancer Res Treat 127(1):133–142
101. van't Veer LJ, Dai H, van de Vijver MJ, He YD, Hart AA, Mao M, Peterse HL, van der Kooy K, Marton MJ, Witteveen AT, Schreiber GJ, Kerkhoven RM, Roberts C, Linsley PS, Bernards R, Friend SH (2002) Gene expression profiling predicts clinical outcome of breast cancer. Nature 415(6871):530–536
102. van de Vijver MJ, He YD, van't Veer LJ, Dai H, Hart AA, Voskuil DW, Schreiber GJ, Peterse JL, Roberts C, Marton MJ, Parrish M, Atsma D, Witteveen A, Glas A, Delahaye L, van der Velde T, Bartelink H, Rodenhuis S, Rutgers ET, Friend SH, Bernards R (2002) A gene-expression signature as a predictor of survival in breast cancer. N Engl J Med 347(25):1999–2009
103. Dowsett M, Sestak I, Lopez-Knowles E, Sidhu K, Dunbier AK, Cowens JW, Ferree S, Storhoff J, Schaper C, Cuzick J (2013) Comparison of PAM50 risk of recurrence score with oncotype DX and IHC4 for predicting risk of distant recurrence after endocrine therapy. J Clin Oncol 31(22):2783–2790. doi:10.1200/JCO.2012.46.1558, Epub 2013 Jul 1

Chapter 9
Preclincial Models for Studying Breast Cancer

9.1 Introduction

Although the use of cell lines in tissue culture is a widely accepted method for testing drugs [1, 2] , it has been observed that not all the data obtained in the *in vitro* system are translational to the human situation. Therefore, the study in animals is extremely important in all the preclinical studies. As an example the data presented in this chapter summarizes data obtained using xenotransplantation model of breast cancer in which the final malignant phenotype of tumorigenesis has been tested. Tumorigenesis in heterologous hhots is a reliable marker to determine that the transformation of human breast epithelial cells (HBEC) treated in vitro with different carcinogenic agents are really transforming one. For this purpose we will describe different examples on how to use the model of xenotransplantation to verify the malignancy of cell lines, or for detecting the transforming abilities, of oncogenes, chemical carcinogens or hormones. We will also describe the metastatic model most often used in breast cancer research and finally we will describe the development of a unique preclinical model for the triple negative breast cancer.

9.2 Xenotransplantation

The process of transplanting a living tissue, cells, or organs from one species to another is known as xenotransplantation and, the tissue, cells, or organs transplanted are known as xenografts. This process most commonly involves implantation of human tumor cells into immunodeficient mice and is used to test the efficiency of compounds and their interactions with pathways within the body. Because human tumor cells can relatively easily be injected into mice and tumor

J. Russo, *The Pathobiology of Breast Cancer*,
DOI 10.1007/978-3-319-40815-6_9

growth can be routinely established, these models are commonly used for testing compounds. An important criterion of malignancy is the ability of transformed cells to grow in an adequate heterotransplantation system [3]. Inmunologically depressed athymic mice *(nu/nu)* [4–10] have the striking capability of discriminating between normal and neoplastic cells. Normal cells do not induce tumors [6], whereas malignant cells do. In an historical experiment performed by Russo et al. [8, 10] was the demonstration of the tumorigenic capability of the MCF-7 cells that is one the first breast cancer cells established and widely used wide world with more the 24,000 citations in PubMed. The method basically consists in *culturing MCF7 cells as previously described* [8], *and removed from the culture vessel by trypsinization, suspended in phosphate buffer saline (PBS) (1 × 10^6 cells per 0.05 ml) and transplanted in 21-day-old Balb/c (nu/nu) mice into the mammary gland fat pad which was cleared according to the method of DeOme* [9]. *The first experiment demonstrated that MCF 7 cells were unable to growth neither in female nor in male athymic mice* [10]. *The gross examination of the area of cell inoculation and the histological study revealed a complete absence of the inoculated cells; only disorganization of the fat and some fibrosis were observed. Only those mice that have received a transplant of pituitary glands or ovaries from syngenic mice induced the growth of MCF-7 cells. Nine of the eleven (82 %) inoculated female mice that received pituitary grafts developed palpable tumors within 12–18 days after inoculation. The tumors adhered to the skin and underlying muscle. No macroscopic metastatic growths were observed. Eight of the thirteen (61.5 %) inoculated female mice that received ovary grafts developed palpable tumors within 12–18 days. Tumors were attached to the skin and underlying muscles; no metastatic growths were observed. The tumors were small, oblong masses of 1.5–2.5 mm at their largest diameter. They adhered to the dermis of the skin and to the muscle of the abdominal wall. The tumors were firm, of a rubbery consistency, and presented resistance to sectioning. The tumor's vascular bed was well developed. The area of the tumor was easily distinguished from the scar produced by the cauterization and the incision made during the transplant procedure. The histological pattern of the 17 tumors studied was identical. The tumors were composed of nests of cells arranged in either clusters or single or double-row strands. The inoculated epithelial cells were surrounded by a dense stroma formed by collagen fibers and fibroblasts. Blood vessels were scarce in the central portion of the tumor and more abundant in the periphery and in areas of invasion. The cells presented a considerable degree of pleomorphism and atypia. The nucleus was oval with few indentations. The nucleoplasm was pale, and a thin layer of heterochromatin was observed on the inner side of the nuclear envelope. More than two nucleoli per nucleus were frequently observed. Intracellular lumina with cellular detritus within were present in some cells (for more details see* [10]*). When stained with toluidine blue, the cytoplasm of most cells appeared strongly basophilic. A few cells with pale cytoplasm were also observed. Similar epithelial cells were also observed in the dermis of the skin overlying the inoculation site and among muscular fibers of the abdominal wall. The intense fibrous* *reaction observed at the inoculation site and in the dermis was not observed around cells invading skeletal muscle. Mitoses were frequently observed in areas of invasion. No metastases were found in any of the tissues studied; however, clusters of cells attached to the adventitia of blood vessels or adjacent to the perineurium were observed in the periphery of the tumor. Invasion of blood vessels or nerves by neoplastic cells was not observed in serial sections. The tumors observed in mice isografted with pituitary glands or ovaries were indistinguishable.*

The successful heterotransplantation of human tumors [7, 11] and cultured human malignant cells [4, 5] into nude mice has proven to be an excellent model for the study of neoplastic tissue and an effective diagnostic tool for differentiating malignant from benign cells [6]. The growth of MCF-7 cells as tumors in nude mice might be predicted by the malignant nature of the tumor of origin and by the demonstration of several transformation markers [10]. However, MCF-7 cells did not form tumors in all inoculated mice but only in those receiving pituitary or ovarian grafts, thus suggesting a hormone dependency for in vivo growth. The fact that more tumors were observed in mice receiving pituitary grafts (82 %) than in those receiving ovarian grafts (61.5 %) suggested that some pituitary hormone could be involved in the development of these tumors. The inoculation of MCF-7 cells into nude mice induces tumors morphologically similar to the tumor of origin. This property of malignant cells has been described for other cell lines maintained for almost 100 passages in *vitro* [6] and transplanted into nude mice, and for human tumors transplanted into the anterior chamber of the guinea pig eye [12]. MCF-7 cells develop a histological pattern in the nude mice similar to that observed in the tumor of origin. The tumor of origin was an infiltrative ductal carcinoma with productive fibrosis. This same pattern of epithelial cells surrounded by a dense stroma is observed in the mouse, suggesting that it is the neoplastic epithelial cell that elicits a stromal response in the host. This observation was also supported by results obtained in an experimental model developed for the study of infiltrating ductal carcinoma no otherwise specified [13]. The absence of tumors in untreated animals could be explained by an inadequate hormonal milieu for the growth of MCF-7 cells. The fact that the original tumor from which MCF-7 cells were derived was responsive to hormones and that MCF-7 cells still retain specific high-affinity estradiol and progesterone receptors after more than 160 passages in culture, supports this explanation. The utilization of hormonal supplementation in the growth of MCF-7 cells in 1976 [14], suggested the replacing the isografts by hormone pellets [15]. We found out that the use of castrated male, estrogen supplemented, was also suitable for the growth of MCF 7 cells [10]. The removal of the uterus and supplementation with estradiol either as pellets or silastic tube containing 5 mg of 17-β–estradiol in female mice is also a standard procedure [10]. The removal of the uterus avoids the swelling and accumulation of fluid in this organ due to the estrogenic stimulation.

9.2.1 *Xenografts for Testing the Tumorigenicity of Chemically Transformed Cells*

In the experimental system of human breast epithelial cells (HBEC) transformed with chemical carcinogens the SCID mice as heterologous hosts for testing tumorigenicity has been used. SCID mice have an autosomal dominant recessive defect that impairs the rearrangement of antigen receptor genes in both T and B lymphocyte progenitors. SCID mice also lack functional T and B cells, which make these animals a suitable host for hetero-transplantation. Tumorigenesis in heterologous hosts is considered to be the final reliable criterion for assessing complete transformation of human cells [12, 16], even though the validity of this model has been questioned, since many human malignancies or cell lines derived from them are not tumorigenic in the nude mouse [17]. The adequacy of SCID mice for testing tumorigenicity has been validated by the observations that 100 % of the mice inoculated with either T24 or MCF-7 cells developed tumors with a short latency period. SCID mice proved to be more adequate as a host for MCF-7 cells than nude mice, since in our model they did not require estrogen supplementation [14]. The tumors developed in inoculated SCID mice were proven to be of human origin by determination of Alu sequences [18], which showed that they were derived from the cells inoculated and not a host reaction.

Transformation in vitro of MCF10A, an spontaneously immortalized human breast epithelial cell line (for more detail see [10]), with a chemical carcinogen like Benz(a)pyrene (BP) resulted in the formation of tumor in SCID mice. One of the cells derived from the BP transformed MCF A cells was the BPIE expressing tumorigenesis in SCID mice [10], also exhibited the highest colony formation in agar-methocel. BP1-E cells also exhibited a higher colony efficiency and , larger colony size and higher colony number than the other non-tumorigenic clones, which indicated that these phenotypes need to be sequentially expressed prior to the manifestation of the tumorigenic phenotype. Anchorage independent growth is considered to be a predictor of tumorigenesis in other cell systems [12, 19]. Nonetheless, in the experimental model neither anchorage independence nor any of the other parameters served, when considered individually, as an indicator of cell tumorigenic potential. These observations are supported by results on human breast epithelial cells obtained from milk, which have been reported to acquire anchorage independent growth and immortalization after SV40 infection, but do not elicit tumorigenesis in nude mice [12, 19–21] and by the observation that the extended lifespan induced in human breast epithelial cells treated with B[a]P *in vitro* was not accompanied by the expression of anchorage independence or tumorigenicity [22]. Tumorigenesis in a heterologous host emerges in chemically treated immortal cells as a consequence of clonal expansion, in which phenotypes indicative of neoplastic transformation are cumulatively expressed through successive processes of selection over long periods of time. During the process of neoplastic transformation no chromosomal changes were detected. The molecular events that are operational in each phase of the transformation process indicate that each carcinogen induces different degrees of point mutations in codons 12 and 61 of the c-Ha-ras oncogene [23] as an event detectable by the 10th passage post-carcinogen treatment. Other genes such as p53, Rb and erbB2 were differently expressed by the various clones derived from carcinogen treated cells [24, 25].

9.2.2 The Oncogene C-HA-RAS Induces a Tumorigenic Phenotype in Human Breast Epithelial Cells

The levels and localization of *ras* expression in normal and malignant breast tissues have been examined and quantitated by analyzing breast tissue samples for the expression of *ras* related mRNA and p2l ras protein which has been found to be expressed in biopsies of both normal and malignant breast tissues [26]. However, whether the *ras* oncogene is a causative agent of human breast cancer have not been proved as yet. Therefore, one way to evaluate the contribution of *ras* genes in the development of the tumorigenic phenotype is to introduce this gene into suitable acceptor cells. Transfection of the non tumorigenic cell lines, previously treated with DMBA, or BP such as clones D3-1 and BP1 with the cHa-ras oncogene not only enhanced colony formation in agar-methocel and invasiveness but induced tumorigenicity with a short latency period in SCID mice [10].

The MCF-10F cells, DMBA or BP-treated cells and the clones D3-1 and BPI did not exhibit tumorigenicity in SCID mice. We have already shown that the subclone BP1-E derived from BP1, expressed the tumorigenic phenotype after 101 days of inoculation whereas the MCF-10F-Tras had lower tumorigenicity because it took 99 days to induce tumors in 3/14 animals, the clones D3-1-Tras and BPI-Tras were highly tumorigenic and the tumors appeared between 47 and 60 days post-inoculation, in 4 out of 4 animals and 11 out of 14 animals, respectively [10]. All the tumors derived from D3-1-Tras and BP1-Tras cells were poorly differentiated adenocarcinomas. They were immunocytochemically positive for keratin whereas the human milk fat globule membrane antigen (HMFGMA) was frankly expressed only in tumors induced by BP1-E cells; tumors derived from *c-Ha-ras* transfected cells showed either a notably reduced expression of this antigen, as observed in D3-1-Tras induced tumors, in which only 10 % of the tumor cells were positive, or complete abolishment of HMFGMA reactivity, as in the BP1 -Tras induced tumor cells. Tumor cell lines derived from the tumors thus originated has been an important resource for understanding the molecular basis of mammary carcinogenesis [10].

9.2.3 Tumorigenicity of 17-β–Estradiol Transformed Human Breast Epithelial Cells

We have developed an in vitro-in vivo model of cells transformation using 17 beta estradiol as a carcinogenic agent and MCF10F as the host cells (see details in [10]). The invasive bsMCF cells that was obtained by treating MFCF10F cells with 70nM of 17-β–estradiol were plated at low density and cell colonies were isolated using cloning rings. The cells were cultured in Dulbecco's modified Eagle medium (DMEM): F12 medium containing 1.05 mM calcium, antimycotics, hormones, growth factors, and equine serum [27]. After trypsinization and plating, the clones obtained were identified as bcMCF-1, bcMCF-2, bcMCF-3, bcMCF-4, bcMCF-5, bcMCF-6 and bcMCF-7. The tumorigenic ability of the bcMCF subclones was tested by injecting them into the mammary fat pad of 45-day-old female SCID mice (see details in [10])

MCF-10F cells that after treatment with 17β-estradiol (E_2) expressed high colony efficiency (CE) and loss of ductulogenic capacity in collagen-matrix represented the first level of *in vitro* transformation. Cells expressing these two parameters were classified as transformed (trMCF), which after further selection for invasiveness in a Matrigel invasion chamber originated the second level of transformation: the invasive (bsMCF) and the cloned (bcMCF) cells [28]. The bsMCF cells formed tumors in SCID mice from which four cell lines, caMCF, were derived. By ring cloning, seven subclones were isolated from the invasive bsMCF cells: bcMCF-1, bcMCF-2, bcMCF-3, bcMCF-4, bcMCF-5, bcMCF-6 and bcMCF-7. All the bcMCF subclones produced invasive poorly differentiated tumors in SCID mice with different morphological phenotypes: spindle cell type (bcMCF-1 and bcMCF-4), epithelial cell type (bcMCF-2, bcMCF-6 and bcMCF-7) and, with mix features of spindle and epithelial type (bcMCF-3 and bcMCF-5) (Table 8.1). As it was previously reported, MCF-10F cells were seeded on Boyden chamber as control; cells that passed through the membrane were selected, expanded and injected in SCID mice; these cells did not produced tumors [27].

9.3 The Labeling of Cancer Cells for an *In Vivo* Imaging System

Many times, it is necessary to follow the growth and migration of cancer cells while the animal is still living [29]. There is a plethora of reporter systems that are used to introduce and select for gene target in cells. These systems have been historically used to study protein expression, interaction between proteins, and function of the proteins in the cells. These reporter genes confer drug resistance, bioluminescence or fluorescence properties into the cells they are introduced. Typical reporter studies link reporter genes directly to a promoter region of interest, the function of which can be monitored by the reporter activity. Tagging of fused proteins is used to detect intracellular localization, degradation, protein-protein interactions, etc. The most common fusion tags used in research are fluorescent proteins (e.g. eGFP) or small protein epitopes (e.g. FLAG, Myc HA) which can be detected by fluorescence FACS or western blots [30]. One method of *in vivo* imaging involves in vitro labeling of cancer cells prior to injection in animals. The labeling of cells can be done using a vector that encodes the luciferase reporter gene isolated from *Photinus pyralis*. This gene has been optimized for mammalian expression. These vectors often contain selection markers for prokaryotes and eukaryotes. Cells can be transfected with the plasmid through different techniques: electroporation, lipid based transfection, calcium phosphate transfection, etc. Once transfected, cells that incorporated the vector are selected with the appropriate antibiotic. Because the gene is incorporated into the genome, as cells divide, the daughter cells with also contain the gene allowing for complete tracking of tumor and cancer cell growth (For details see reference Book). Luciferase positive cells are then injected in animals. Non-invasive *in vivo* study of experimental tumor formation and/or metastasis with bioluminescent imaging allows the study of tumor distribution, growth, and regression in individual animals. This technology permits repeated measurements over an extended time period in living animals without the need to sacrifice animals at pre-set time points [31].

All imaging is performed using an imaging system, such as the "Perkin-Elmer IVIS Spectrum in vivo imaging system". This system is designed to image bioluminescent or fluorescent signals in animals. To visualize injected cancer cells *in vivo*, animals are anesthetized and then transferred to the imaging system, where the animals are imaged. For animals bearing tumors expressing firefly luciferase, an injection of the appropriate luciferin substrate dissolved in phosphate buffered saline at a physiological concentration is injected intraperitoneally. Then the animals are placed in the imaging system, and the light emitted by the tumors is measured. Animals bearing tumors that produce fluorescent markers (enhanced green fluorescent protein, red fluorescent protein, etc.) may be imaged, but injection of substrates is not needed. However, animals will be illuminated at the excitation frequency of the fluorophore for brief periods (generally a few seconds). Since the animals are under a general anesthetic they suffer no stress or discomfort due to the substrate injections or illumination (For more details see reference Book)

We have used MDB.MA-231 cells transfected by Lipofectamine/Plus Reagent from Life Technologies. The plasmid used for the transfection was pGL4.51 [luc2/CMV/Neo] from Promega. Transfected cells were selected occurred over a period of 10–14 days using Neomycin (Invitrogen). After selection, cells were allowed to expand. To ensure the presence of luciferase in the cells, a Luciferase Assay (Promega) was performed using the EnVision Workstation plate reader. Two million (2×10^6) cells suspended in PBS were injected into the lateral tail vein using a 26 gauge needle. The animals were followed over a period of 3 weeks to determine the location of the MDA-MB-231 cells. Weekly inspection of the animals allow to follow up closely the process of nesting the metastatic cells (for more details see [1]).

For small animal the procedure basically consist in injecting 2×10^6 cells suspended in 200ul Dulbecco's PBS (Calcium and Magnesium Free) per animal. The animals must be weight and marked with an individual identifying mark. For making the vein tail injection must be considered that the mouse tail is shaped like a cube, with a vein on each side. Use one of the lateral veins, using a 26 gauge needle. If the operator is in the vein the injected material will be seen go down the length of the vein. If the operator misses the vein it'll get a "blib". In some cases the cells remain there and the tumor is formed in the tail of the animals. The largest volume that should be injected is 5cc but the mouse may go into shock.

When using mice as a model for metastasis assays, the site of injection is important for proper absorption or distribution of additive. For intra venous (IV) injections, the Laboratory Animal Sciences Program of the National Cancer Institute recommends the use of the lateral tail vein. Although this injection can prove difficult because of the small size of the area of injection, it is readily used in metastasis assays and shown to be an effective means cell administration and this is the standard model for metastatic assays [32, 33]. Another option for injection site to observe metastasis is a intra-ventricular injection. This method is used less in literature due to the location of the injection site [33]. In order to inject drugs or molecules to be tested in the lateral ventricles of the brain of experimental animals, a cannula is implanted several days before the experiments, under anesthesia. In most of the protocols, the 3-D location of the cannula is reported (antero-posterior, lateral, and depth). The drug or vehicle is injected at a pre-set schedule using a microinjector at a high rate or at a low rate. Special attention should be paid to be volume of fluid injected, as it will directly depend on the type of animal used and the age of the animal. For example, Passini et al. injected a volume of 2ul into each lateral ventricle of neonate mice with a finely drawn glass micropipette [34]. When an experiment calls for multiple injections into the lateral ventricle/s of the brain, subcutaneously implanted osmotic mini-pump connected to an implanted indwelling lateral ventricle cannula similar to that used for the acute infusion studies can be used. At the end of the infusion, the cannula is closed, the infusion pump removed under anesthesia, and the animals remain alive until the end of the experiment, time at which they are sacrificed and the tissue dissected and processed [35].

Evaluation of metastatic foci in the lungs and liver is a relatively simple, yet a laborious procedure. After dissection, the lungs and liver are washed in PBS and are placed in Bouin's fixative for a minimum of 24 hours. Metastatic foci will appear white whereas the normal tissue will stain yellow. Quantification of surface metastasis can be done by counting the number of surface metastatic foci under a stereo microscope or by the naked eye. The number of foci can be expressed as a total sum, tumor volume, percentage, or metastatic score [36–40]. The presence of microscopic metastatic foci can be histologically examined by embedding the organ in a paraffin block and staining histological sections with Hematoxylin and Eosin (Fig. 9.8a and 9.8b) [36–38].

9.3.1 A Model for Triple Negative Breast Cancer

Triple negative breast cancer (TNBC) tends to have a much poorer prognosis than other subtypes of breast cancer, accounts for 10–15 % prevalence of cases. TNBC is a highly diverse group of cancers. These tumors are of higher histological grade, affect more young women, are more likely to recur early and metastasize to distant sites [41]. Treatment of patients with TNBC has been challenging due to the heterogeneity of the disease and absence of well-defined molecular targets [42]. TNBC cell lines and related animal models are essential tools to develop therapeutics for TNBC. Of seventeen TNBC cell lines listed in American Type Culture Collection (ATCC) TNBC panel 3, seven cell lines including BT-20, BT-549, DU4475, HCC1806, MDA-MB-157, MDA-MB-231, and MDA-MB-468 are described to be tumorigenic in mice [43–48]. Besides these seven cell lines, two other TNBC cell lines, Sum149 and Sum159, are also widely used for *in vivo* studies [49, 50]. Compared to the diversity of TNBC, the number of available TNBC cell lines that can be used for *in vivo* studies is limited. In addition, these cell lines are usually established from the primary or metastatic tumors and lack parental cell lines at early stages. The transformation of normal cells to malignant cells is a multistep process that involves the accumulation of genetic and epigenetic changes. The use of a cell model in which normal cells are progressively transformed into malignant cells facilitates the identification and characterization of genes and pathways responsible for the progression thus providing new insights for the treatment. We have developed a unique cell model consisting of a series of cell lines and which presents with EMT during the progression [21, 27, 28, 51]. The baseline cell of this model is MCF10F, a spontaneously immortalized normal-like triple negative human breast epithelial cell line [21]. MCF10F cell line treated with 17-β estradiol for two weeks exhibited features of transformation and was named trMCF. The trMCF cells were then plated in Boyden chambers, and the invaded cells were selected and named bsMCF. The bsMCF cell line showed characteristics of EMT; it was highly invasive in Matrigel chamber, and tumorigenic in SCID mice [27]. bsMCF cells were also metastatic in SCID mice when injected into the tail vein. However the development of lung metastases required injection of over 2×10^6 cells/mouse which killed some mice during injection. Here, we report the development and characterization of two additional cell lines with high tumorigenic and metastatic capabilities. The two new cell lines, named as XtMCF and LmMCF, were derived from xenograft tumor and lung metastasis of luciferase transfected bsMCF cells, respectively. Moreover, we demonstrated that XtMCF and LmMCF cells have undergone EMT, show $CD24^{weak}/CD44^{+}/EpCAM^{+}$ CSC properties, and the EGF like domain of EpCAM in these cells is cleaved off.

9.4 Development of XtMCF and LmMCF Cell Lines

Our laboratory has established a cell model consisting of cell line MCF10F, trMCF, and bsMCF [27, 28], which represents the initiation and transformation of TNBC. The bsMCF cell line has undergone EMT completely; it is tumorigenic, and metastatic in CB17/SCID mice. Due to the large number of cells per injection required to develop lung metastases which, is inconvenient for *in vivo* studies, we sought to develop new derivative cell lines with higher tumorigenic and metastatic capacity. We established xenograft and model by injecting bsMCF-luc cells into MFP or tail vein of CB17/SCID mice. For xenograft model, all five mice injected with 3×10^6 bsMCF-luc cells developed xenografts (Fig. 9.1). Xenografts from two mice were used to derive cell lines. Two cell lines were developed with no difference in cell morphology and expression of E-cadherin and vimentin, thus one of two cell lines was chosen to use in the following study and referred as XtMCF cell line. For lung metastatic model, all five mice injected with 2×10^6 bsMCF-luc cells developed lung metastases. Two lung tumors from two individual mice were used to develop cell lines. Only one cell line was established and was named LmMCF cell line (Fig. 9.2a).

MTT assay showed trMCF, bsMCF-luc, XtMCF, and LmMCF cell lines had roughly similar growth speeds (Fig. 9.2b). Morphologically, trMCF cells grew as inter-connected colonies of polygonal cells. bsMCF cells were polygonal cells which grew as fibroblast-like cells. There was no difference in morphology between bsMCF and bsMCF-luc cells. XtMCF cells were very similar to bsMCF-luc cells. LmMCF cell size was smaller than bsMCF-luc, showing multiple elongated filopodia (usually more than four filopodia) and enlarged spin head at the tip of filopodia (Fig. 9.2c).

To check the chromosomal abnormalities, XtMCF and LmMCF cells at passage 10 were used for karyotype analysis. Both cell lines were aneuploidy female (the karyotypes are shown in Fig. 9.2d). For XtMCF, modal number was 76 to 80 (4n), range was 71 to 95. The karyotype was presented as: 76 ~ 80:

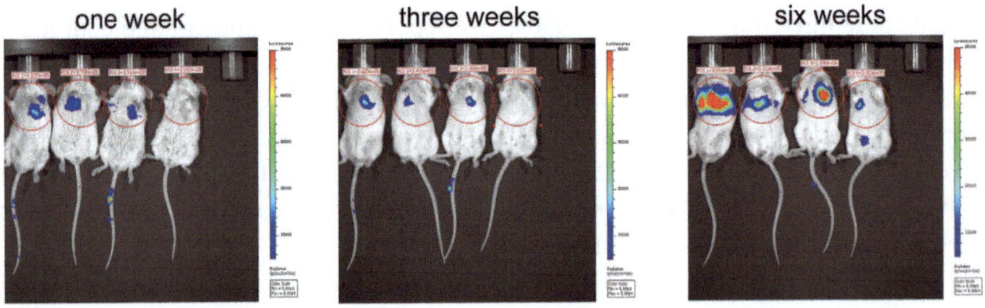

Fig. 9.1 Tumor growth curves. CB17/SCID mice received a single injection of 3×10^6 cells to MFP; tumor growth was monitored twice a week. trMCF was not tumorigenic. bsMCF-luc and MDA-MB-231 had similar tumor growth dynamic in first 6 weeks and then bsMCF-luc exceeded MDA-MB-231

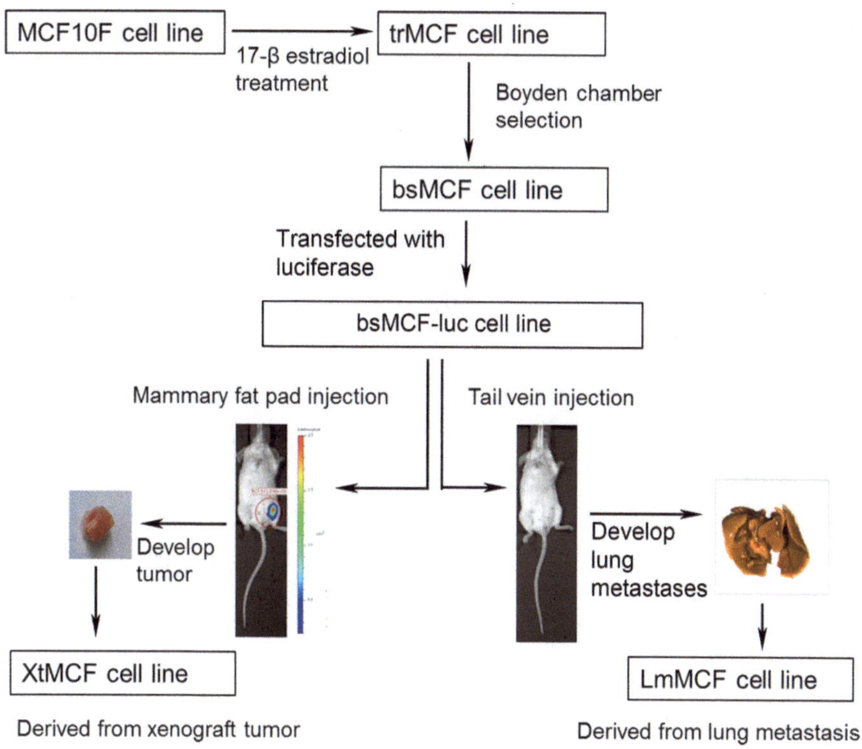

Fig. 9.2 Development of two new TNBC cell lines. (**a**) Schematic representation of establishment of a TNBC model. (**b**) Cell growth curve by MTT assay. Cells were plated in 96-well plate at 500 cells/well, the proliferation was measured in four consecutive days starting from one day post plating. (**c**) Morphological images taken by phase contrast microscope. *Arrow* indicates the filopodium and enlarged head shown by LmMCF cells. Magnification, 200×. (**d**) Karyotype analysis of XtMCF and LmMCF cell line

XX,-X,-X,add(1)(p36.2),add(2)(q21?),-2,der(3)t(3;?)(q11;?),del(3) (p13),-7,-8,-8,-9,-9,+11,-16,-18,-20,-21[cp20]. For LmMCF, modal number was 79 to 83 (4n), range was 68 to 83. The karyotype was presented as: 79 ~ 83: XX, -X,del(X)(q26),add(1)(p36.2)*x*2,-2,der(3)t(3;?)(q11;?),der(9)t(1;9)(p11;q34),-12,+14,-18,-20,-21,-22,der(22)t(1;22)(q10;p11) [cp20]. Both of XtMCF and LmMCF had the addition of chromosome 1p36.2, which was one of the characteristic changes in MCF10F cell line, and was present in xenografts of bsMCF cells and cell lines derived from these xenografts [27].

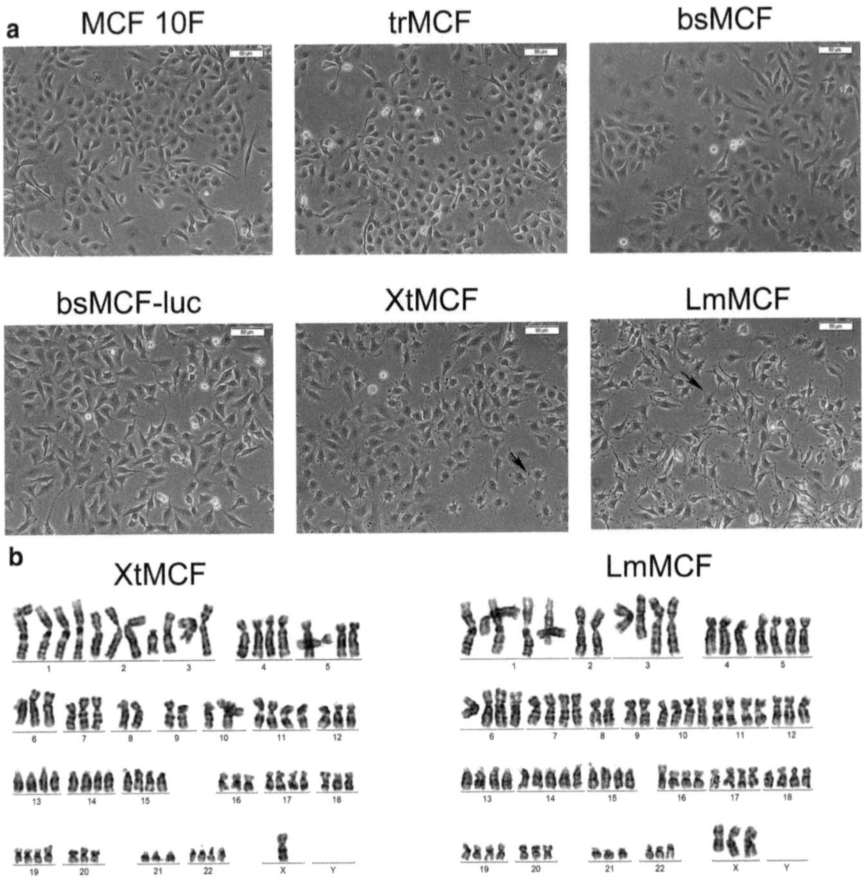

Fig. 9.3 Molecular classification of XtMCF and LmMCF cell lines. (**a**) IF staining of ER alpha, PgR, HER2 and CK5, CK19 and CK18. Magnification: 400×. The staining was overlapped with DAPI (blue) to show nuclei. bsMCF-luc, XtMCF, and LmMCF are classified to basal B cell lines. (**b**) WB analysis of CK18 and EMT markers. (**c**) IF staining of EMT markers. Magnification, 400×. bsMCF-luc, XtMCF, and LmMCF underwent a complete EMT process. CK18 fluorescence was exposed for 20 milliseconds for all cell lines and then 100 milliseconds (long exposure) to show the expression in bsMCF-lus, XtMCF and LmMCF cells

9.4.1 Molecular Characterization of XtMCF and LmMCF Cells

The two new cell lines were characterized using antibodies frequently used for the classification of breast cancer. XtMCF and LmMCF cells were TNBC cells (Fig. 9.3a). CK5 was positive in 100 % of trMCF cells and decreased by 54.3 % ± 4.1 % in bsMCF-luc cells. XtMCF and LmMCF were totally negative for CK5. CK18 was significantly reduced in bsMCF-luc, XtMCF, and LmMCF cells compared to trMCF cells; CK19 was negative in all four cell lines (Fig. 9.3a). The assessment of EMT status showed E-cadherin was positive in trMCF cells but undetectable by IF staining in bsMCF-luc, XtMCF and LmMCF cells. Vimentin was positive only in 10.1 % ± 1.3 % of trMCF cells; in contrast, it was present in all bsMCF-luc, XtMCF, and LmMCF cells (Fig. 9.3c). The expression of CK18, E-cadherin, and vimentin were confirmed by WB (Fig. 9.3b). Based on these results, bsMCF-luc, XtMCF, and LmMCF cell lines were classified as basal B TNBC cell lines [52].

9.4.2 *XtMCF and LmMCF Cells Are Differed from bsMCF-luc Cells in Migration, Solid Masses Formation, and Colony Formation Capacity*

We next examined if XtMCF and LmMCF showed different phenotypes from bsMCF-luc *in vitro*. Cell migration was investigated using wound healing assay (Fig. 9.4a). Quantification of cell movement over 17 hours showed XtMCF cells migrated faster than bsMCF-luc and LmMCF cells. There was no difference in migration between bsMCF-luc and LmMCF cells (Fig. 9.4b).

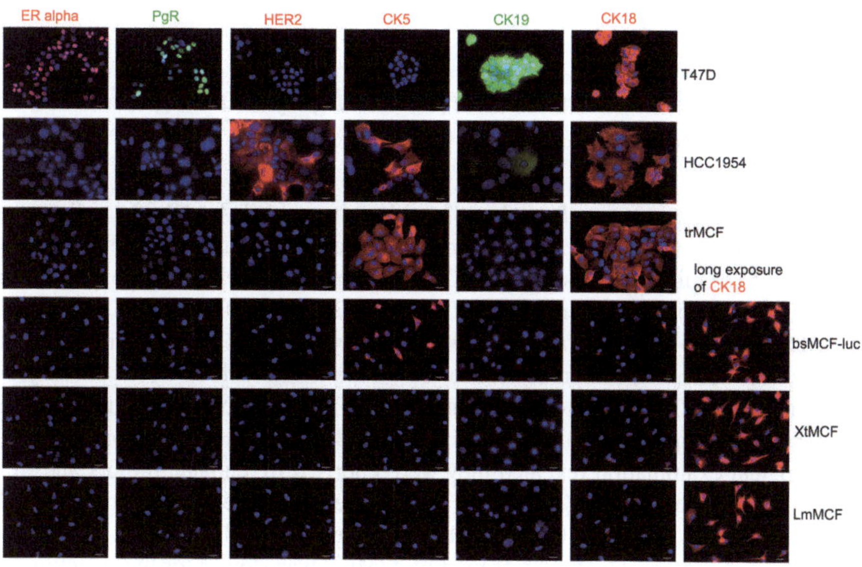

Fig. 9.4 *In vitro* phenotypes of XtMCF and LmMCF cells. (**a**) Wound healing assay over 17 hours of culture. Magnification, 100×. (**b**) Quantification of wound healing assay shows that XtMCF cells move faster than bsMCF-luc and LmMCF cells. * indicates $p<0.05$. (**c**) 3D culture in bovine type I collagen. Cells were mixed with collagen and plated onto pre-coated 24-well plate at 1500 cells/well; pictures of structures formed in collagen were acquired after 6 days of culture. (**d**) XtMCF and LmMCF cells form fewer colonies than bsMCF-luc cells in methylcellulose. ** indicates $p<0.01$ compared to bsMCF-luc cells. (**e**) Anchorage independent growth in methylcellulose. Single cell suspension was mixed with methylcellulose and plated onto agar coated 24-well plate at 1500 cells/well. Pictures were taken after 10 days of culture. (**f**) XtMCF cells form larger colonies than bsMCF-luc and LmMCF, whereas LmMCF cells form smaller colonies than other cell lines in methylcellulose. ** indicates $p<0.01$ compared to bsMCF-luc cells

We previously reported that MCF10F cells can form ducts in collagen which indicated these cells are well differentiated [53]. bsMCF (or C5) loses the ability to form ducts, but instead forms solid mass and grows in clusters in the collagen [54, 55]. No difference was observed for the growth pattern of bsMCF and bsMCF-luc cells in collagen. The masses were usually tightly packed with protrusions invading into the surrounding collagen reminiscing the growth and invasion of primary tumors *in vivo*. Strikingly, XtMCF and LmMCF grew in clusters only; the formation of mass was very rare (Fig. 9.4c).

Anchorage independent growth of cells in agar has been closely related to the process of *in vivo* carcinogenesis. Our results showed that XtMCF and LmMCF cells formed significantly less number of colonies compared to bsMCF-luc (Fig. 9.4d, e). The morphology of the colonies also varied among the three cell lines. Colonies of bsMCF-luc and LmMCF cells were more circular, whereas about half of colonies of XtMCF cells were oval-like (Fig. 9.4e). Notably, budding from the surface of colonies (arrows in Fig. 9.4e) was frequently observed in colonies from XtMCF and LmMCF, but not bsMCF-luc cells. Besides that, LmMCF cells also grew as cell clusters or aggregates (arrow head in Fig. 9.4e) in methylcellulose. Quantification showed XtMCF formed larger colonies, whereas LmMCF formed smaller colonies compared to bsMCF-luc cells (Fig. 9.4f).

9.4.3 XtMCF and LmMCF Cells Are Highly Tumorigenic and Metastatic In Vivo

Preliminary experiments where 3×10^6 bsMCF-luc or MDA-MB-231 cells were injected into the MFP of CB17/SCID mice showed the formation of xenografts in 100 % (5/5) of mice. Tumors reached 10 mm in diameter seven weeks post injection (Fig. 9.1). Compared to xenografts of bsMCF-luc and MDA-MB-231 cell lines, XtMCF and LmMCF xenografts grew significantly faster even with the injection of only 2×10^6 cells. The tumors started to grow exponentially three weeks post injection and reached 10 mm in diameter in 30 days (Fig. 9.5a, b). These tumors were highly invasive; half of the tumors invaded to the skin or muscles of abdominal wall. Histological examination revealed they were poorly differentiated tumors and frequently invaded to muscles (Fig. 9.5d). To further evaluate the tumorigenic potential of these two cell lines, 1×10^5 and 5×10^4 cells were injected into MFP, and 100 % mice formed tumors. The tumor growth of XtMCF was slower for the injection of 1×10^5 and 5×10^4 cells compared to the tumor growth when 2×10^6 cells were injected. However tumor growth of LmMCF cells was almost the same for all three cell concentrations (Fig. 9.5c). Tumor weights at sacrifice are shown in Fig. 9.5e.

The metastatic capacity of the developed cells was evaluated by tail vein injection. With the injection of 1×10^6 cells, the whole lung surface was filled with tumors and it was difficult to count tumor foci at sacrifice 25 days post injection (Fig. 9.5a). The left lobe was more affected than the right lobes in all mice. Histological examination showed metastases present both on the lung surface and

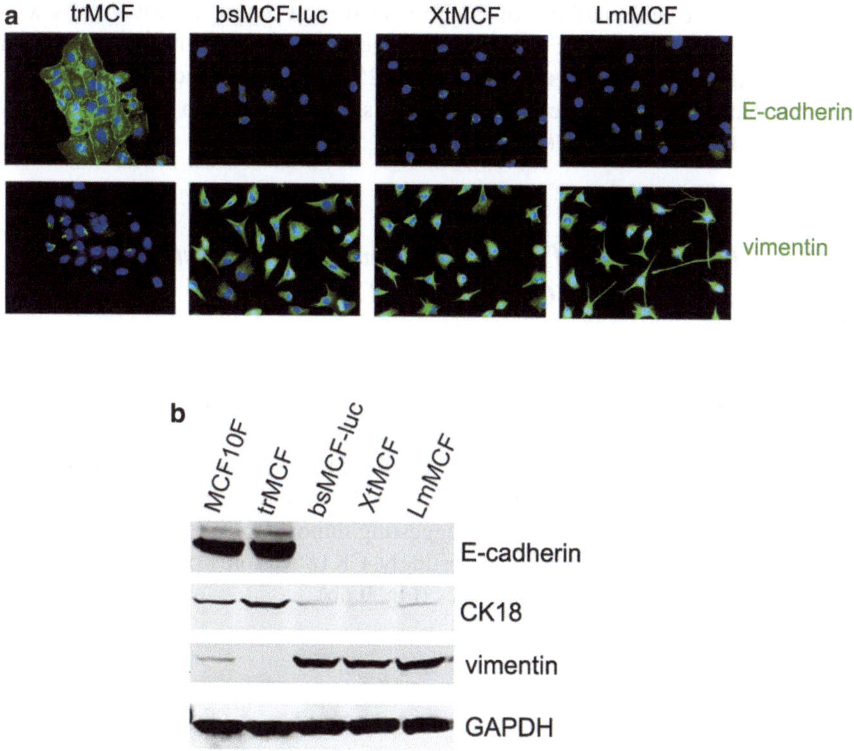

Fig. 9.5 XtMCF and LmMCF cells display high tumorigenic and metastatic potential. (**a**) Pictures of xenografts and lungs fixed with Buin's solution. Magnification, 0.8×. (**b**) Tumor growth curves. CB17/SCID mice received a single injection of 2×10^6 cells to MFP; tumor growth was monitored twice a week. Y-axis: days after injection. (**c**) Tumor growth curves. CB17/SCID mice received a single injection of 1×10^5 or 5×10^4 cells to MFP. (**d**) H&E staining of xenografts, *arrows* indicate the muscle cells. Magnification, 400×. XtMCF and LmMCF tumor cells are highly invasive. (**e**) Evaluation of tumorigenicity. Cell number, days post-injection at sacrifice, tumor frequency (tumor formed/injected mice), and tumor weight at sacrifice are shown in table. (**f**) Evaluation of metastatic potential. Cell number, days post-injection at sacrifice, lung metastasis frequency (number of mice with lung metastasis/injected mice), and tumor foci on lung surface at sacrifice are shown in table. (**g**) H&E staining of lungs from the injection of 1×10^6 cells into tail vein. LmMCF cells are more metastatic than XtMCF cells. *Arrows* indicate the metastases. Magnifications are shown in figure. (**h**) H&E staining of liver and heart from the mice injected with tumor cells into tail vein. *Arrows* indicate metastases. (**i**) Bouin's solution fixed Liver and heart from tail vein injection of 1×10^6 LmMCF cells. *Arrows* indicate metastases. Magnification, 0.8×

inside of the lung (Fig. 9.5f, g). The amount of metastasis foci was higher in the LmMCF model than in XtMCF model. Besides the metastases in the lung, 1/5 of mice in XtMCF model and 1/6 of mice in LmMCF model revealed metastases in the liver (Fig. 9.5h, i). Worth mentioning is the fact that 1/6 of mice in LmMCF model also showed metastasis to pericardium, one of the common sites for breast cancer metastasis [56] (Fig. 9.5h, i). Strikingly, only 6×10^5 XtMCF cells were able to form micro-metastasis in the lungs of 5/6 mice 18 days post injection.

Remarkably, the lung surfaces of mice injected with 5×10^5 LmMCF cells were filled with tumors even just 18 days post injection. We then injected mice with 8×10^4 LmMCF cells and sacrificed mice 25 days post injection. The results showed that even this reduced number of cells was sufficient to form lung metastases in 100 % (5/5) of mice (Fig. 9.5a, f).

9.4.4 Classification of Xenografts and Lung Metastases Formed by XtMCF and LmMCF Cells

IHC staining was performed to classify the xenografts and lung metastases. Consistent to the *in vitro* data (Fig. 9.3), these xenografts and lung metastases were TNBC. The tumor cells were highly proliferative, Ki67 positive cells was 34.6±9.1 % and 40±12.3 % for XtMCF and LmMCF xenografts, respectively, and 21.6±4.7 % for LmMCF lung metastases (Fig. 9.6a). CK5, CK14, and CK19 were negative and vimentin was positive, suggesting tumors from XtMCF and LmMCF cell lines were basal-like TNBC. Interestingly, CK18 was only positive in the xenografts but not present in lung metastases (Fig. 9.6b).

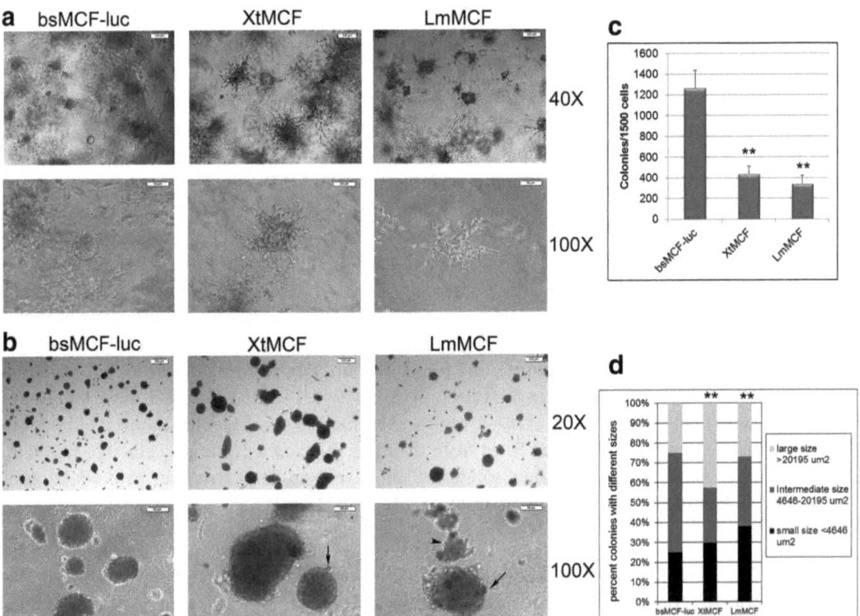

Fig. 9.6 Tumors developed by XtMCF and LmMCF cells injection in CB17/SCID mice are basal B TNBC. Xenografts from the mammary fat pad injection of 2×10^6 cells and lungs from the tail vein injection of 1×10^6 cells were used to construct TMA block. Human breast cancer samples were included in the same block for staining control. Paraffin sections were stained with antibodies indicated. Representative images are shown. Magnification, 400×

9.4.5 XtMCF and LmMCF Cells Present CD24^{weak}/CD44^{+}/EpCAM^{+} Cancer Stem Cells Properties, EGF-Like Domain of EpCAM Is Cleaved Off

We performed tumorsphere formation assay, which is widely used to enrich and expand potential CSC [57, 58]. Under non-adherent conditions, all three cell lines generated tumorspheres, demonstrating their stem-like properties. Consistent to *in vivo* tumorigenicity data, both XtMCF and LmMCF cells produced more tumorspheres than bsMCF-luc cells. Of note, LmMCF cells formed more spheres than XtMCF (Fig. 9.7a, b).

We next evaluated CSC markers CD24 and CD44 by IF staining. CD24 was weakly expressed in all cell lines of this model. CD44 was weakly positive in some of MCF10F and trMCF cells. In contrast, all bsMCF-luc, XtMCF, and LmMCF cells were positive for CD44 (Fig. 9.7c), confirming that EMT contributes to generate CD24^{weak}/CD44^{+} cells [59]. Quantification of CD44 fluorescent intensity did not show significant difference among these three cell lines. Next, we examined the expression of epithelial cell adhesion molecule (EpCAM). Initially, the EpCAM antibody we used was from Abbiotech. This antibody recognizes amino acid (AA) 55 to 150 which corresponds to thyroglobulin repeat-like domain and a part of cysteine poor region (Fig. 9.7e). IF staining showed that the staining mainly located in cytoplasm and nuclei, was weak in MCF10F and trMCF cells, whereas it exhibited increasing staining from bsMCF-luc to XtMCF and LmMCF cells (Fig. 9.7c, d). This staining pattern was observed in the study using the antibody to the thyroglobulin repeat-like domain (2G8) or cysteine poor region (311-K1) [60]. As this antibody was not suitable for WB, the antibody EpCAM[VU1D9] which recognizes the EGF-like domain was used. A strong 38 KDa band was detected in epithelial cell lines MCF7, trMCF, and MCF10F, and a weak band in MDA-MB-231. It was not detectable in mesenchymal-like cell lines bsMCF-luc, XtMCF, LmMCF, and Sum159pt (Fig. 9.7e). IF staining of EpCAM[VU1D9] revealed the staining was located to the cell surface, and was only shown in epithelial cells but not in mesenchymal-like cells (Fig. 9.7c). To confirm the expression of EpCAM in mesenchymal-like cells, the antibody EpCAM[E144], which recognizes the cytoplasmic tail, was used. The reactivity of EpCAM[E144] was not satisfactory, only epithelial cell lines showed a weak 38 KDa band under both reducing and non-reducing condition (Fig. 9.7f). By IF staining, EpCAM[E144] detected both cell surface and intracellular EpCAM in epithelial cell lines, but only intracellular EpCAM in mesenchymal-like cell lines (Fig. 9.7c). We next validated our observations in more breast cancer cell lines. Consistently, EpCAM[VU1D9] was not detected in mesenchymal-like cell lines (Figs. 9.7g and 9.8), EpCAM(Abbiotech) and EpCAM [E144] were detected in all cell lines with the stronger expression in luminal cell lines by IF staining. These data indicate that EGF-like domain at N-terminal EpCAM is cleaved off in cells which have undergone EMT.

We then assessed the expression of CSC markers in xenografts and metastases. Consistent to what we have already reported, CD24 was weak in the xenografts and

a

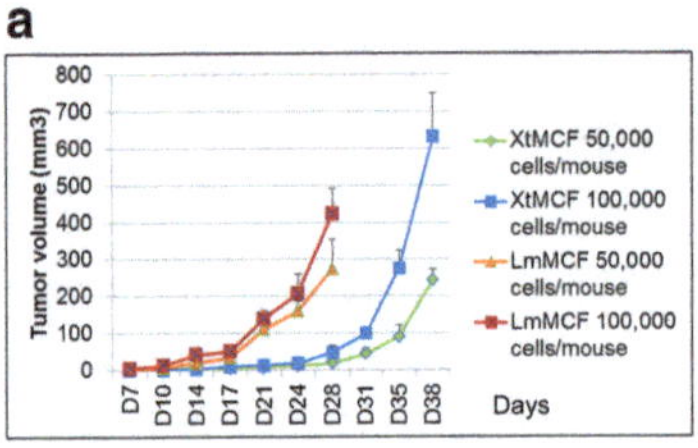

b

Cell line	Cells injection	Days post-injection	Tumor frequency	Tumor weight (mg)
XtMCF	2×10⁶	30	4/4	340 ± 32.5
LmMCF	2×10⁶	30	4/4	389.5 ± 105.5
XtMCF	1×10⁵	38	4/4	740.0 ± 364.8
LmMCF	1×10⁵	28	4/4	383.3 ± 12.5
XtMCF	5×10⁴	38	4/4	269.3 ± 98.4
XtMCF	5×10⁴	34	5/5	143.4 ± 90.3
LmMCF	5×10⁴	28	4/4	246.0 ± 77.7
LmMCF	5×10⁴	25	5/5	138.0 ± 45.4

c

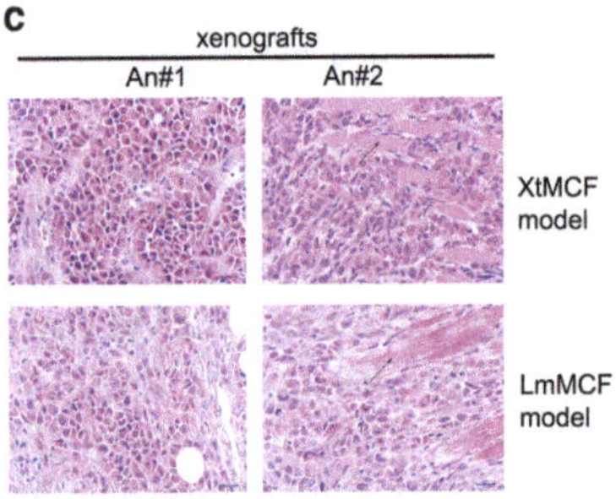

d

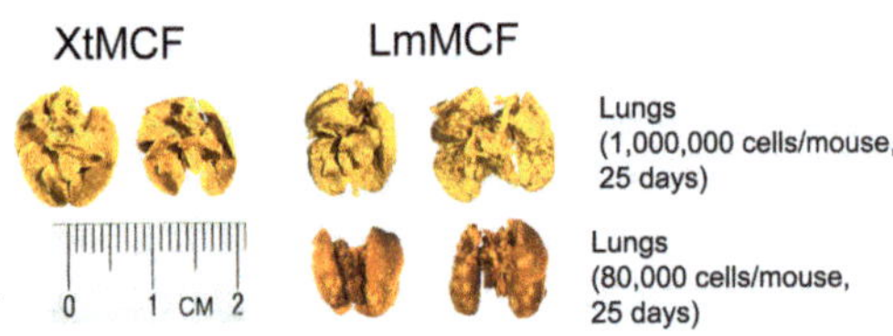

e

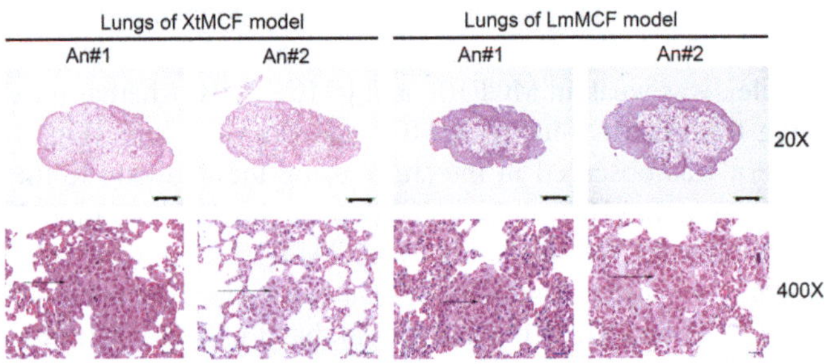

f

Cell line	Cells injection	Days post-injection	Lung metastasis frequency	Tumor foci on lung surface
XtMCF	1×10⁶	25	5/5	uncountable
XtMCF	6×10⁵	18	5/6	Micro-metastases by IHC staining
LmMCF	1×10⁶	25	6/6	uncountable
LmMCF	5×10⁵	18	5/5	uncountable
LmMCF	8×10⁴	25	5/5	107±34

Fig. 9.7 XtMCF and LmMCF cells present CD24^{weak}/CD44^{+}/EpCAM^{+} cancer stem cell properties; EGF-like domain at N-terminal EpCAM is cleaved off. (**a**) XtMCF and LmMCF are capable of forming tumorspheres cultured under non-adherent condition. Representative images of tumorspheres are shown. Magnification, 20×. Scale bar: 200 μm. (**b**) XtMCF and LmMCF produce more tumorspheres than parental bsMCF-luc cells. LmMCF produces more tumorspheres than XtMCF cells. *$p<0.05$ when compare to bsMCF-luc, or when compare LmMCF to XtMCF. (**c**) bsMCF-luc, XtMCF, and LmMCF cells have undergone EMT process, are CD24^{weak}/CD44^{+}/EpCAM+, the EpCAM does not have EGF-like domain. Magnification, 400×. (**d**) EpCAM level is up-regulated in XtMCF and LmMCF cell lines compared to bsMCF-luc. The quantification was performed on

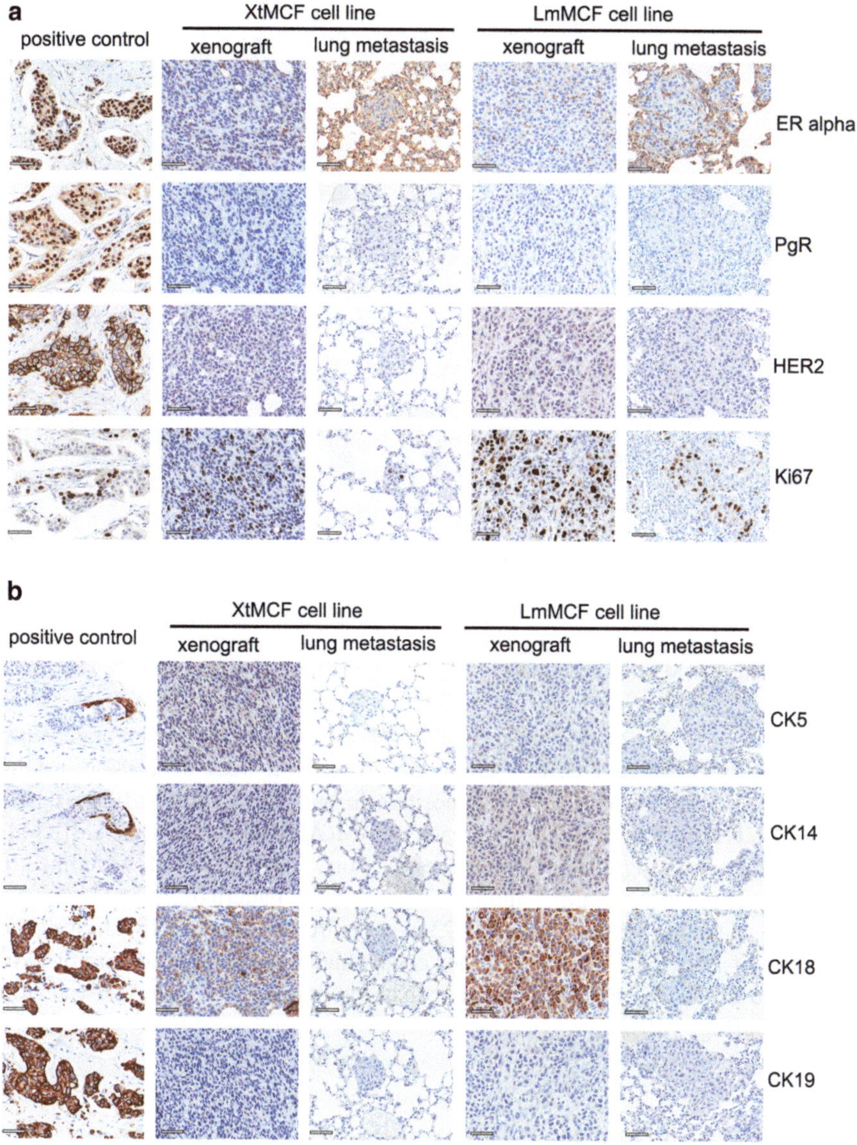

Fig. 9.8 MDA-MB-231, Sum159pt, and Hs578t cells have undergone EMT, are $CD24^{weak}/CD44^{+}/EpCAM^{+}$. The EpCAM in these cells does not have EGF-like domain. Magnification, 400×

Fig. 9.7 (continued) the staining using EpCAM (Abbiotech) antibody. *$p<0.05$ when compared to bsMCF-luc, or when LmMCF was compared to XtMCF. (**e**) EpCAM molecule and antibody map. The locations of epitopes of antibody VU1D9, Abbiotech, and E144 are indicated. Arrow: N terminal cleavage site between Arg-80/Arg-81. (**f**) bsMCF-luc, XtMCF, LmMCF, MDA-MB-231, and Sum159pt cells have undergone EMT, do not have full length EpCAM expression examined by WB. (**g**) Full length EpCAM expression is observed only in epithelial breast cancer cell lines but not mesenchymal-like cell lines by WB

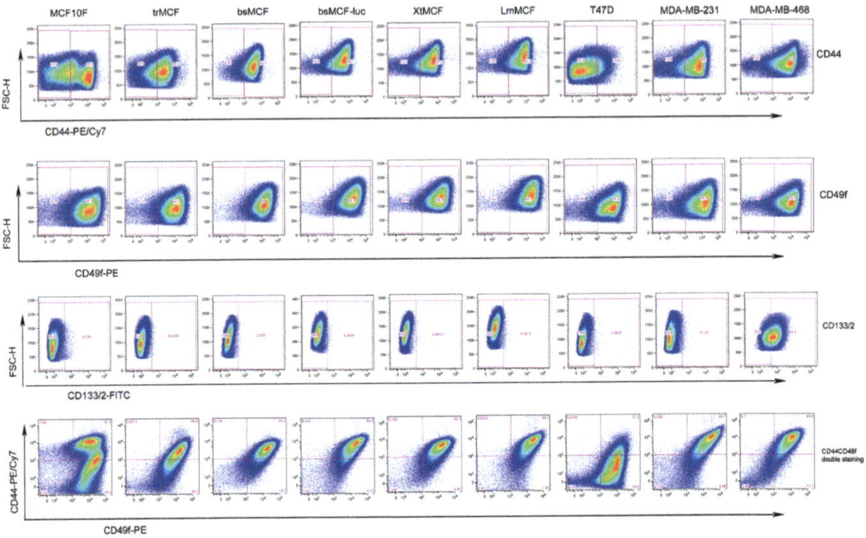

Fig. 9.9 Tumor cells of xenografts and lung metastases from XtMCF and LmMCF cells injection in CB17/SCID mice have undergone EMT process, are $CD24^{-/weak}/CD44^{+}/EpCAM^{+}$. There is no full length EpCAM in these xenografts and metastases. Expression of CD44, cleaved EpCAM, and vimentin are stronger in LmMCF model than XtMCF model. Sections from TMA were stained by IHC for evaluation. Magnification, 400×

lung metastases from XtMCF cells (Fig. 9.9). For LmMCF cell model, there was a moderate expression of CD24 in xenografts, whereas a weak expression in lung metastases was observed. CD44 was positive in both xenografts and metastases. EpCAM[VU1D9] was negative, whereas EpCAM(Abbiotech) and EpCAM[E144] were positive in xenografts and metastases. Of note, CD44, EpCAM, and vimentin all were stronger in the LmMCF model. The expression pattern of EpCAM was also evaluated and confirmed in MDA-MB-231 xenografts. In human TNBC tissues, the reactivity of EpCAM[VU1D9] in tumor cells which have lost E-cadherin or have undergone EMT, was significantly lower than tumor cells which show E-cadherin expression (Fig. 9.10).

9.4.6 Relevance of the Triple Negative Breast Cancer Model

Triple negative cancer (TNBC) represents a heterogeneous group of cancers. Cluster analysis of human TNBC identified six subtypes displaying unique gene expression and ontologies [48]. Approximately 80 % of TNBC show features of basal like cancers [61]. Transcriptional profile analysis assigned twenty one TNBC cell lines into three clusters [62, 63]: luminal, basal A and basal B. HCC2185 is the only TNBC cell line in luminal cluster [52, 63]. Basal A contains cell lines such as BT-20, Sum149, and MDA-MB-468, which preferentially expresses genes such as *CK5/6*,

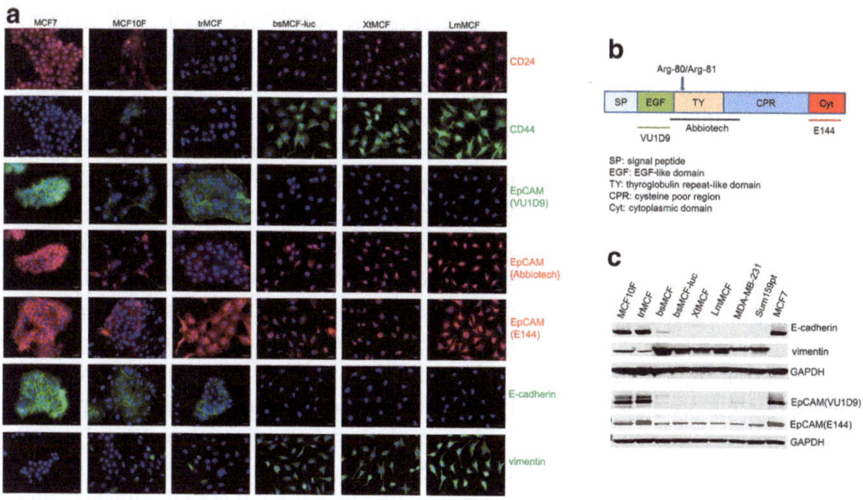

Fig. 9.10 IHC staining of EpCAM and EMT markers in MDA-MB-231 xenografts and human primary TNBC. EpCAM in MDA-MB-231 xenografts does not have EGF-like domain. TNBC tumor cells that lost E-cadherin or underwent EMT show reduced EpCAM expression. Magnification, 400×

CK14, and *EGFR*. Basal B contains cell lines such as MDA-MB-231, Sum159pt, and Hs578t, which preferentially expresses genes such as *CD44*, *MSN*, *CAV1/2*, *AXL*, *VIM* and *SNAI2*, and exhibits a stem-cell like profile [62]. This classification of TNBC cell lines is closely associated to cell morphology and invasive potential. Basal B cells have a more mesenchymal-like appearance, are less differentiated and much more invasive compared to the other two clusters. Analysis of the relationship between TNBC cell lines and tumor subtypes showed most of basal A and basal B cell lines resemble basal-like tumors [55], indicating that TNBC cell lines are suitable for investigations of subtype specific cancer cell biology.

Although there are over twenty TNBC cell lines, MDA-MB-231 is the most widely used *in vitro* and *in vivo*. In BALB/CAJCI-nu/nu mice, it took five weeks to form a xenograft around 6.5 mm in diameter with the subcutaneous injection of 5×10^6 MDA-MB-231 cells [64]. MDA-MB-468 cells had a growth speed similar to MDA-MB-231 in the same mouse strain [64]. The growth speed of MDA-MB-231 xenograft in CB17/SCID was almost the same as in nude mice, while BT-549 cells grew a little bit slower than MDA-MB-231 cells in CB17/SCID mice [65]. Sum149 and Sum159 are two highly tumorigenic cell lines, it was reported the injection of 1×10^5 cells in nonobese diabetic SCID mice could produce tumors in 3/4 and 5/6 mice, respectively [66]. Although Sum149 cells are of high tumorigenesis, the growth speed was not that fast, MFP injection of 5×10^6 cells to nude mice MFP formed tumors around 6 mm in diameter in three weeks [9]. Regarding Sum159, Flanagan et al. reported that the tumor growth kinetic of orthotopic Sum159 was similar to those observed in the MDA-MB-231 cells [67].

Considering the heterogeneity and complexity of TNBC, more cell lines and animal models are needed. We here report the establishment of a progressive TNBC model consisting of normal MCF10F, transformed cell line trMCF, and tumorigenic cell lines bsMCF, XtMCF and LmMCF. Compared to the nine tumorigenic TNBC cell lines mentioned in the introduction and in this discussion, XtMCF and LmMCF cell lines are the most tumorigenic and metastatic. The expression of CK18 confirmed the epithelial origin of this cell model. We observed that CK18 was down-regulated in bsMCF cell line and its derivatives. Furthermore, CK18 was lost in the lung metastases, whereas still present in the xenografts of both XtMCF and LmMCF cells, suggesting down-regulation of CK18 may be critical to breast tumor progression. This idea is supported by Woelfle's study, which found that down-regulation of CK18 was significantly correlated to advanced tumor stage and high grade [68]. Of interest, CK5 was progressively down-regulated in our cell model. Our previous study showed CK5 positive cell number was inversely correlated to clinical stage of TNBC [69]. Aguiar et al. also found CK5/6 expression was negatively associated with the probability of invasion [70], suggesting that our cell model reflects features of TNBC progression.

The EMT process is not only closely related to cancer invasion and metastasis [71] but also conferred to the generation of CSC [59, 72]. As bsMCF-luc, XtMCF, and LmMCF have undergone EMT, we evaluated their CSC properties and showed that the number of tumorspheres was progressively increasing from bsMCF-luc to XtMCF and LmMCF cells, consistent to *in vivo* tumorigenic and metastatic potential. This result contradicted the colony formation assay result, which showed fewer colonies were produced by XtMCF and LmMCF cells, indicating the number of colonies in agar may not be always associated with tumorigenicity. The growth pattern of colonies in agar should also be considered, for example, budding from the surface of colonies was frequently observed in XtMCF and LmMCF colonies, which may indicate its aggressiveness and tumorigenicity. In addition, XtMCF and LmMCF grew in collagen in clusters only, not forming solid masses, suggesting these two new cell lines may be more aggressive and lose the potential to grow connectively with other cells.

We postulated the evaluation of CSC markers would give us a rationale for the high tumorigenic and metastatic potential of these two cell lines. As expected, the bsMCF-luc, XtMCF and LmMCF cells were $CD24^{weak}/CD44^{+}$. EpCAM was an important marker for isolating CSC from breast [73], colon [74], and pancreatic cancer [75]. Lin et al. showed EpCAM induces expressions of reprogramming factors (*OCT4*, *SOX2*, *NANOG*, and *c-MYC*) and EMT genes, regulates EMT progression and tumorigenesis [76]. In addition, EpCAM can be cleaved at several sites, the nuclear translocation of cytoplasmic domain (EpCID) associates with Wnt pathway, promotes cell proliferation and tumor formation in mice [77, 78]. One of EpCAM cleavage sites between two arginine residues (AA80 and AA81) was detected and described in the late 1980s soon after the cloning of EpCAM, but the functional consequence is still unknown [79]. Since EGF-like domain mediates lateral and reciprocal interactions of EpCAM molecules [60], the loss of this domain may affect cell connections. Interestingly, we observed the EGF-like domain of

EpCAM was absent in mesenchymal-like cells, suggesting the EGF-like domain might be cleaved off from the cleavage site between AA80 and AA81. This was supported by Keller and collaborators, who showed that mesenchymal-like cells do not express EpCAM using EpCAM[VU1D9] [80]. Gorges et al. also showed that circulating tumor cells escape from EpCAM based detection due to EMT [81]. The majority of commercial antibodies for EpCAM react with overlapped or partly overlapped epitope at EGF-like domain [60]. This may result in failing detection of EpCAM in cells which have undergone EMT. Our study indicates the EGF-like domain cleaved off EpCAM may also be important for tumor initiation. Furthermore, although the total level of EpCAM is low in mesenchymal-like cells, the subcellular localization might be more important to the EpCAM nuclear signaling. ALDH1 is another important CSC marker for breast cancer [82]. There was no detectable expression of this protein in bsMCF, XtMCF and LmMCF cells and tumors (data not shown).

Another interesting observation was the nuclear E-cadherin staining in xenografts of LmMCF and MDA-MB-231 cell lines, which was consistent to the finding that nuclear translocation of cleaved cytoplasmic domain of E-cadherin plays oncogenic roles [83, 84].

Taken together, we present the development and characterization of two highly tumorigenic and metastatic basal B TNBC cell lines, XtMCF and LmMCF. To best of our knowledge, they are the most tumorigenic and metastatic TNBC cell lines compared to all reported cell models used for TNBC studies. In addition, the normal (MCF10F) and early stage (trMCF) counterparts of these two cell lines are also available. Altogether, these cell lines can be used to study the evolution of TNBC, investigate molecular pathways at different stages of transformation and progression in a relatively constant genetic background, and most importantly, identify new treatments for TNBC. In addition, XtMCF and LmMCF cell lines present CSC properties and can be used for developing CSC targeted therapy. The finding that the EGF-like domain of EpCAM is cleaved off in cancer cells which have undergone EMT also provides new insights in research of EMT and CSC, two important fields in cancer biology.

References

1. Russo J, Russo IH (2014) Methodological approaches to breast cancer research. Springer, New York
2. Baselga J, Norton L, Albanell J, Mee Kim Y, Mendelsohn J (1998) Recombinant humanized anti-HER2 antibody (Herceptin™) enhances the antitumor activity of paclitaxel and doxorubicin against HER2/*neu* overexpressing human breast cancer xenografts. Cancer Res 58:2825–2831
3. Sanford KK (1974) Biologic manifestation of oncogenesis in vitro: a critique. J Natl Cancer Inst 53:1481–1485
4. Giovanella BC, Stehlin JS, Williams LJ Jr (1972) Development of invasive tumors in the "nude" mouse after injection of cultured human melanoma cells. J Natl Cancer Inst 48:1531–1533

5. Giovanella BC, Stehlin JS (1973) Heterotransplantation of human malignant tumors in "nude" thymusless mice. I. Breeding and maintenance of "nude" mice. J Natl Cancer Inst 51:615–619
6. Giovanella BC, Stehlin JS, Williams LJ Jr (1974) Heterotransplantation of human malignant tumors in "nude" thymusless mice. II. Malignant tumors induced by injection of cell cultures derived from human solid tumors. J Natl Cancer Inst 52:921–930
7. Povlsen CO, Fialkow PJ, Klein E (1973) Growth and antigenic properties of a biopsy-derived Burkitt's lymphoma in thymusless (nude) mice. Int J Cancer 11:30–39
8. Russo J, Soule HD, McGrath C, Rich MA (1976) Re-expression of the original tumor pattern by a human breast carcinoma cell line (MCF-7) in sponge cultures. J Natl Cancer Inst 56:279–282
9. DeOme KB, Faulkin LJ Jr, Bern HA, Blair PB (1959) Development of mammary tumors from hyperplastic alveolar nodules transplanted into gland-free mammary fat pads of female C3H mice. Cancer Res 19:350–359
10. Russo J, Russo IH (2004) Biological and molecular basis of breast cancer. Springer-Verlag, Heidelberg
11. Rygaard J, Povlsen CO (1969) Heterotransplantation of a human malignant tumor to "nude" mice. Acta Pathol Microbiol Scand 77:758–760
12. Greene HSN (1952) The significance of the heterologous transplantability of human cancer. Cancer 5:24–44
13. Russo J, McGrath CM (1975) Scirrhous carcinoma in the mouse: a model for human mammary carcinoma. Excerpta Medica, Amsterdam, p 488
14. Russo J, McGrath CM, Russo IH, Rich MA (1976) Tumoral growth of a human breast cancer cell line (MCF-7) in athymic mice. In: Nieburgs HE (ed) III Int. Symp. on detection and prevention of cancer, New York, pp 617–626
15. Shafie SM, Giartham FH (1981) Role of hormones in the growth and regression of human breast cancer cells (MCF-7) transplanted into athymic mice. J Natl Cancer Inst 67:51–56
16. Ura H, Bonfil RD, Reich R et al (1989) Expression of type IV collagenase and procollagen genes and its correlation with the tumorigenic, invasive and metastatic abilities of oncogene-transformed human bronchial epithelial cells. Cancer Res 49:4615–4621
17. Smith HS, Wolman SR, Hackett AJ (1984) The biology of breast cancer at the cellular level. Biochim Biophys Acta 738:103–123
18. Cooper CS, Blair DG, Oskarsson MK, Tainsky MA, Eader LA, Vande Woude GF (1984) Characterization of human transforming genes from chemically transformed teratocarcinoma, and pancreatic carcinoma cell lines. Cancer Res 44:1–10
19. Strange R, Aguilar-Cordova E, Young UT, Billey HT, Dandekar S, Cardiff R (1989) Harvey-ras mediated neoplastic development in the mouse mammary gland. Oncogene 4:309–315
20. Ozzello L (1971) Ultrastructure of the human mammary gland. In: Summers SC (ed) Pathology annual. Appleton, New York., pp 1–59
21. Soule HD, Maloney TM, Wolman SR, Peterson WD, Brenz R, McGrath CM, Russo J, Pauley RJ, Jones RF, Brooks SC (1990) Isolation and characterization of a spontaneously irmno human breast epithelial cell line, MCF-10. Cancer Res 50:6075–6086
22. Stampfer MR, Bartley JC (1984) Induction of transformation and continuous cell lines from normal human mammary epithelial cells after exposure to benzo[a]pyrene. Proc Natl Acad Sci U S A 82:2394–2398
23. Zhang PL, Calaf G, Russo J (1992) Point mutation in codons 12 and 61 of the c-Ha-ras gene in carcinogen-treated human breast epithelial cells (HBEC). Proc Am Assoc Cancer Res 33:669a
24. Abarca-Quinones J, Calaf G, Estrada S, Barnabas-Sohi N, Zhang PL, Garcia M, Russo J (1992) Phenotypic progression of human breast epithelial cells HBEC transformed with chemical carcinogen. Proc Am Assoc Cancer Res 33:670a
25. Calaf G, Russo J (1992) Emergence of progressive neoplastic phenotypes of human breast epithelial (HBEC) treated with chemical carcinogens in vitro. Proc Am Assoc Cancer Res 33:1141a

26. Rochlitz CF, Scott GK, Dodson JM, Liu E, Dollbaum CH, Smith HS, Benz CH (1989) Incidence of activating ras oncogene mutations associated with primary and metastatic human breast cancer. Cancer Res 49:357–360
27. Russo J, Fernandez SV, Russo PA et al (2006) 17-Beta-estradiol induces transformation and tumorigenesis in human breast epithelial cells. FASEB J 20:1622–1634
28. Huang Y, Fernandez S, Goodwin S, Russo PA, Russo IH, Sutter T, Russo J (2007) Epithelial to mesenchymal transition in human breast epithelial cells transformed by 17beta-estradiol. Cancer Res 67:11147–11157
29. Zelmer A, Ward TH (2013) Noninvasive fluorescence imaging of small animals. J Microsc. doi:10.1111/jmi.12063
30. Liu C, Billadeau DD, Abdelhakim H, Leof E, Kaibuchi K, Bernabeu C, Bloom GS, Yang L, Boardman L, Shah VH, Kang N (2013) IQGAP1 suppresses TbetaRII-mediated myofibroblastic activation and metastatic growth in liver. J Clin Invest 123(3):1138–1156. doi:10.1172/JCI63836
31. Edinger M, Cao YA, Hornig YS, Jenkins DE, Verneris MR, Bachmann MH, Negrin RS, Contag CH (2002) Advancing animal models of neoplasia through in vivo bioluminescence imaging. Eur J Cancer 38(16):2128–2136
32. Vines DC, Green DE, Kudo G, Keller H (2011) Evaluation of mouse tail-vein injections both qualitatively and quantitatively on small-animal PET tail scans. J Nucl Med Technol 39(4):264–270. doi:10.2967/jnmt.111.090951
33. Ray DE, Holton JL, Nolan CC, Cavanagh JB, Harpur ES (1998) Neurotoxic potential of gadodiamide after injection into the lateral cerebral ventricle of rats. AJNR Am J Neuroradiol 19:1455–1462
34. Passini MA, Watson DJ, Vite CH, Landsburg DJ, Feigenbaum AL, Wolfe JH (2003) Intraventricular brain injection of adeno-associated virus type 1 (AAV1) in neonatal mice results in complementary patterns of neuronal transduction to AAV2 and total long-term correction of storage lesions in the brains of β-glucuronidase-deficient mice. J Virol 77: 7034–7040
35. Allen RM, Uban KA, Atwood EM, Albeck DS, Yamamoto DJ (2007) Continuous intracerebroventricular infusion of the competitive NMDA receptor antagonist, LY235959, facilitates escalation of cocaine self-administration and increases break point for cocaine in Sprague-Dawley rats. Pharmacol Biochem Behav 88(1):82–88
36. Becker JC et al (1996) Eradication of human hepatic and pulmonary melanoma metastases in SCID mice by antibody-interleukin 2 fusion proteins. Proc Natl Acad Sci U S A 93(7): 2702–2707
37. Bijian K et al (2012) Synthesis and biological activity of novel organoselenium derivatives targeting multiple kinases and capable of inhibiting cancer progression to metastases. Eur J Med Chem 48:143–152
38. Bugge TH et al (1997) Growth and dissemination of Lewis lung carcinoma in plasminogen-deficient mice. Blood 90(11):4522–4531
39. Gillies SD et al (1998) Antibody-IL-12 fusion proteins are effective in SCID mouse models of prostate and colon carcinoma metastases. J Immunol 160(12):6195–6203
40. Louie E et al (2010) Identification of a stem-like cell population by exposing metastatic breast cancer cell lines to repetitive cycles of hypoxia and reoxygenation. Breast Cancer Res 12(6):R94
41. Dent R et al (2007) Triple-negative breast cancer: clinical features and patterns of recurrence. Clin Cancer Res 13(15):4429–4434
42. Metzger-Filho O et al (2012) Dissecting the heterogeneity of triple-negative breast cancer. J Clin Oncol 30(15):1879–1887
43. Price JE et al (1990) Tumorigenicity and metastasis of human breast carcinoma cell lines in nude mice. Cancer Res 50(3):717–721
44. van Slooten HJ, Bonsing BA, Hiller AJ, Colbern GT, van Dierendonck JH, Cornelisse CJ, Smith HS (1995) Outgrowth of BT-474 human breast cancer cells in immune-deficient mice: a new in vivo model for hormonedependent breast cancer. Br J Cancer 72(1):22–30

45. Madsen MW, Briand P (1990) Relationship between tumorigenicity, in vitro invasiveness, and plasminogen activator production of human breast cell lines. Eur J Cancer 26(7):793–797
46. Langlois A et al (1979) Morphological and biochemical properties of a new human breast cancer cell line. Cancer Res 39(7 Part 1):2604–2613
47. Lacroix M, Leclercq G (2004) Relevance of breast cancer cell lines as models for breast tumours: an update. Breast Cancer Res Treat 83(3):249–289
48. Lehmann BD et al (2011) Identification of human triple-negative breast cancer subtypes and preclinical models for selection of targeted therapies. J Clin Invest 121(7):2750–2767
49. Zhang D et al (2009) Epidermal growth factor receptor tyrosine kinase inhibitor reverses mesenchymal to epithelial phenotype and inhibits metastasis in inflammatory breast cancer. Clin Cancer Res 15(21):6639–6648
50. Ginestier C et al (2010) CXCR1 blockade selectively targets human breast cancer stem cells in vitro and in xenografts. J Clin Invest 120(2):485–497
51. Russo J et al (2006) The concept of stem cell in the mammary gland and its implication in morphogenesis, cancer and prevention. Front Biosci 11:151–172
52. Kao J et al (2009) Molecular profiling of breast cancer cell lines defines relevant tumor models and provides a resource for cancer gene discovery. PLoS One 4(7):e6146
53. Kocdor H et al (2009) Human chorionic gonadotropin (hCG) prevents the transformed phenotypes induced by 17 β-estradiol in human breast epithelial cells. Cell Biol Int 33(11): 1135–1143
54. Kocdor M et al (2013) Progressive increase of glucose transporter-3 (GLUT-3) expression in estrogen-induced breast carcinogenesis. Clin Transl Oncol 15(1):55–64
55. Tiezzi DG, Fernandez SV, Russo J (2007) Epithelial mesenchymal transition during the neoplastic transformation of human breast epithelial cells by estrogen. Int J Oncol 31(4): 823–828
56. Reynen K, Köckeritz U, Strasser R (2004) Metastases to the heart. Ann Oncol 15(3):375–381
57. Ciliberto G (2010) Mammosphere-forming cells from breast cancer cell lines as a tool for the identification of CSC-like and early progenitor-targeting drugs. Cell Cycle 9(14):2878–2887
58. Grimshaw MJ et al (2008) Mammosphere culture of metastatic breast cancer cells enriches for tumorigenic breast cancer cells. Breast Cancer Res 10(3):R52
59. Morel A-P et al (2008) Generation of breast cancer stem cells through epithelial-mesenchymal transition. PLoS One 3(8):e2888
60. Balzar M et al (2001) Epidermal growth factor-like repeats mediate lateral and reciprocal interactions of Ep-CAM molecules in homophilic adhesions. Mol Cell Biol 21(7):2570–2580
61. Tan DS et al (2008) Triple negative breast cancer: molecular profiling and prognostic impact in adjuvant anthracycline-treated patients. Breast Cancer Res Treat 111(1):27–44
62. Neve RM et al (2006) A collection of breast cancer cell lines for the study of functionally distinct cancer subtypes. Cancer Cell 10(6):515–527
63. Grigoriadis A et al (2012) Molecular characterisation of cell line models for triple-negative breast cancers. BMC Genomics 13(1):619
64. Yunokawa M et al (2012) Efficacy of everolimus, a novel mTOR inhibitor, against basal-like triple-negative breast cancer cells. Cancer Sci 103(9):1665–1671
65. Tate CR et al (2012) Targeting triple-negative breast cancer cells with the histone deacetylase inhibitor panobinostat. Breast Cancer Res 14(3):R79
66. Fillmore CM, Kuperwasser C (2008) Human breast cancer cell lines contain stem-like cells that self-renew, give rise to phenotypically diverse progeny and survive chemotherapy. Breast Cancer Res 10(2):R25
67. Flanagan L et al (1999) SUM-159PT cells: a novel estrogen independent human breast cancer model system. Breast Cancer Res Treat 58(3):193–204
68. Woelfle U et al (2004) Down-regulated expression of cytokeratin 18 promotes progression of human breast cancer. Clin Cancer Res 10(8):2670–2674
69. Russo J, Russo IH (2014) Techniques and methodological approaches in breast cancer research. Springer, New York

70. Aguiar FN et al (2013) Basal cytokeratin as a potential marker of low risk of invasion in ductal carcinoma in situ. Clinics 68(5):638–643
71. Xue C et al (2003) The gatekeeper effect of epithelial-mesenchymal transition regulates the frequency of breast cancer metastasis. Cancer Res 63(12):3386–3394
72. Mani SA et al (2008) The epithelial-mesenchymal transition generates cells with properties of stem cells. Cell 133(4):704–715
73. Al-Hajj M et al (2003) Prospective identification of tumorigenic breast cancer cells. Proc Natl Acad Sci 100(7):3983–3988
74. Dalerba P et al (2007) Phenotypic characterization of human colorectal cancer stem cells. Proc Natl Acad Sci 104(24):10158–10163
75. Li C et al (2007) Identification of pancreatic cancer stem cells. Cancer Res 67(3):1030–1037
76. Lin C-W et al (2012) Epithelial cell adhesion molecule regulates tumor initiation and tumorigenesis via activating reprogramming factors and epithelial-mesenchymal transition gene expression in colon cancer. J Biol Chem 287(47):39449–39459
77. Schnell U, Kuipers J, Giepmans BN (2013) EpCAM proteolysis: new fragments with distinct functions? Biosci Rep 33(2):e00030. doi:10.1042/BSR20120128
78. Maetzel D et al (2009) Nuclear signalling by tumour-associated antigen EpCAM. Nat Cell Biol 11(2):162–171
79. Thampoe IJ, Ng JS, Lloyd KO (1988) Biochemical analysis of a human epithelial surface antigen: differential cell expression and processing. Arch Biochem Biophys 267(1):342–352
80. Keller PJ et al (2010) Mapping the cellular and molecular heterogeneity of normal and malignant breast tissues and cultured cell lines. Breast Cancer Res 12(5):R87
81. Gorges TM et al (2012) Circulating tumour cells escape from EpCAM-based detection due to epithelial-to-mesenchymal transition. BMC Cancer 12:178
82. Ginestier C et al (2007) ALDH1 is a marker of normal and malignant human mammary stem cells and a predictor of poor clinical outcome. Cell Stem Cell 1(5):555–567
83. Ferber EC et al (2008) A role for the cleaved cytoplasmic domain of E-cadherin in the nucleus. J Biol Chem 283(19):12691–12700
84. David JM, Rajasekaran AK (2012) Dishonorable discharge: the oncogenic roles of cleaved E-cadherin fragments. Cancer Res 72(12):2917–2923

Chapter 10
Biological Basis of Breast Cancer Prevention

10.1 Introduction

Epidemiological, clinical and pathological studies have uncovered novel aspects regarding the complexity of breast cancer [1–11], and reproductive history has been found to be a common denominator for risk [7, 8, 11]. Increased breast cancer incidence and mortality were associated with nulliparity as early as the 1700s, as reported by Bernardino Ramazzini, who attributed the phenomenon to the childlessness of nuns in Italian convents [12]. MacMahon et al. [8] reported that pregnancy exerted a protective effect in women who bore children from their early teen years to their mid-twenties, relative to a risk of 100 for nulliparous women. Numerous studies have confirmed these results and have additionally reported that multiple pregnancies significantly decrease the risk of developing breast cancer after age 50 [8], whereas full-term pregnancy later in life increases a woman's breast cancer risk, reaching the same levels observed in nulliparous women when it occurs between 30 to 34 years of age, increasing even further after 35 years [7, 8]. An understanding of the mechanisms that determine whether a pregnancy would prevent breast cancer or increase its risk requires taking into consideration the age at the first pregnancy [13–15].

10.2 Pregnancy as a Physiological Process That Prevent Breast Cancer

Pregnancy itself is a complex process, that only succeeds when a woman's ovaries are fully functional and secrete estrogen and progesterone, hormones that are essential for the maintenance of pregnancy. The ovaries work under the control of the

J. Russo, *The Pathobiology of Breast Cancer*,
DOI 10.1007/978-3-319-40815-6_10

hypothalamic-pituitary-gonadal (HPG) axis [16, 17], which synchronizes the ovarian secretions with those of pituitary and placental hormones for stimulating breast development in preparation for milk production [17, 18]. Primiparous women younger than 25 years of age who have elevated serum levels of hCG during their first trimester have 33 % decreased risk of a breast cancer diagnosis after age 50, whereas estrogen concentrations have been positively associated with risk of breast cancer before age 40, supporting the role of this or other pregnancy hormones in the development of breast cancer [19–23].

In experiments performed in rats, pregnancy, the gold standard for the induction of mammary gland differentiation, needs to be completed to prevent mammary cancer. In rats it has been shown that when their first pregnancy was interrupted 12 days after conception and they received DMBA 21 days later [24] the tumor incidence and number of tumors per animal in pregnancy-interrupted rats and age-matched virgin rats were similar, whereas rats that completed their pregnancy had a significantly reduced tumorigenic response. Completion of the first pregnancy results in full differentiation of the mammary gland that culminates in the secretion of milk, which persists during the length of the lactational period [17, 22]. At post-weaning the lobular structures regress and the remaining cells exhibit a marked reduction in proliferative rate, lengthening in the G_1 phase of the cell cycle, greater capabilities to repair DNA damaged by the carcinogen and lower affinity for binding DMBA to DNA [22]. These structural, functional and molecular changes persist in the mammary gland, resulting in a significant reduction of mammary cancer incidence that is evident in various strains of rats and mice [25], in spite of histopathological differences in tumor type between these species. Blakely et al. [26] have confirmed that in four genetically distinct inbred strains of rats (Lewis, Wistar-Furth, Fischer 344, and Copenhagen) and in mice pregnancy and lactation induce similar structural and genomic changes in mammary glands studied by microarray analysis. Gene analysis identified a genomic signature that has sufficed for distinguishing nulliparous from parous animals and explain the almost total refractoriness of the parous rat mammary gland to develop carcinomas after carcinogen administration [26, 27]. These observations indicate that when the development of the mammary gland has been completed by an early pregnancy, steroid hormone- or hCG treatment of virgin animals the period of maximal susceptibility to cancer (PMSC) or Stem Cells with euchromatin nucleus (EUN) has completed a first cycle of differentiation under specific hormonal influences, becoming a Cell with heterochromatin nucleus (HTN) [28] a suggestion of chromatin remodeling, which is resistant to transformation by a carcinogen (Fig. 10.1). Although more differentiated, the HTN cells have retained the capacity to regenerate the complete lobular system required by subsequent pregnancies. This concept has been further demonstrated in transgenic WAP-driven Cre and Rosa 26-fl-stop-fl-LacZ mice in which parity-induced mammary epithelial cells (PI-MEC) originated from differentiated cells during pregnancy, survived post lactational involution and increased their percentage with successive pregnancies [29]. PI-MEC, like the HTN cells show capacity for self-renewal and contribute to mammary outgrowth in transplanta-

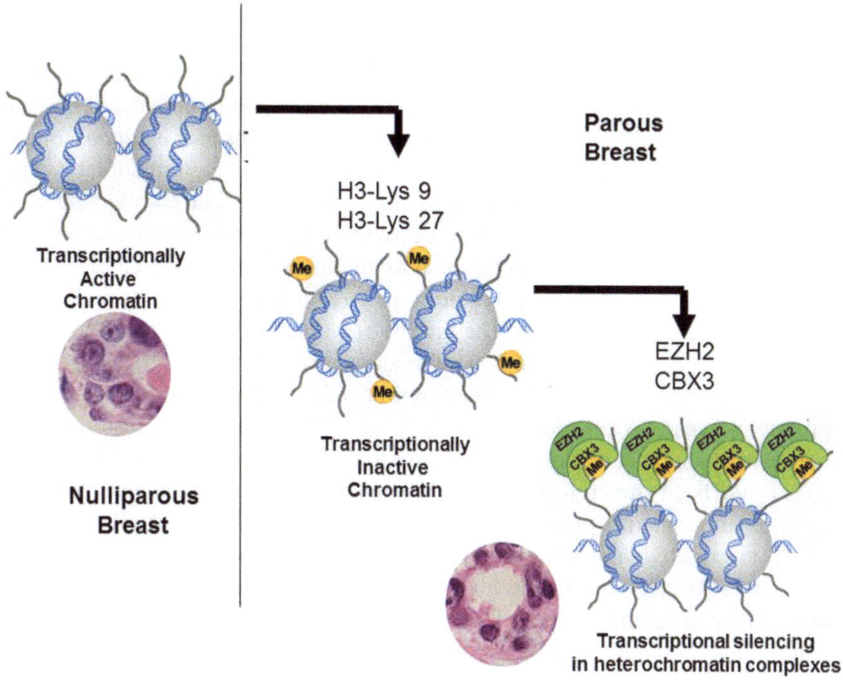

Fig. 10.1 The development of the mammary gland has been completed by an early pregnancy, steroid hormone- or hCG treatment. The period of maximal susceptibility to cancer or Stem Cells with euchromatic nucleus (EUN) in the nulliparous breast are more transcriptionally active, but when they have completed a first cycle of differentiation under specific hormonal influences, becoming a cell with heterochromatin nucleus (HTN) meaning that a chromatin remodeling has been taken place which makes the cells of the parous breast resistant to transformation by a carcinogen

tion studies. PI-MEC can function as alveolar progenitors in subsequent pregnancies, and it is thought that they would be related to differences in response to hormonal stimulation and carcinogenic agents observed between nulliparous and parous females [30–32].

The relevance of the findings that the first full term pregnancy occurring during the high risk susceptibility window (HRSW) (see Chap. 1) but before exposure to a carcinogen prevents cancer initiation is equivalent to the well demonstrated protective effect of an early first full term pregnancy (FTP) in women. A first FTP initiated approximately two weeks after carcinogen exposure, on the other hand, results in a high incidence of mammary cancer, a phenomenon that could explain the increased cancer risk observed in women first parous after age 30, supporting the assumption that during that lengthened HRSW the breast has been exposed to carcinogenic stimuli before pregnancy. These data emphasize the importance of discriminating whether the first pregnancy would produce protection by inducing complete differentiation of the breast activating the same mechanisms that hormonal treatments do, or would increase breast cancer risk as a consequence of genotoxic or epigenetic exposures during the HRSW.

The development of the breast is a continuous process initiated by the fourth week of intrauterine life that progresses under the influence of maternal, placental and environmental factors until birth and by diet and by environmental exposures after weaning. During these periods the maturation of the hypothalamic gonadal (HPG) axis [16, 17, 33] and endogenous hormone secretions play essential roles on the development of the breast at puberty, which is driven by the initiation of ovulation and the establishment of regular menstrual cycles [34]. The architecture of the breast of normally cycling women has been widely described as composed of three main lobular structures that are classified on the basis of their degree of development into lobules type 1 (Lob 1), lobules type 2 (Lob 2) and lobules type 3 (Lob 3) [22, 35, 36] (see also Sect. 10.3). The breasts of women who have never conceived a child remain composed of Lob 1, with moderate formation of Lob 2 with successive menstrual cycles; Lob 3 become present only occasionally during the early reproductive years. After menopause the breast further regresses, resulting in an increase in the number of Lob 1 in response to the decline in Lob 2 and Lob 3 with aging. It has been shown that the breast parenchyma of postmenopausal nulliparous women contains predominantly euchromatin nucleus (EUN) cells [28] (Fig. 10.1), which do not achieve the most differentiated stage of heterochromatin nucleus (HTN) cells due to the absence of pregnancy, therefore retaining their susceptibility to be transformed. Therefore, a carcinogenic insult or an inappropriate hormonal stimulus, such as hormone replacement therapy [37], would transform the EUN cells into a cancer stem cell. This concept has been confirmed in a systematic study [38] comparing mammographic density with histology in women receiving or not receiving menopausal hormone therapy (HT). Noncancerous tissue from mastectomy specimens was examined histologically to quantitate the content of fibrous stroma, ducts, and lobule types 1, 2, and 3. Tissue samples were also evaluated for estrogen receptor, progesterone receptor, and Ki67 activity in the ducts and lobules. Breast density was quantified by digitizing the contralateral mammogram and computer-assisted interactive thresholding. High breast density in women using HT was correlated with greater fibrous stroma ($P=0.020$) and increase in the number of lobule type 1 ($P=0.016$). Breast density also correlated with Ki67 activity in the ducts ($P=0.031$) and lobules ($P=0.023$) for both groups combined. Estrogen and progesterone receptors did not correlate with either breast density or HT use. Increased fibrous stroma and lobule type 1 are associated with increasing mammographic density in women using HT, independent of estrogen and progesterone receptor up-regulation. The increase in breast cancer risk with HT use may be due to an increase in target lobule type 1 cells that is associated with increase breast density [38]. This data clearly shows that the breast at menopause can respond to hormones and activating ductal growth that will generate the Lobules type 1 or terminal ductal lobular unit where carcinoma start explaining why HRT increased the incidence of breast cancer and when women stopped the use of HRT the incidence get lower [39–45]. At difference of the menopausal breast the pubertal girl have a significant fibrous stroma associated with greater number of lobules type one (Lob1) [35, 36] that are the target of carcinogenesis. Therefore it is predictable that any substance that affects breast density like increase in the number of Lob 1 or in the case of the rat increase in the number of TEB will be a risk factor. On the other hand a decrease density that is associated with an increase in mammary fat will reduce risk and in the rat will be associated with less number of TEB [22].

10.3 Breast Development and Differentiation as the Biological Clue in Cancer Prevention

The development of the breast from birth to puberty follows a general pattern common for all normally cycling women, with the formation of Lob 1, Lob 2 and Lob 3 [35, 36]. The progression of lobular development under the cyclic influence of ovarian hormones is rapidly accelerated during the first pregnancy, which to be successful requires the timely fertilization of an oocyte followed by its uterine implantation. The embryo drives a process that establishes a collaboration of the newly formed placenta with the maternal environment [46]. The placenta alone elaborates a myriad of proteins, glycoproteins, steroid hormones, growth factors, tumor suppressor factors and cytokines that control the local environment of the fetus and regulate the metabolic activities of both the mother and the fetus [47]. In addition to estrogen and progesterone, newly secreted hormones, such as human growth hormone (hGH), hCG, human placental lactogen (hPL), and inhibin stimulate breast development and differentiation [48, 49]. Elevated serum levels of Metastin (KISS1) have been detected during pregnancy [50], but the role of this hormone in breast development has not been identified as of yet. LH, progesterone and hCG are the main hormones driving the initial phase of growth, followed by the secretion of the pituitary hormone prolactin (PRL) that stimulates milk secretion and contributes to the development of the fully differentiated Lob 4 during the last trimester of pregnancy and lactation. After weaning, Lob 4 regresses to Lob 3, which persists in the breast as long as women continue cycling. At peri-menopause the number of Lob 3 progressively decrease due to their involution to Lob 2 and Lob 1 [22].

The morphological, physiological and genomic changes resulting from pregnancy and hormonally-induced differentiation of the breast and their influence on breast cancer risk have been addressed above and in the literature [51–56]. The observations that during the post-menopausal years the breasts of both parous and nulliparous women contain predominately Lob 1, and the fact that nulliparous women are at higher risk of developing breast cancer than parous women, indicate that Lob 1 in these two groups of women either differ biologically, or exhibit different susceptibility to carcinogenesis [54]. Novel markers showing changes in cell types and increases in chromatin condensation define the concept of differentiation in the adult breast and further clarify this concept [28]. These findings confirm the universality of the histone 3 methylation in lysines 9 and 27 during differentiation, since a similar phenomenon has been described to occur during embryonic stem cell (ESC) differentiation [57]. The observed chromatin changes in parous epithelial cells are complemented by the expression of genes related to increasing cell

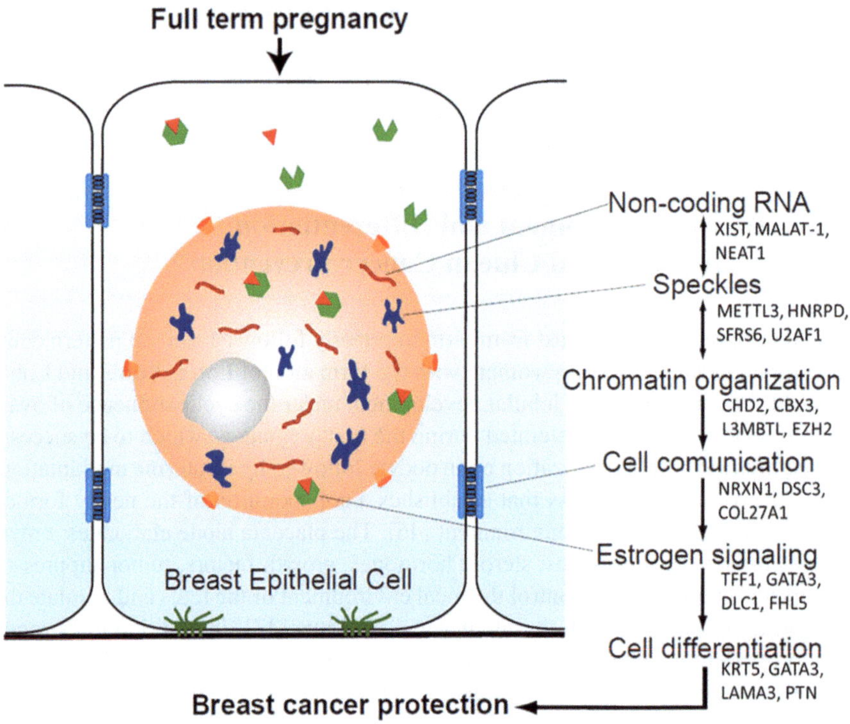

Fig. 10.2 The genomic signature of prevention induced by full term pregnancy is characterized by the expression of genes related to differentiation and cell proliferation, cell communication, spliceosomes, lncRNA and estrogen signaling

adhesion, such as NRXN1, DSC3. COL27A1, PNN, COL4A6, LAMC2, COL7A1, COL16A1,and LAMA3, and differentiation, that include MGP KRT5 GATA3 and LAMA3 [28, 56] (Fig. 10.2).

Genes that are regulated downstream by ER-α were found to be up regulated in the parous breast, supporting a parity mediated protective effect evident in younger parous women [58] but lasting until menopause. Among the ER-α downstream regulated genes was GATA3, which encodes a protein that belongs to the GATA family of transcription factors that regulates T lymphocyte differentiation and maturation. GATA3 is crucial to mammary gland morphogenesis and differentiation of progenitor cells and a putative tumor suppressor [59]. Induction of GATA 3 expression in GATA3-negative undifferentiated carcinoma cells is sufficient to induce tumor differentiation and inhibition of tumor dissemination [60]. Therefore, the observation that genes involved in the estrogen receptor regulated pathways are up-regulated in the parous breast in spite of the lack of transcriptomic differences in this receptor's levels between parous and nulliparous postmenopausal breast tissues suggests that they could be under permanent transcriptional modification as a manifestation of a higher degree of cell differentiation.

Studies of breast development under the influence of parity in women and in animal models are in agreement on the pregnancy-induced differentiation of the breast, a process that ultimately becomes manifested as a specific genomic signature in the mammary gland [51–53, 55, 58, 61, 62]. Although variations in gene expression among different studies and species are expected, an increase in immune activity, including overexpression of lipopolysaccharide binding protein (LBP/Lbp) has been reported in the post-pregnancy breast of premenopausal women [58] and in the mammary gland of four different strains of rats [61]. Interestingly, this response was observed in both recently pregnant in distant pregnant groups but not in the postmenopausal group. These discrepancies might indicate that the up regulation of inflammation/immune response–related genes persists during post-partum involution, but wanes after menopause sets in (see next section).

Importantly, there has been a reported shift in the cell population of the postmenopausal breast as a manifestation of the reprogramming of the organ after pregnancy [28] (Fig. 10.1). These observations are in agreement with what is observed in the rat mammary gland, which also contains two types of luminal epithelial cells, designated dark (DC) and intermediate (IC) cells, in addition to the myoepithelial cells [63]. The DC and IC are equivalent to the HTN and EUN cells described in the parous breast [28]. DCs increase after pregnancy and lactational involution; whereas the ICs significantly outnumber the DCs in ductal hyperplasias and ductal carcinomas [63, 64]. The analysis of nuclear ultrastructural and morphometric parameters of rodent ICs have allowed to differentiate the mammary progenitor stem cell from the cancer stem cells [54, 63, 64]. Nuclear morphometric analysis of breast and ovarian carcinomas has confirmed the predictive value of nuclear grade on the progression of premalignant lesions to invasiveness [65–67]. The findings of a significant decrease in the number of EUN cells with a subsequent increase in the number of HTN cells expressing specific biomarkers identified at the chromatin and transcriptional levels support the value of morphometric analysis as an adjuvant to molecular studies. The data clearly indicate [28] that there are morphological indications of chromatin remodeling in the parous breast, such as an increase in the number of epithelial cells with condensed chromatin and increased reactivity with anti-H3K9me^2 and H3K27me^3 antibodies (Fig. 10.1). Histone methylation is a major determinant for the formation of active and inactive regions of the genome and is crucial for the proper programming of the genome during development [68]. In the parous breast there is up regulation of transcription factors and chromatin remodeling genes such as CHD2 or chromodomain helicase DNA binding protein 2 and the CBX3 or Chromobox homolog 3, whose products are required for controlling recruitment of protein/protein or DNA/protein interactions. CBX3 is involved in transcriptional silencing in heterochromatin-like complexes, and recognizes and binds H3 tails methylated at lysine 9, leading to epigenetic repression. Two other important genes related to the polycomb group (PcG) protein that are up regulated in the parous breast are the L3MBTL gene or l(3)mbt-like and the histone-lysine

N-methyltransferase or EZH2. Members of the PcG form multimeric protein complexes that maintain the transcriptional repressive state of genes over successive cell generations. EZH2 is an enzyme that acts mainly as a gene silencer, performing this role by the addition of three methyl groups to lysine 27 of histone 3, a modification that leads to chromatin condensation to occur [57, 69, 70].

RNA molecules recruit PcG complexes to the locus of transcription or to sites located elsewhere in the genome. An important role has been attributed to noncoding RNAs (ncRNAs) [71]. It has been postulated [28] that the increased chromatin condensation in the parous breast could have been initiated by ncRNAs, a postulate supported by the observed up regulation of several ncRNAs that included nuclear paraspeckle assembly transcript 1 (NEAT1), MALAT-1 (NEAT2), and X inactive specific transcript (XIST) (Fig. 10.2) [72] all critical components of the speckles. There is a relationship between the chromatin remodeling process and post transcriptional control maintained by the spliceosome machinery that is stored in nuclear speckles. Among the components of the spliceosome machinery that are up-regulated in the parous breast are the heterogeneous nuclear ribonucleoproteins HNRPA3, HNRPA2B1, HNRPD and HNRPU (Fig. 10.3). The functional role of these HNRPs in the postmenopausal breast could be implicated in the regulation of mRNA stability, other functions like mammary gland involution, acting as negative regulators of telomere length maintenance [73] or regulating the trafficking of mRNA molecules [74]. Other members of the spliceosome complex are the small nuclear ribonucleoproteins (snRNPs), which function as suppressors of tumor cell growth and may have major implications as cancer therapeutic targets. Among these we have found that the transcripts regulated by the genes SF3B1, SFRS2, SFRS7, SFRS8, SFRS14, SFRS16, SNRP70, SNRPB, SNRPA1, PRF3 and PHF5A are over expressed in the parous breast [28]. Other members of the splicing factor compartment that are localized in the nuclear speckles are CCNL1 and CCNL2 (Fig. 10.3). It has been demonstrated that CCNL2 protein is overexpressed in the nucleus of epithelial cells composing the Lob 1 of the parous breast [28]. CCNL1 and CCNL2 are transcriptional regulators that participate in the pre-mRNA splicing process and the expression of critical factors leading to cell apoptosis, possibly through the Wnt signal transduction pathway [75, 76], which we found to be down regulated in the parous breast.

Another component of the spliceosome complex that regulates genes involved in the apoptotic process is the RNA binding motif protein 5 (RBM5). The over expression of RBM5 retards ascites associated tumor growth and enhances p53-mediated inhibition of cell growth and colony formation [77, 78] mechanisms that could also be operational in the parous breast. The spliceosome plays a critical role in differentiating mouse ESC and self-renewal, pluripotency and tissue lineage specification of human ESC [79]. Post-transcriptional modifications of RNA, including packaging into the nuclear speckles of the breast epithelial cells and recognition by RNA-binding proteins and/or microRNAs are crucial processes in differentiating breast epithelial cells. Data discussed here emphasize the importance of post-transcriptional regulatory mechanisms as a critical component underlying the differentiation of the breast (Figs. 10.2 and 10.3).

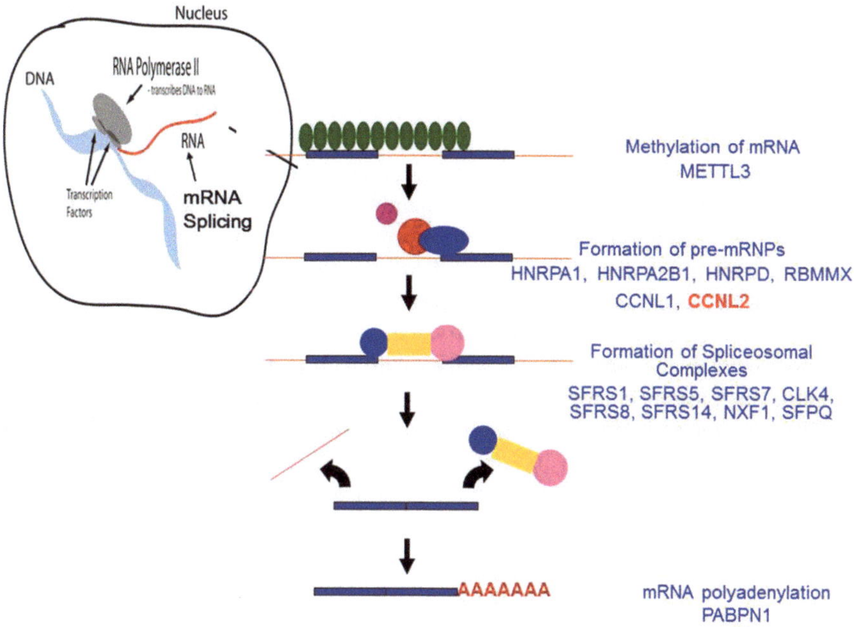

Fig. 10.3 Among the components of the spliceosome machinery that are up-regulated in the parous breast are the heterogeneous nuclear ribonucleoproteins and the functional role of these HNRPs in the postmenopausal breast could be implicated in the regulation of mRNA stability, other functions like mammary gland involution, acting as negative regulators of telomere length maintenance or regulating the trafficking of mRNA molecules. Other members of the splicing factor compartment that are localized in the nuclear speckles are CCNL1 and CCNL2. CCNL1 and CCNL2 are transcriptional regulators that participate in the pre-mRNA splicing process and the expression of critical factors leading to cell apoptosis, possibly through the Wnt signal transduction pathway, which has been found to be down regulated in the parous breast

10.4 Basis of the Dual Effect of Late Pregnancy in the Increase Risk of Breast Cancer

Differences in gene expression in the breast of parous versus nulliparous healthy premenopausal women has been reported [80] by Santucci-Pereira and colleagues. The authors used Affymetrix Human Genome U133 Plus 2.0 microarrays, and analyzed the gene expression profile of breast tissue from 30 nulliparous (NP) and 79 parous (P) premenopausal volunteers between the ages of 30 and 47 years who were free of breast pathology. Because of the known short-term increase in breast cancer risk preceding the long-term protective effect of FTP, the authors also examined gene expression differences in P vs. NP women as a function of time since last FTP. Through multiple regression analysis, controlling for confounders, they found 416 probesets differentially expressed (fold-change ≥ 1.2 and false discovery rate $< 10\,\%$) comparing all P vs. all NP, and/or, P women whose last FTP was less than 5 years before biopsy vs. all NP women. Among these, 352 probesets, representing 238 genes, were up regulated, while 64 probesets, representing 48 genes, were down regulated in the parous compared to nulliparous breast. Of interest is that among the up regulated genes, they observed three expression patterns designated transient, long term changing and long term constant.

The transient genes up regulated after FTP but whose expression levels rapidly returned to nulliparous level were genes mainly related to immune response (CCL5, CD48, IL7R). Among other genes of the immune response that are also upregulated are those controlling dendritic cells and T killer cells and in Fig. 10.4 is an interpretation on how these transcripts could be protective in the parous women, but in those women in which the pathway is not activated, like those that have cancer during the first 5 years after pregnancy, could well explain their increase risk by an absence of immune response mechanism that eliminate the transformed cells.

The long-term changing genes up regulated following FTP, whose expression levels decreased with increasing time since last FTP, but did not return to nulliparous levels, are genes related to immune response (CD38, CXCL10) and develop-

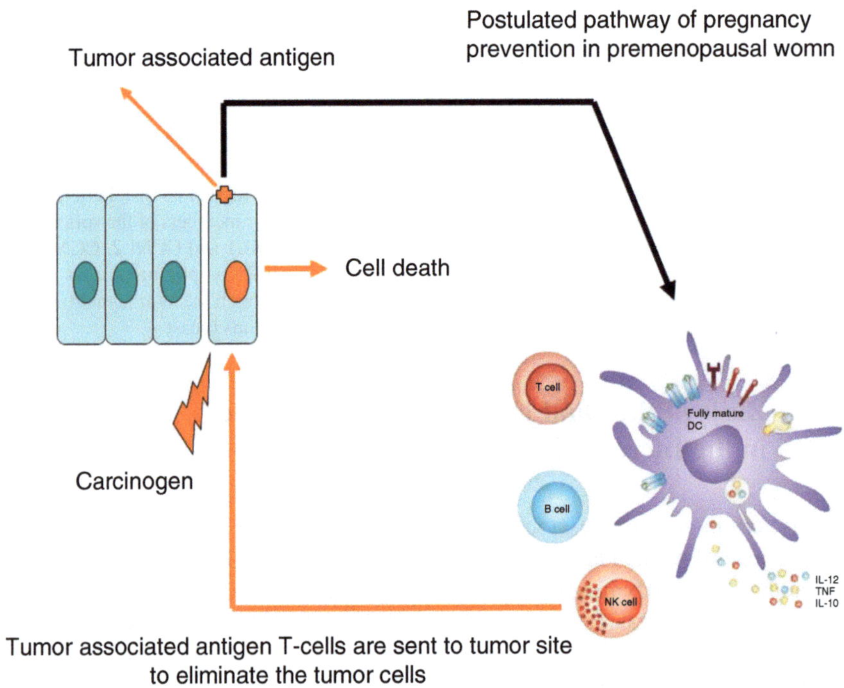

Fig. 10.4 Among the genes of the immune response that are activated during the firsts five years post pregnancy in the premenopausal women are those controlling dendritic cells and T killer cells. These transcripts could be protective in the parous women, but in those women in which the pathway is not activated, like those that have cancer during the first 5 years after pregnancy, could well explain their increase risk by an absence of immune response mechanism that eliminate the transformed cells

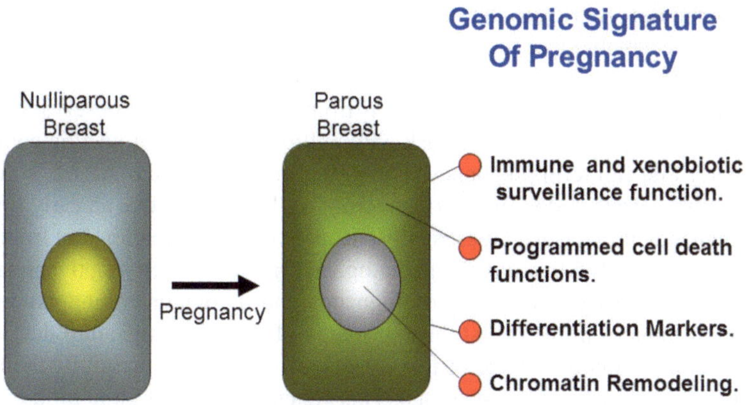

Fig. 10.5 Full term pregnancy induces long-term expression changes in genes related to the processes of immune and xenobiotic surveillance function, programmed cell death, differentiation and chromatin remodeling

ment (DKK3, LAMA2). The long-term constant genes that remained up regulated in the parous compared to nulliparous breast, independent of time since last FTP, were mainly involved in developmental processes (BHLHE22, FZD8, KRT5), cell differentiation (RASGRP1, DSC3) and chromatin remodeling (NAP1L2).

The Santucci-Pereira study [80] shows that a first full term pregnancy induces long-term expression changes in genes related to the processes of development, cell differentiation and chromatin remodeling as has also be found in the parous postmenopausal breast [28] (Fig. 10.5). Additionally, the transiently activated genes related to immune response during the first five years after FTP may play a role in the short-term increase of breast cancer risk following FTP. A better understanding of the molecular effects of parity on the breast may help the development of novel strategies for preventing breast cancer [80].

In brief the genomic profile of nulliparous and parous women in the premenopausal and postmenopausal period has shown that there are genes which are only activated during the first five years after pregnancy that may contribute to the increased risk experienced by certain women after pregnancy [28, 55, 56, 80]. At the same time pregnancy induces a long lasting genomic signature that starts after pregnancy, explaining its preventive effect. The molecular mechanism related to prevention revolves around the chromatin remodeling process [28] (Fig. 10.1).

10.5 Current Strategies in Breast Cancer Prevention

Current strategies to prevent breast cancer focused on a unique feature of this disease, its endocrine, namely estrogen, dependence, which can be manipulated to control growth or prevent tumor development utilizing either selective estrogen receptor modulators (SERMs), such as tamoxifen [8, 81–86], or aromatase inhibitors (AI's), such as Arimidex, Letrozole and Exemestane [83, 85]. However, these strategies are not widely acceptable to a majority of treated women who would not have developed breast cancer even if untreated. Therefore what is needed is a new strategy for breast cancer prevention that has emerged from epidemiological observations of a direct association of breast cancer risk with nulliparity and of protection conferred by an early first full term pregnancy [8, 86–91]. Important to emphasize is that the novelty of this strategy does not germane from the knowledge that an early first full term pregnancy protects the breast against neoplastic transformation, but from studies [48, 49, 63, 92–99] that unveil the biological principle underlying the protection conferred by an early first full-term pregnancy and by demonstrating experimentally that it induces in the breast the expression of a specific signature that results from the completion of a cycle of this organ's differentiation driven by the reproductive process (see previous sections). This signature (Fig. 10.5), in turn, is a biomarker associated with lifetime decreased breast cancer risk. More importantly, the biological principle has been harnessed by demonstrating in an experimental model that a short treatment with human chorionic gonadotropin (hCG), a placental hormone secreted during pregnancy, induces the same genomic signature than pregnancy, inhibiting not only the initiation but also the progression of mammary carcinomas, stopping the development of early lesions, such as intraductal proliferations, and carcinomas in situ. These observations indicate that hCG administered for a very short period of time has significant potential as a chemopreventive agent, protecting the normal cell from becoming malignant [24, 48, 51, 56, 57, 100–103]. This new biological concept also implies that when the genomic signature of protection or refractoriness to carcinogenesis is acquired, the hormonal treatment with hCG is no longer required. This is a novel concept that contra poses the current knowledge that a chemopreventive agent needs to be given for a long period to suppress a metabolic pathway or abrogate the function of an organ.

10.5.1 Experimental Data Supporting the New Strategy in Prevention

The direct association of breast cancer risk with the prolongation in the period encompassed between menarche and the first full term pregnancy, as well as the protection afforded by pregnancy has been explained by experimental studies performed in laboratory animals [24, 48, 49, 59, 103–106]. It has been demonstrated that mammary cancer in rodents can be induced with the polycyclic hydrocarbon 7,12-dimethylbenz(a)anthracene (DMBA) preferentially when the carcinogen is administered to young nulliparous females [49]. Those females that have completed a full term pregnancy prior to carcinogen exposure fail to develop carcinomas [24, 104, 105, 107, 108]. Altogether these studies have revealed that the susceptibility of the mammary gland to be transformed by a chemical carcinogen is modulated by specific biological conditions of the host and of the target organ [109–111]. Tumor incidence and number of tumors per animal, which are the biological endpoints when evaluating tumorigenic response, are maximal when the carcinogen is administered to young but cycling virgin rats. Cancer incidence is directly proportional to the number of terminal end buds (TEBs) that are at their peak of cell proliferation [109–112]. Stimulation of the development and differentiation of the gland, resulting in profuse lobular development and depression of DNA synthesis, such as it occurs during pregnancy, or after completion of a 21 day-treatment of virgin rats with hCG, reduce the susceptibility of the mammary epithelium to be transformed by the carcinogen. The reduction in cancer incidence is permanent, as demonstrated by the similar degree of reduction when DMBA is administered after a delay of 21, 42, or 63 days after termination of hCG treatment.

Pregnancy alone or followed by lactation, induces in the mammary gland a permanent protective effect from chemically induced carcinogenesis, since administration of a carcinogen to parous rats when the glands have regressed to a resting stage either fails to induce carcinomas or considerably lowers their incidence [109, 113], whereas mammary glands showing gestational or lactational hyperplasia are moderately refractory to DMBA induced carcinogenesis [114, 115]. This indicates that it is not the transient hormonal status occurring during pregnancy and lactation that protects the mammary gland, but the permanent changes induced in the gland structure and in the biological properties of the glandular epithelium by the reproductive phenomenon [116].

The observation that pregnancy before carcinogen administration seems to be the only truly protective factor in chemically-induced mammary gland carcinogenesis, suggests that placental hormones play an important role in mammary growth and development during pregnancy [52, 59, 110, 116–120]. The main placental hormone, human chorionic gonadotropin (hCG) has a stimulatory effect on the mammary gland when administered exogenously, producing either a gestational or a lactational type of mammary development that considerably reduces the incidence of tumors, [59, 115]. The fact that the hormonal changes of pregnancy accelerate DMBA-induced mammary tumor growth when mating occurs after carcinogen administration [112, 114, 115] indicates that *the most important event in determining the role that this hormone plays in either preventing initiation or in promoting tumor growth is the sequence in which it reaches the mammary gland.*

It has been demonstrated that the inhibitory effect of pregnancy on mammary cancer initiation is mediated by hCG, since virgin rats treated for 21 days with a daily intraperitoneal injection of this hormone prior to carcinogen administration exhibit a dose-related reduction in tumor incidence and number of tumors per animal [24, 48, 49, 59, 103, 106]. This phenomenon is in great part mediated by the induction of mammary gland differentiation, inhibition of cell proliferation, increase in the DNA repair capabilities of the mammary epithelium, decrease binding of the carcinogen to the DNA, and activation of genes controlling programmed cell death (PCD) [24, 51, 56, 57, 99–105, 121, 122]. The activation of these genes by hCG is of great relevance because PCD is a physiological and phylogenetically conserved form of active cell death (or apoptosis) that has been associated with specific phases of development that control cell proliferation and differentiation [51].

10.5.2 Clinical Studies Supporting the New Strategy in Prevention

Based on preclinical data that had demonstrated that hCG treatment of virgin rats prevented the initiation and inhibited the progression of DMBA-induced mammary carcinomas, the effect of hCG on primary breast cancer in post-menopausal patients was evaluated [117, 118]. In a double-blind, placebo-controlled study, 25 post-menopausal women with primary operable breast cancer (T1-T3) whose diagnosis was made by core biopsy performed on day 0, received on alternate days for 2 weeks intramuscular injections of either r-hCG (recombinant hCG) (500 μg); n = 20) or placebo (n = 5). Surgery (mastectomy or lumpectomy) was performed on day 15. The tumor tissue obtained in the initial core biopsy and that removed at the time of therapeutic surgery were evaluated to determine the rate of cell proliferation, or proliferative (Ki67) index, inhibin immunoreactivity, and percentage of cells positive for estrogen (ER) and progesterone receptors (PR). The most remarkable effects attributed to this two-week treatment were a significant reduction in Ki67 index from 18 % in the initial biopsy to 4 % in the mastectomy/lumpectomy specimens ($p < 0.00006$), and increased synthesis of inhibin. Serum hormonal levels were those characteristics of post-menopausal women, and remained unchanged during and after the treatment, except for elevation in hCG levels during treatment. Hormone administration was well tolerated by all patients, and no local or systemic side effects were reported at any time. The data clearly indicated that hCG is an inhibitor of cell proliferation independently of the ovarian function (postmenopausal women) and independent of the estrogen and progesterone receptor status of the host tissue. In addition the data indicates that the recombinant form of this hormone does not affect the hormonal milieu of the patient [117, 118].

10.5.3 Pregnancy and HCG Induce Permanent Genomic Imprinting or a Specific Signature of Protection

The genomic signature of the mammary gland induced in virgin animals by exogenous administration of hCG is similar to that induced by pregnancy, and that specific genomic profiles are still manifested by 42 days post termination of treatment [119, 120] (Fig. 10.5). The importance of these specific signatures is highlighted by the fact that administration of carcinogen to hCG-treated or control virgin rats whose mammary glands appear morphologically similar will induce a markedly different tumorigenic response, supporting the concept that the differentiation induced by hCG is expressed at genomic level, and results in a shift of the susceptible stem cell EUN to a refractory stem cell HTN. The permanence of these changes, in turn, makes them ideal surrogate markers for the evaluation of hCG effect as a breast cancer preventive agent.

10.6 Developing a Prevention Clinical Trial Using HCG

Based on preclinical data that have demonstrated that r-hCG exerts a mammary cancer preventive effect that is mediated by the induction of gland differentiation, which results in permanent changes in the genomic signature of this organ the following clinical trial has been designed (Fig. 10.6). The study will evaluate the genomic profile of breast epithelial cells obtained from core biopsies specimens performed in high risk women treated for 90 days with r-hCG. The objective of the proposed study is to characterize the genomic profile of breast epithelial cells obtained from asymptomatic high breast cancer risk nulliparous premenopausal women carriers of *BRCA1* deleterious mutations. Gene expression measurements will be obtained at baseline (time 0), after treatment with r-hCG at 90 days (time 1) and at 270 days from baseline (time 2) (Fig. 10.6). The primary objective of the study is to compare the gene expression profiles of these women across the three time points and identify differentially expressed genes. It is of interest in comparing the expression profiles between all pairs of time points as well as across time. The comparison of profiles before and after treatment with r-hCG, both at 90 and 270 days are of particular interest. The women will receive three-weekly injections of 250 μg r-hCG for a total of 12 weeks. Normal breast tissue specimens will be collected by Spirotome core needle biopsies at the beginning (0 day) and at the end of treatment (90 days) and at six months post-treatment (270 days). Core Needle Biopsies specimens will be primarily utilized for analysis of genomic expression by cDNA microarray. In addition, a series of surrogate intermediate markers such as cytomorphologic evaluation and cell proliferation index will be analyzed.

High breast cancer risk women carriers of BRCA1 and BRCA2 mutations will be invited to participate in this study and the inclusion criteria are: 1) Premenopausal women between the ages of 18–24 years of age; 2) having normal menstrual cycles and intact ovaries; 3) nulliparous, never pregnant (G0P0); 4) carriers of a deleterious mutation on the *BRCA1/2* gene, as determined by testing in a CLIA-certified clinical genetics laboratory; 5) not pregnant and not on oral contraceptives or hormone

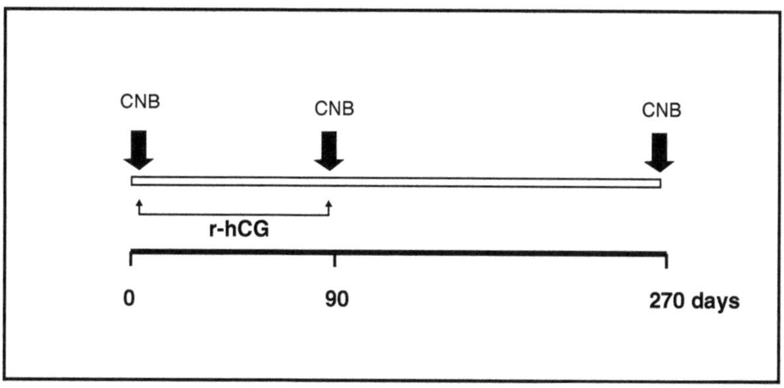

Fig. 10.6 Schematic representation of the clinical protocol for testing rhCG as a preventive agent

replacement therapy; 6) currently not participating in a chemopreventive trial for breast cancer; 7) currently not taking Tamoxifen for chemoprevention; 8) no previous diagnosis of breast or ovarian cancer; 9) normal ovarian size report from pelvic ultrasound; 10) eligible candidates should not be taking oral contraceptives, and if taking them, stopping six weeks prior to the initiation of treatment and the performance of the first CNB, and blood drawing; 11) willingness to self-administer r-hCG for three months and to return for two repeat CNBs.

Among the exclusion criteria are: 1) Prior hypersensitivity to hCG preparations or one of its excipients; 2) prior history of ovarian cancer; 3) ovarian enlargement of undetermined origin at the time of admission; 4) ovarian cysts larger than 2 cm; 5) microcystic ovaries, which have been reported to predispose to the development of ovarian hyperstimulation syndrome (OHSS) under treatment with FSH for assisted reproduction techniques; 6) history of prior cancer other than non-melanoma skin cancer; 7) ever pregnant; 8) taking medications that could interfere with the study protocol such as hormonal contraceptives, androgens, prednisone, thyroid hormones, insulin; 9) severe cognitive deficit, and 10) unable to give informed consent.

Eligible women will be instructed on the purpose of the study; an overview of methods and number of visits involved in the study, and the proposed research outcome. All the volunteers participating in this study will sign an informed consent and the participants will be asked to complete the study questionnaire. Height and weight will be determined and blood will be collected by venipuncture by a phlebotomist. When the results from the blood test and the vaginal ultrasounds are received confirming the eligibility of each candidate, the nurse coordinator will schedule the participant for the Core Needle Biopsy (CNB) and further follow up for drug administration (Fig. 10.7). Four CNB samples will be collected at 0, 90, and 270 days of treatment during the progestational phase of the menstrual cycle (Fig. 10.7). At the time of each CNB, 45 ml of blood will be collected by venipuncture. These blood samples will be separated into three distinct layers: the plasma at the top of the tube, the erythrocytes (red blood cells) at the bottom of the tube, and the lymphocytes or "buffy coat" in a thin white layer between the plasma and red cells. The plasma and red cells will be frozen and saved for possible future evaluations and half of the pooled lymphocytes will be aliquoted for cryopreservation. Four 10 ml Serum Separator Tubes of approximately 7.5 ml each will be collected from each participant for storage for future evaluations. After the blood clots, the tubes will be centrifuged and the serum isolated. The serum will be stored in 1-ml aliquots at −80 °C.

The study drug, r-hCG (Ovidrel, Serono) is an analog of luteinizing hormone (LH) and binds to the LH/hCG receptor of the granulose and theca cells of the ovary. In rodents, hCG obtained from the urine of pregnant women as well as r-hCG act mainly as LH, inducing the ovulation of already existing ovarian follicles and maintaining their respective corpora lutea in a pseudo pregnancy condition in which there is no further ovulation or risk of pregnancy. After cessation of treatment the corpora lutea regress, and the ovaries return to their normal size, with maturation of new follicles for resuming their cyclic activity. HCG has been used clinically for

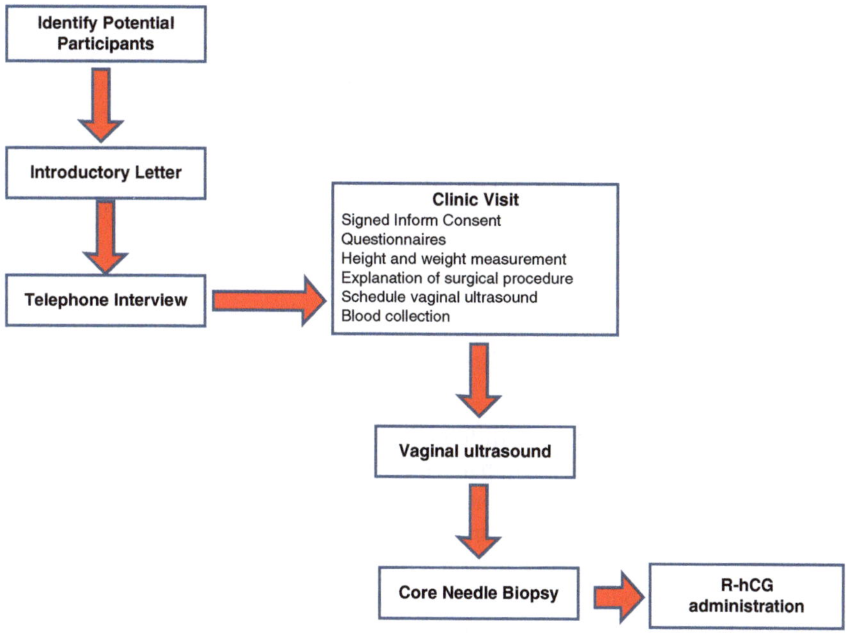

Fig. 10.7 Organigram of the hCG protocol used for prevention

many years for the treatment of male and female infertility, corpus luteum insufficiency, habitual or threatened abortion, hypogonadism and cryptorchidism in the male, and weight reduction [123, 124]. R-hCG has been approved by FDA for its use as a subcutaneous injection and for patient self-administration in Assisted Reproductive Technologies (ART) [125, 126]. The hormone is well tolerated without significant toxicities. Due to the direct effect of hCG on the ovarian follicle it is recommended to monitor women receiving follitropin, a recombinant preparation of follicle stimulating hormone (FSH) prior to the administration of r-hCG in ART for ovarian hyperstimulation syndrome (OHSS). In our knowledge, there is no report in the literature that r-hCG *per se* induces OHSS. Nevertheless, in this protocol, ovarian size will be monitored by ultrasound, as it is recommended for the combined FSH/r-hCG treatments for ovulation induction in ART as a precautionary measure. In addition the r-hCG will be administered in the progestational phase (16–20 of her menstrual cycle--day 1 is the first day of bleeding) to avoid interference with the ovulation process.

Total RNA will be isolated from the CNB samples using Trizol (Invitrogen, Inc.). The concentration and the quality of each RNA will be determined as described [28, 55, 56]. The Gene expression analysis will be performed as previously published by our laboratory [28, 55, 56]. Gene expression profiling will be done using Affimetrix technology [28, 55, 56]. Analysis of differential expression between two time points will involve adjustment for multiple testing in terms of controlling the false discovery rate (FDR). Using a reference design, in order to detect an effect size of 1 (in terms of $\log_2$ ratios, an effect size of 1 corresponds to a 2-fold difference between any two time points being compared) between any two time points, at a significance level of 0.001 with 95 % power, we would need a total of approximately 23 samples or arrays. This is based on a standard deviation (of $\log_2$ expression ratios across samples) of approximately 0.5 [127]. A significance level of 0.001 results in 1 false discovery per 1000 non-differentially expressed genes. Due to the paired nature of the comparisons, we would need 23 patients for each comparison. Assuming a drop-out rate of 20 % at each subsequent time point, we would need a minimum of 18 patients (or a total minimum of 54 arrays across the 3 time points).

10.7 Summary and Conclusions

Breast cancer originates in the undifferentiated terminal duct of the Lob 1 that contains stem cells (Stem cell EUN)] that is the site of origin of ductal carcinomas. The susceptibility of Lob 1 to undergo neoplastic transformation has been attributed to its high rates of cell proliferation and of carcinogen binding to the DNA and low reparative activity. The hormonal milieu of an early full term pregnancy or hCG treatment induces lobular development, completing the cycle of differentiation of the breast. This process induces a specific genomic signature in the mammary gland that is represented by the Stem cell HTN. Even though differentiation significantly reduces cell proliferation in the mammary gland, the mammary epithelium remains capable of responding with proliferation to given stimuli, such as a new pregnancy. The stem cell HTN is able to metabolize the carcinogen and repair the induced DNA damage more efficiently than the stem cell EUN, as it has been demonstrated in the rodent experimental system. There is also evidence that hCG has an effect in the cancer cell by further the differentiation pattern. The finding that differentiation is a powerful inhibitor of cancer initiation provides a strong rationale for identifying the genes that control this process. The basic biological concept is that pregnancy or hCG shifts the stem cell EUN to the stem cell HTN that is refractory to carcinogenesis.

The mechanisms discussed above play a role in the protection exerted by hCG from chemically induced carcinogenesis, and might be even involved in the life-time reduction in breast cancer risk induced in women by full term and multiple pregnancies. The implications of these observations are two-fold: on one hand, they indicate that hCG, as pregnancy, may induce early genomic changes that control the progression of the differentiation pathway, and that these changes are permanently imprinted in the genome, regulating the long-lasting refractoriness to carcinogenesis. The permanence of these changes, in turn, makes them ideal surrogate markers of hCG effect in the evaluation of this hormone as a breast cancer preventive agent.

Based in the knowledge on the pathogenesis of mammary cancer it has been tested the effect of hCG hormone on the early phases of tumor progression, namely from TEBs damaged by DMBA to intraductal proliferation, *in situ* carcinomas and invasive carcinomas and demonstrate that this hormone inhibits the progression of 7,12-dimethylbenz(a)anthracene (DMBA)-induced mammary tumors.

Treatment of young virgin rats with hCG induced a profuse lobular development of the mammary gland, practically eliminating the highly proliferating TEBs, with overall reduction in the proliferative activity of the mammary epithelium, and induction of the synthesis of inhibin, a secreted protein with tumor-suppressor activity.

The hormonal treatment induced differentiation of the mammary gland, which was manifested at morphological, cell kinetic and functional levels. The morphological changes consisted of progressive branching of the mammary parenchyma and lobule formation. They were accompanied by reduction in the rate of cell proliferation.

The functional changes comprised increased synthesis of inhibin, β-casein and other milk-related bioactive peptides. In addition, hCG also increased the expression of the programmed cell death genes inducing as well apoptosis, and down regulation of cyclins. Programmed cell death genes were activated through a p53-dependent process, modulated by c-myc, and with partial dependence on the bcl-2 family-related genes.

Data generated with the new tools provided by the cDNA micro array techniques have allowed to demonstrate that while lobular development regressed after the cessation of hormone administration, programmed cell death genes remained activated, DNA repair genes, chromatin remodeling, transcription factors and immune-surveillance gene transcripts. The genomic signature is specific for pregnancy and hCG and significantly different than the one induced by y other hormones such as estrogen and progesterone [128].

References

1. Botha JL, Bray F, Sankila R, Parkin DM (2003) Breast cancer incidence and mortality trends in 16 European countries. Eur J Cancer 39:1718–1729
2. Clarke CA, Purdie DM, Glaser SL (2006) Population attributable risk of breast cancer in white women associated with immediately modifiable risk factors. BMC Cancer 6:170–181
3. Jemal A, Bray F, Center MM, Ferlay J, Ward E et al (2011) Global cancer statistics. CA Cancer J Clin 61:69–90
4. Althuis MD, Dozier JM, Anderson WF, Devesa SS, Brinton LA (2005) Global trends in breast cancer incidence and mortality 1973-1997. Int J Epidemiol 34:405–412
5. Maskarinec G, Pagano I, Chen Z, Nagata C, Gram IT (2007) Ethnic and geographic differences in mammographic density and their association with breast cancer incidence. Breast Cancer Res Treat 104:47–56
6. Nagata C, Mizoue T, Tanaka K, Tsuji I, Wakai K et al (2006) Tobacco smoking and breast cancer risk: an evaluation based on a systematic review of epidemiological evidence among the Japanese population. Jpn J Clin Oncol 36:387–394
7. Hinkula M, Pukkala E, Kyyronen P, Kauppila A (2001) Grand multiparity and the risk of breast cancer: population-based study in Finland. Cancer Causes Control 12:491–500
8. MacMahon B, Cole P, Lin TM, Lowe CR, Mirra AP et al (1970) Age at first birth and breast cancer risk. Bull World Health Organ 43:209–221
9. Ma H, Henderson KD, Sullivan-Halley J, Duan L, Marshall SF et al (2010) Pregnancy-related factors and the risk of breast carcinoma in situ and invasive breast cancer among postmenopausal women in the California Teachers Study cohort. Breast Cancer Res 12:R35
10. Phipps AI, Chlebowski RT, Prentice R, McTiernan A, Wactawski-Wende J et al (2011) Reproductive history and oral contraceptive use in relation to risk of triple-negative breast cancer. J Natl Cancer Inst 103:470–477

11. Toniolo P, Grankvist K, Wulff M, Chen T, Johansson R et al (2010) Human chorionic gonadotropin in pregnancy and maternal risk of breast cancer. Cancer Res 70:6779–6786
12. Mustacchi P (1961) Ramazzini and Rigoni-Stern on parity and breast cancer. Clinical impression and statistical corroboration. Arch Intern Med 108:639–642
13. Brinton LA, Sherman ME, Carreon JD, Anderson WF (2008) Recent trends in breast cancer among younger women in the United States. J Natl Cancer Inst 100:1643–1648
14. Kroman N, Melbye M, Mouridsen HT (2002) Prognostic influence of age at diagnosis in premenopausal breast cancer patients. Scand J Surg 91:305–308
15. Kroman N, Mouridsen HT (2003) Prognostic influence of pregnancy before, around, and after diagnosis of breast cancer. Breast 12:516–521
16. Downs JL, Wise PM (2009) The role of the brain in female reproductive aging. Mol Cell Endocrinol 299:32–38
17. Russo IH, Medado J, Russo J (1989) Endocrine influences on mammary gland structure and development. In: Jones TC, Mohr U, Hunt RD (eds) Integument and mammary gland of laboratory animals. Springer-Verlag, Berlin, pp 252–266
18. Chen T, Lundin E, Grankvist K, Zeleniuch-Jacquotte A, Wulff M et al (2010) Maternal hormones during early pregnancy: a cross-sectional study. Cancer Causes Control 21:719–727
19. Albrektsen G, Heuch I, Thoresen S, Kvale G (2006) Clinical stage of breast cancer by parity, age at birth, and time since birth: a progressive effect of pregnancy hormones? Cancer Epidemiol Biomarkers Prev 15:65–69
20. Beatson G (1896) On the treatment of inoperable cases of carcinoma of the mammary. Suggestions for new method of treatment with illustrative cases. Lancet 2:104–107
21. Lukanova A, Surcel HM, Lundin E, Kaasila M, Lakso HA et al (2011) Circulating estrogens and progesterone during primiparous pregnancies and risk of maternal breast cancer. Int J Cancer 130:910–920
22. Russo J, Russo IH (eds) (2004) Molecular basis of breast cancer: prevention and treatment. Springer-Verlag, Berlin, p 447
23. Russo J, Russo IH (2006) The role of estrogen in the initiation of breast cancer. J Steroid Biochem Mol Biol 102:89–96
24. Russo J, Russo IH (1980) Susceptibility of the mammary gland to carcinogenesis. II. Pregnancy interruption as a risk factor in tumor incidence. Am J Pathol 100:497–512
25. Medina D, Smith GH (1999) Chemical carcinogen-induced tumorigenesis in parous, involuted mouse mammary glands. J Natl Cancer Inst 91:967–969
26. Blakely CM, Stoddard AJ, Belka GK, Dugan KD, Notarfrancesco KL et al (2006) Hormone-induced protection against mammary tumorigenesis is conserved in multiple rat strains and identifies a core gene expression signature induced by pregnancy. Cancer Res 66:6421–6431
27. Russo J, Balogh GA, Heulings R, Mailo DA, Moral R et al (2006) Molecular basis of pregnancy-induced breast cancer protection. Eur J Cancer Prev 15:306–342
28. Russo J, Santucci-Pereira J, de Cicco RL, Sheriff F, Russo PA et al (2012) Pregnancy-induced chromatin remodeling in the breast of postmenopausal women. Int J Cancer 131:1059–1070
29. Wagner KU, Boulanger CA, Henry MD, Sgagias M, Hennighausen L et al (2002) An adjunct mammary epithelial cell population in parous females: its role in functional adaptation and tissue renewal. Development 129:1377–1386
30. Booth BW, Boulanger CA, Smith GH (2008) Selective segregation of DNA strands persists in long-label-retaining mammary cells during pregnancy. Breast Cancer Res 10:R90
31. Boulanger CA, Wagner KU, Smith GH (2005) Parity-induced mouse mammary epithelial cells are pluripotent, self-renewing and sensitive to TGF-beta1 expression. Oncogene 24:552–560
32. Russo J, Tay LK, Ciocca DR, Russo IH (1983) Molecular and cellular basis of the mammary gland susceptibility to carcinogenesis. Environ Health Perspect 49:185–199
33. Blackshaw S, Scholpp S, Placzek M, Ingraham H, Simerly R et al (2010) Molecular pathways controlling development of thalamus and hypothalamus: from neural specification to circuit formation. J Neurosci 30:14925–14930

34. Hendriks AE, Laven JS, Valkenburg O, Fong SL, Fauser BC et al (2011) Fertility and ovarian function in high-dose estrogen-treated tall women. J Clin Endocrinol Metab 96:1098–1105
35. Russo J, Russo IH (1987) Development of the human mammary gland. In: Neville MC, Daniel C (eds) The mammary gland. Plenum, New York, pp 67–93
36. Russo J, Rivera R, Russo IH (1992) Influence of age and parity on the development of the human breast. Breast Cancer Res Treat 23:211–218
37. Howell A, Evans GD (2011) Hormone replacement therapy and breast cancer. Recent Results Cancer Res 188:115–124
38. Harvey JA, Santen RJ, Petroni GR, Bovbjerg VE, Smolkin ME, Sheriff FS, Russo J (2008) Histologic changes in the breast with menopausal hormone therapy use: correlation with breast density, estrogen receptor, progesterone receptor, and proliferation indices. Menopause 15(1):67–73
39. Gurney EP, Nachtigall MJ, Nachtigall LE, Naftolin F (2014) The Women's Health Initiative trial and related studies: 10 years later: a clinician's view. J Steroid Biochem Mol Biol 142:4–11
40. Ghazal S, Pal L (2013) Perspective on hormone therapy 10 years after the WHI. Maturitas 76(3):208–212. doi:10.1016/j.maturitas.2013.08.014
41. Manson JE, Chlebowski RT, Stefanick ML, Aragaki AK, Rossouw JE, Prentice RL, Anderson G, Howard BV, Thomson CA, LaCroix AZ, Wactawski-Wende J, Jackson RD, Limacher M, Margolis KL, Wassertheil-Smoller S, Beresford SA, Cauley JA, Eaton CB, Gass M, Hsia J, Johnson KC, Kooperberg C, Kuller LH, Lewis CE, Liu S, Martin LW, Ockene JK, O'Sullivan MJ, Powell LH, Simon MS, Van Horn L, Vitolins MZ, Wallace RB (2013) Menopausal **hormone therapy** and health outcomes during the intervention and extended poststopping phases of the Women's Health Initiative randomized trials. JAMA 310(13):1353–1368
42. Obiorah I, Jordan VC (2013) Scientific rationale for postmenopause delay in the use of conjugated equine estrogens among postmenopausal women that causes reduction in **breast cancer** incidence and mortality. Menopause 20(4):372–382
43. Santen RJ (2014) Menopausal **hormone therapy** and **breast cancer**. J Steroid Biochem Mol Biol 142:52–61
44. Chlebowski RT, Anderson GL, Gass M, Lane DS, Aragaki AK, Kuller LH, Manson JE, Stefanick ML, Ockene J, Sarto GE, Johnson KC, Wactawski-Wende J, Ravdin PM, Schenken R, Hendrix SL, Rajkovic A, Rohan TE, Yasmeen S, Prentice RL, WHI Investigators (2010) Estrogen plus progestin and **breast cancer** incidence and mortality in postmenopausal women. JAMA 304(15):1684–1692. doi:10.1001/jama.2010.1500
45. Chlebowski RT, Kuller LH, Prentice RL, Stefanick ML, Manson JE, Gass M, Aragaki AK, Ockene JK, Lane DS, Sarto GE, Rajkovic A, Schenken R, Hendrix SL, Ravdin PM, Rohan TE, Yasmeen S, Anderson G, WHI Investigators (2009) **Breast cancer** after use of estrogen plus progestin in postmenopausal women. N Engl J Med 360(6):573–587. doi:10.1056/NEJMoa0807684
46. Lathi RB (2006) Implantation and placental physiology in early human pregnancy: the role of the maternal decidua and the trophoblast. In: De Groot L, Jameson LJ (eds) Endocrinology. Elsevier Saunders, Philadelphia, pp 3341–3351
47. Parry S, Strauss F (2006) Placental hormones. In: De Groot L, Jameson L (eds) Endocrinology. Elsevier Saunders, Philadelphia, pp 3353–3367
48. Alvarado M, Ho T-Y, Russo J, Russo I (1994) Human chorionic gonadotropin regulates the synthesis of inhibin in the ovary and the mammary gland of rats. Endocrine 2:1107–1114
49. Alvarado MV, Alvarado NE, Russo J, Russo IH (1994) Human chorionic gonadotropin inhibits proliferation and induces expression of inhibin in human breast epithelial cells in vitro. In Vitro Cell Dev Biol Anim 30A:4–8
50. Horikoshi Y, Matsumoto H, Takatsu Y, Ohtaki T, Kitada C et al (2003) Dramatic elevation of plasma metastin concentrations in human pregnancy: metastin as a novel placenta-derived hormone in humans. J Clin Endocrinol Metab 88:914–919
51. Russo IH, Koszalka M, Russo J (1991) Comparative study of the influence of pregnancy and hormonal treatment on mammary carcinogenesis. Br J Cancer 64:481–484

52. Russo J, Russo IH (1997) Role of differentiation in the pathogenesis and prevention of breast cancer. Endocr Relat Cancer 4:7–21
53. Russo J, Balogh GA, Russo IH (2008) Full-term pregnancy induces a specific genomic signature in the human breast. Cancer Epidemiol Biomarkers Prev 17:51–66
54. Russo IH, Russo J (2011) Pregnancy-induced changes in breast cancer risk. J Mammary Gland Biol Neoplasia 16:221–233
55. Belitskaya-Levy I, Zeleniuch-Jacquotte A, Russo J, Russo IH, Bordas P et al (2011) Characterization of a genomic signature of pregnancy identified in the breast. Cancer Prev Res 4:1457–1464
56. Peri S, de Cicco RL, Santucci-Pereira J, Slifker M, Ross EA et al (2012) Defining the genomic signature of the parous breast. BMC Med Genomics 5:46–57
57. Golob JL, Paige SL, Muskheli V, Pabon L, Murry CE (2008) Chromatin remodeling during mouse and human embryonic stem cell differentiation. Dev Dyn 237:1389–1398
58. Asztalos S, Gann PH, Hayes MK, Nonn L, Beam CA et al (2010) Gene expression patterns in the human breast after pregnancy. Cancer Prev Res 3:301–311
59. Wilson BJ, Giguere V (2008) Meta-analysis of human cancer microarrays reveals GATA3 is integral to the estrogen receptor alpha pathway. Mol Cancer 7:49–57
60. Chou J, Provot S, Werb Z (2010) GATA3 in development and cancer differentiation: cells GATA have it! J Cell Physiol 222:42–49
61. D'Cruz CM, Moody SE, Master SR, Hartman JL, Keiper EA et al (2002) Persistent parity-induced changes in growth factors, TGF-beta3, and differentiation in the rodent mammary gland. Mol Endocrinol 16:2034–2051
62. Medina D (2004) Breast cancer: the protective effect of pregnancy. Clin Cancer Res 10:380S–384S
63. Russo IH, Russo J (1996) Mammary gland neoplasia in long-term rodent studies. Environ Health Perspect 104:938–967
64. Russo J, Tait L, Russo IH (1983) Susceptibility of the mammary gland to carcinogenesis. III. The cell of origin of rat mammary carcinoma. Am J Pathol 113:50–66
65. Bussolati G, Marchio C, Gaetano L, Lupo R, Sapino A (2008) Pleomorphism of the nuclear envelope in breast cancer: a new approach to an old problem. J Cell Mol Med 12:209–218
66. Palmer JE, Sant Cassia LJ, Irwin CJ, Morris AG, Rollason TP (2008) The prognostic value of nuclear morphometric analysis in serous ovarian carcinoma. Int J Gynecol Cancer 18:692–701
67. Tan PH, Goh BB, Chiang G, Bay BH (2001) Correlation of nuclear morphometry with pathologic parameters in ductal carcinoma in situ of the breast. Mod Pathol 14:937–941
68. Cao R, Wang L, Wang H, Xia L, Erdjument-Bromage H et al (2002) Role of histone H3 lysine 27 methylation in Polycomb-group silencing. Science 298:1039–1043
69. Kubicek S, Schotta G, Lachner M, Sengupta R, Kohlmaier A et al (2006) The role of histone modifications in epigenetic transitions during normal and perturbed development. Ernst Schering Res Fdn Workshop 57:1–27
70. Lin W, Dent SY (2006) Functions of histone-modifying enzymes in development. Curr Opin Genet Dev 16:137–142
71. Guenther MG, Young RA (2010) Transcription. Repressive transcription. Science 329: 150–151
72. Erwin JA, Lee JT (2010) Characterization of X-chromosome inactivation status in human pluripotent stem cells. Curr Protoc Stem Cell Biol (Chapter 1: Unit 1B 6). doi:10.1002/9780470151808.sc01b06s12
73. Huang PR, Hung SC, Wang TC (2010) Telomeric DNA-binding activities of heterogeneous nuclear ribonucleoprotein A3 in vitro and in vivo. Biochim Biophys Acta 1803:1164–1174
74. Han SP, Friend LR, Carson JH, Korza G, Barbarese E et al (2010) Differential subcellular distributions and trafficking functions of hnRNP A2/B1 spliceoforms. Traffic 11:886–898
75. Loyer P, Trembley JH, Grenet JA, Busson A, Corlu A et al (2008) Characterization of cyclin L1 and L2 interactions with CDK11 and splicing factors: influence of cyclin L isoforms on splice site selection. J Biol Chem 283:7721–7732

76. Yang L, Li N, Wang C, Yu Y, Yuan L et al (2004) Cyclin L2, a novel RNA polymerase II-associated cyclin, is involved in pre-mRNA splicing and induces apoptosis of human hepatocellular carcinoma cells. J Biol Chem 279:11639–11648
77. Fushimi K, Ray P, Kar A, Wang L, Sutherland LC et al (2008) Up-regulation of the proapoptotic caspase 2 splicing isoform by a candidate tumor suppressor, RBM5. Proc Natl Acad Sci U S A 105:15708–15713
78. Kobayashi T, Ishida J, Musashi M, Ota S, Yoshida T et al (2011) p53 transactivation is involved in the antiproliferative activity of the putative tumor suppressor RBM5. Int J Cancer 128:304–318
79. Salomonis N, Schlieve CR, Pereira L, Wahlquist C, Colas A et al (2010) Alternative splicing regulates mouse embryonic stem cell pluripotency and differentiation. Proc Natl Acad Sci U S A 107:10514–10519
80. Santucci-Pereira J, Zeleniuch-Jacquotte A, Afanasyeva Y, Zhong H, Peri S, et al. (2014) Gene expression profile induced by pregnancy in the breast of premenopausal women. Proc Am Assoc Cancer Res. 23:2360a
81. King MC, Wieand S, Hale K, Lee M, Walsh T, Owens K, Tait J, Ford L, Dunn BK, Costantino J, Wickerham L, Wolmark N, Fisher B (2001) Tamoxifen and breast cancer incidence among women with inherited mutations in BRCA1 and BRCA2: National Surgical Adjuvant Breast and Bowel Project (NSABP-P1) Breast Cancer Prevention Trial. JAMA 286:2251–2256
82. Narod SA, Brunet JS, Ghadirian P, Robson M, Heimdal K, Neuhausen SL, Stoppa-Lyonnet D, Lerman C, Pasini B, de los Rios P, Weber B, Lynch H (2000) Tamoxifen and risk of contralateral breast cancer in BRCA1 and BRCA2 mutation carriers: a case-control study. Hereditary Breast Cancer Clinical Study Group. Lancet 356:1876–1881
83. Baum M (2001) The ATAC (Arimidex, Tamoxifen, Alone or in Combination) adjuvant breast cancer trial in post-menopausal women. Breast Cancer Res Treat 69:210
84. Mouridsen H, Gershanovich M, Sun Y, Perez-Carrion R, Boni C, Monnier A et al (2001) Superior efficacy of letrozole versus tamoxifen as first-line therapy for postmenopausal women with advanced breast cancer: results of a phase III study of the International Letrozole Breast Cancer Group. J Clin Oncol 19:2596–2606
85. Robertson JF, Nicholson RI, Bundred NJ, Anderson E, Rayter Z, Dowsett M et al (2001) Comparison of the short-term biological effects of 7alpha-[9-(4,4,5,5,5-pentafluoropentylsulfinyl)-nonyl]estra-1,3,5, (10)-triene-3,17beta-diol (Faslodex) versus tamoxifen in postmenopausal women with primary breast cancer. Cancer Res 61:6739–6746
86. Trapido EJ (1983) Age at first birth, parity and breast cancer risk. Cancer 51:46–948
87. Chie WC, Hsieh C, Newcomb PA, Longnecker MP, Mittendorf R, Greenberg ER, Clapp RW, Burke KP, Titus-Ernstoff L, Trentham-Dietz A, MacMahon B (2000) Age at any full-term pregnancy and breast cancer risk. Am J Epidemiol 151:715–722
88. Holmberg E, Holm LE, Lundell M, Mattsson A, Wallgren A, Karlsson P (2001) Excess breast cancer risk and the role of parity, age at first childbirth and exposure to radiation in infancy. Br J Cancer 85:362–366
89. Vessey MD, McPherson K, Roberts MM, Neil A, Jones L (1985) Fertility and the risk of breast cancer. Br J Cancer 52:625–628
90. Kelsey JL, Horn-Ross PL (1993) Breast cancer: magnitude of the problem and descriptive epidemiology. Epidemiol Rev 15:7–16
91. Lambe M, Hsieh CC, Chan HW, Ekbom A, Trichopoulos D, Adami HO (1996) Parity, age at first and last birth, and risk of breast cancer: a population-based study in Sweden. Breast Cancer Res Treat 38:305–311
92. Russo IH, Koszalka M, Russo J (1990) Effect of human chorionic gonadotropin on mammary gland differentiation and carcinogenesis. Carcinogenesis 11:1849–1855
93. Russo J, Saby J, Isenberg W, Russo IH (1977) Pathogenesis of mammary carcinoma induced in rats by 7,12-dimethylbenz(a)anthracene. J Natl Cancer Inst 59:435–445
94. Russo IH, Russo J (1994) Role of hCG and inhibin in breast cancer (Review). Int J Oncol 4:297–306

95. Srivastava P, Russo J, Russo IH (1998) Chorionic gonadotropin inhibits rat mammary carcinogenesis through activation of programmed cell death. Carcinogenesis 18:1799–1808
96. Mgbonyebi OP, Tahin Q, Russo J, Russo IH (1996) Serum levels of chorionic gonadotropin in treated female rats during the progression of DMBA-induced tumorigenesis. Proc Am Assoc Cancer Res 37(1564a)
97. Tahin Q, Mgbonyebi OP, Russo J, Russo IH (1996) Influence of hormonal changes induced by the placental hormone chorionic gonadotropin on the progression of mammary tumorigenesis. Proc Am Assoc Cancer Res 37(1622a)
98. Russo J, Russo IH (2000) Human chorionic gonadotropin in breast cancer prevention. In: Ethier SP (ed) Endocrine oncology. Humana Press, Totowa, pp 121–136
99. Russo J, Tay LK, Russo IH (1982) Differentiation of the mammary gland and susceptibility to carcinogenesis. Breast Cancer Res Treat 2:5–37
100. Russo IH, Koszalka M, Russo J (1990) Human chorionic gonadotropin and rat mammary cancer prevention. J Natl Cancer Inst 82:1286–1289
101. Russo J, Wilgus G, Russo IH (1979) Susceptibility of the mammary gland to carcinogenesis. I. Differentiation of the mammary gland as determinant of tumor incidence and type of lesion. Am J Pathol 96:721–734
102. Russo J, Russo IH (1980) Influence of differentiation and cell kinetics on the susceptibility of the mammary gland to carcinogenesis. Cancer Res 40:2677–2687
103. Russo J, Russo IH (1994) Toward a physiological approach to breast cancer prevention. Cancer Epidemiol Biomarkers Prev 3:353–364
104. Russo IH, Russo J (1998) Role of pregnancy and chorionic gonadotropin in breast cancer prevention. In: Birkhauser MH, Rozenbaum H (eds) Proc. IV. European congress on menopause. Editions ESKA, Paris, pp 133–142
105. Russo IH, Russo J (1978) Developmental stage of the rat mammary gland as determinant of its susceptibility to 7,12-dimethylbenz[al anthracene. J Natl Cancer Inst 61:1439
106. Russo J (1983) Basis of cellular autonomy in susceptibility to carcinogenesis. Toxicol Pathol 11:149
107. Russo IH, Pokorzynski T, Russo J (1986) Contraceptives as hormone-preventive agents in mammary carcinogenesis. Proc Am Assoc Cancer Res 27(912a)
108. Dao TL (1964) Carcinogenesis of mammary gland in rat. Prog Exp Tumor Res 5:157
109. Russo J, Russo IH (1978) DNA-labeling index and structure of the rat mammary gland as determinants of its susceptibility to carcinogenesis. J Natl Cancer Inst 61:1451
110. Dao TL, Bock FG, Greiner MJ (1960) Mammary carcinogenesis by 3methylcholanthrene 11: inhibitory effect of pregnancy and lactation on tumor induction. J Natl Cancer Inst 25:991
111. Huggins C, Grand LC, Brillantes F (1959) Critical significances of breast structure in the induction of mammary cancer in the rat. Proc Natl Acad Sci 45:1294
112. Ciocca DR, Parente A, Russo J (1982) Endocrinologic milieu and susceptibility of the rat mammary gland to carcinogenesis. Am J Pathol 109:74
113. Tay LK, Russo J (1983) Metabolism of 7,12-dimethylbenz(a)anthracene by rat mammary epithelial cells in culture. Carcinogenesis 4:733–738
114. Russo J, Tay LK, Russo IH (1982) Differentiation of the mammary gland and susceptibility to carcinogenesis. Breast Cancer Res Treat 2:5–75
115. Tay LK, Russo J (1981) 7,12-dimethylbenz [a] anthracene-induced DNA binding and repair synthesis in susceptible and non-susceptible mammary epithelial cells in culture. J Natl Cancer Inst 67:155
116. Russo J, Miller J, Russo IH (1982) Hormonal treatment prevents DMBA-induced rat mammary carcinoma. Proc Am Assoc Cancer Res 23(348a)
117. Russo J, Janssens J, Russo IH (2000) Recombinant human chorionic gonadotropin (r-hCG) significantly reduces primary tumor cell proliferation in patients with breast cancer. Breast Cancer Res Treat 64(161a)
118. Janssens JP, Russo J, Russo IH, Michiels L, Donders G, Verjans M, Riphagen I, Van den Bossche T, Deleu M, Sieprath P (2007) Human chorionic gonadotropin (hCG) and prevention of breast cancer. Mol Cell Endocrinol 269:93–98

119. Russo J, Mailo D, Hu Y-F, Balogh GA, Sheriff F, Russo IH (2005) Breast differentiation and its implication in cancer prevention. Clin Cancer Res 11:931s
120. Mailo D, Russo J, Sheriff F, Hu YF, Tahin Q, Mihaila D, Balogh G, Russo IH (2002) Genomic signature induced by differentiation in the rat mammary gland. Proc Am Assoc Cancer Res 43:5368A
121. Rao DN, Ganesh B, Desai PB (1994) Role of reproductive factors in breast cancer in a low-risk area: a case-control study. Br J Cancer 70:129–152
122. Russo IH, Russo J (1993) Chorionic gonadotropin: a tumoristatic and preventive agent in breast cancer. In: Teicher BA (ed) Drug resistance in oncology. Marcel Dekker, New York, pp 537–560
123. Kliesch S, Behre HM, Nieschlag E (1994) High efficacy of gonadotropin or pulsatile gonadotropin-releasing hormone treatment in hypogonadotropic hypogonadal men. Eur J Endocrinol 131:347–354
124. Chen C, Jones WR, Fern B, Forde C (1985) Monitoring embryos after in vitro fertilization using early pregnancy factor. In: Seppala M, Edwards RG (ed) In vitro fertilization and embryo transfer, vol 442. The New York Academy of Science, p 428
125. The European Recombinant Human Chorionic Gonadotrophin Study Group (2000) Induction of final follicular maturation and early luteinization in women undergoing ovulation induction for assisted reproduction treatment--recombinant HCG versus urinary HCG. Hum Reprod 15:1446–1451
126. Chang P, Kenley S, Burns T, Denton G, Currie K, DeVane G, O'Dea L (2001) Recombinant human chorionic gonadotropin (rhCG) in assisted reproductive technology: results of a clinical trial comparing two doses of rhCG (Ovidrel) to urinary hCG (Profasi) for induction of final follicular maturation in in vitro fertilization-embryo transfer. Fertil Steril 76:67–74
127. Simon RM, Dobbin K (2003) Experimental design of DNA microarray experiments. Biotechniques 34:S16–S21
128. Russo J, Russo IH (2012) The trasncriptoma of the human breast. Springer, New York

ERRATUM TO

The Pathobiology of Breast Cancer

Jose Russo

J. Russo, *The Pathobiology of Breast Cancer*,
DOI 10.1007/978-3-319-40815-6

DOI 10.1007/978-3-319-40815-6_11

The original version of this book is revised. The previous version of this book was inadvertently published without the Preface, Acknowledgement and the Author's profile/bio.

The updated online version of the original book can be found at
http://dx.doi.org/10.1007/978-3-319-40815-6

J. Russo, *The Pathobiology of Breast Cancer*,
DOI 10.1007/978-3-319-40815-6_11

MIX
Papier aus verantwortungsvollen Quellen
Paper from responsible sources
FSC® C105338

If you have any concerns about our products,
you can contact us on
ProductSafety@springernature.com

In case Publisher is established outside the EU,
the EU authorized representative is:
Springer Nature Customer Service Center GmbH
Europaplatz 3, 69115 Heidelberg, Germany

Printed by Libri Plureos GmbH
in Hamburg, Germany